Related titles on radar:

Advances in Bistatic Radar Willis and Griffiths
Airborne Early Warning System Concepts, 3rd Edition Long
Bistatic Radar, 2nd Edition Willis
Design of Multi-Frequency CW Radars Jankiraman
Digital Techniques for Wideband Receivers, 2nd Edition Tsui
Electronic Warfare Pocket Guide Adamy
Foliage Penetration Radar: Detection and characterisation of objects under trees Davis
Fundamentals of Ground Radar for ATC Engineers and Technicians Bouwman
Fundamentals of Systems Engineering and Defense Systems Applications Jeffrey
Introduction to Electronic Warfare Modeling and Simulation Adamy
Introduction to Electronic Defense Systems Neri
Introduction to Sensors for Ranging and Imaging Brooker
Microwave Passive Direction Finding Lipsky
Microwave Receivers with Electronic Warfare Applications Tsui
Phased-Array Radar Design: Application of radar fundamentals Jeffrey
Pocket Radar Guide: Key facts, equations, and data Curry
Principles of Modern Radar, Volume 1: Basic principles Richards, Scheer and Holm
Principles of Modern Radar, Volume 2: Advanced techniques Melvin and Scheer
Principles of Modern Radar, Volume 3: Applications Scheer and Melvin
Principles of Waveform Diversity and Design Wicks *et al.*
Pulse Doppler Radar Alabaster
Radar Cross Section Measurements Knott
Radar Cross Section, 2nd Edition Knott *et al.*
Radar Design Principles: Signal processing and the environment, 2nd Edition
Nathanson *et al.*
Radar Detection DiFranco and Rubin
Radar Essentials: A concise handbook for radar design and performance Curry
Radar Foundations for Imaging and Advanced Concepts Sullivan
Radar Principles for the Non-Specialist, 3rd Edition Toomay and Hannan
**Test and Evaluation of Aircraft Avionics and Weapons
Systems** McShea
Understanding Radar Systems Kingsley and Quegan
Understanding Synthetic Aperture Radar Images Oliver and Quegan
Radar and Electronic Warfare Principles for the Non-Specialist, 4th Edition Hannen
Inverse Synthetic Aperture Radar Imaging: Principles, algorithms and applications
Chen and Martorella
Stimson's Introduction to Airborne Radar, 3rd Edition Griffiths, Baker and Adamy
Test and Evaluation of Avionics and Weapon Systems, 2nd Edition McShea
Angle-of-Arrival Estimation Using Radar Interferometry: Methods and applications Holder
Biologically-Inspired Radar and Sonar: Lessons from nature Balleri,
Griffiths and Baker
The Impact of Cognition on Radar Technology Farina, De Maio and Haykin
**Novel Radar Techniques and Applications, Volume 1: Real aperture array radar,
imaging radar, and passive and multistatic radar** Klemm, Nickel, Gierull, Lombardo,
Griffiths and Koch
**Novel Radar Techniques and Applications, Volume 2: Waveform diversity and
cognitive radar, and target tracking and data fusion** Klemm, Nickel, Gierull, Lombardo,
Griffiths and Koch
Radar and Communication Spectrum Sharing Blunt and Perrins
Systems Engineering for Ethical Autonomous Systems Gillespie
Shadowing Function from Randomly Rough Surfaces: Derivation and applications
Bourlier and Li
Photo for Radar Networks and Electronic Warfare Systems Bogoni, Laghezza and Ghelfi
Multidimensional Radar Imaging Martorella
Radar Waveform Design Based on Optimization Theory Cui, De Maio, Farina and Li

Micro-Doppler Radar and its Applications

Edited by
Francesco Fioranelli, Hugh Griffiths,
Matthew Ritchie and Alessio Balleri

The Institution of Engineering and Technology

Published by SciTech Publishing, an imprint of The Institution of Engineering and Technology, London, United Kingdom

The Institution of Engineering and Technology is registered as a Charity in England & Wales (no. 211014) and Scotland (no. SC038698).

© The Institution of Engineering and Technology 2020

First published 2020

The Institution of Engineering and Technology
Michael Faraday House
Six Hills Way, Stevenage
Herts, SG1 2AY, United Kingdom

www.theiet.org

British Library Cataloguing in Publication Data
A catalogue record for this product is available from the British Library

ISBN 978-1-78561-933-5 (hardback)
ISBN 978-1-78561-934-2 (PDF)

Typeset in India by MPS Limited

Contents

About the editors   xi
List of reviewers   xiii
Preface   xv

1 Multistatic radar micro-Doppler   1
Matthew Ritchie, Francesco Fioranelli and Hugh Griffiths

  1.1 Introduction   2
  1.2 Bistatic and multistatic radar properties   4
  1.3 Description of the radars and experimental setups   6
  1.4 Analysis of multistatic radar data   9
    1.4.1 Micro-Doppler signatures   9
    1.4.2 Feature diversity for multistatic radar systems   12
    1.4.3 Multiple personnel for unarmed/armed classification   14
    1.4.4 Simple neural networks for multistatic radar data   17
    1.4.5 Personnel recognition using multistatic radar data   19
    1.4.6 Multistatic drone micro-Doppler classification   22
  1.5 Concluding remarks   31
  Acknowledgements   32
  References   32

2 Passive radar approaches for healthcare   35
Karl Woodbridge, Kevin Chetty, Qingchao Chen and Bo Tan

  2.1 Introduction   36
    2.1.1 Ambient assisted living context   36
    2.1.2 Sensor technologies in healthcare   37
  2.2 Passive radar technology for human activity monitoring   38
    2.2.1 Current radar technology research with application to healthcare   38
    2.2.2 Passive radar technologies for activity monitoring   39
  2.3 Human activity recognition using Wi-Fi passive radars   42
    2.3.1 Fundamentals of Wi-Fi passive radars   42
    2.3.2 Micro-Doppler activity recognition   44
    2.3.3 Using Wi-Fi CSI informatics to monitor human activities   51

2.4   The future for passive radar in healthcare   59
2.5   Conclusions   60
References   62

**3   Sparsity-driven methods for micro-Doppler detection and
classification   75**
Gang Li and Shobha Sundar Ram

3.1   Introduction   75
3.2   Fundamentals of sparse signal recovery   76
      3.2.1   Signal model   76
      3.2.2   Typical algorithms for sparse signal recovery   77
      3.2.3   Dictionary learning   79
3.3   Sparsity-driven micro-Doppler feature extraction for dynamic
      hand gesture recognition   81
      3.3.1   Measurement data collection   82
      3.3.2   Sparsity-driven dynamic hand gesture recognition   84
      3.3.3   Experimental results   89
3.4   Sparsity-based dictionary learning of human micro-Dopplers   94
      3.4.1   Measurement data collection   95
      3.4.2   Single channel source separation of micro-Dopplers from
              multiple movers   95
      3.4.3   Target classification based on dictionaries   97
3.5   Conclusions   98
References   99

**4   Deep neural networks for radar micro-Doppler signature
classification   103**
Sevgi Zubeyde Gurbuz and Moeness G. Amin

4.1   Radar signal model and preprocessing   105
      4.1.1   2D Input representations   106
      4.1.2   3D Input representations   108
      4.1.3   Data preprocessing   108
4.2   Classification with data-driven learning   110
      4.2.1   Principal component analysis   111
      4.2.2   Genetic algorithm-optimised frequency-warped cepstral
              coefficients   112
      4.2.3   Autoencoders   113
      4.2.4   Convolutional neural networks   115
      4.2.5   Convolutional autoencoders   115
4.3   The radar challenge: training with few data samples   116
      4.3.1   Transfer learning from optical imagery   116
      4.3.2   Synthetic signature generation from MOCAP data   117
      4.3.3   Synthetic signature generation using adversarial learning 118

4.4   Performance comparison of DNN architectures   126
4.5   RNNs and sequential classification   128
References   129

5   **Classification of personnel for ground-based surveillance**   **137**
Ronny Harmanny, Lorenzo Cifola, Roeland Trommel and Philip van Dorp

5.1   Introduction   137
5.2   Modelling and measuring human motion   139
       5.2.1   Human gait   139
       5.2.2   Measuring the signature of human gait   141
       5.2.3   Visualization of the micro-Doppler signature   144
       5.2.4   Motion capturing activities on human motion   145
5.3   Model-driven classification   147
       5.3.1   Introduction to model-driven classification methods   148
       5.3.2   Results of model-based classification   149
       5.3.3   Particle-filter-based techniques for human gait classifica-
                tion and parameter estimation   152
5.4   Data-driven classification   160
       5.4.1   Deep supervised learning for human gait classification   160
       5.4.2   Deep unsupervised learning for human gait classification   167
       5.4.3   Conclusion   175
5.5   Discussion   175
References   178

6   **Multimodal sensing for assisted living using radar**   **181**
Aman Shrestha, Haobo Li, Francesco Fioranelli and Julien Le Kernec

6.1   Sensing for assisted living: fundamentals   182
       6.1.1   Radar signal processing: spectrograms   182
       6.1.2   Wearable sensors: basic information   185
6.2   Feature extraction for individual sensors   186
       6.2.1   Features from radar data   187
       6.2.2   Features from wearable sensors   190
       6.2.3   Classifiers   191
6.3   Multi-sensor fusion   192
       6.3.1   Principles and approaches of sensor fusion   192
       6.3.2   Feature selection   196
6.4   Multimodal activity classification for assisted living: some exper-
       imental results   198
       6.4.1   Experimental setup   198
       6.4.2   Heterogeneous sensor fusion: classification results for
                multimodal sensing approach   202
6.5   More cases of multimodal information fusion   208
       6.5.1   Sensor fusion with same sensor: multiple radar sensors   208
6.6   Conclusions   210
References   212

7   Micro-Doppler analysis of ballistic targets   **217**
Adriano Rosario Persico, Carmine Clemente and John Soraghan

    7.1   Introduction   217
    7.2   Radar return model at radio frequency   219
        7.2.1   Cone   224
        7.2.2   Cylinder   228
        7.2.3   Sphere   229
    7.3   Laboratory experiment   230
    7.4   Classification framework   233
        7.4.1   Feature vector extraction   236
        7.4.2   Classifier   240
    7.5   Performance analysis   241
        7.5.1   ACVD approach   242
        7.5.2   pZ Moments approach   242
        7.5.3   Kr Moments approach   243
        7.5.4   2D Gabor filter approach   245
        7.5.5   Performance in the presence of the booster   245
        7.5.6   Average running time   249
    7.6   Summary   254
    References   254

8   Signatures of small drones and birds as emerging targets   **257**
Børge Torvik, Daniel Gusland and Karl Erik Olsen

    8.1   Introduction   257
        8.1.1   Classes and configuration of UAVs   258
        8.1.2   Literature review   260
    8.2   Electromagnetic predictions of birds and UAVs   263
        8.2.1   Electromagnetic properties, size and shape   263
        8.2.2   RCS predictions   264
    8.3   Target analysis using radar measurements   269
        8.3.1   Body RCS   269
        8.3.2   Rotor RCS   270
        8.3.3   Maximum CPI   272
        8.3.4   Micro-Doppler analysis   273
        8.3.5   Micro-Doppler analysis on polarimetric data   277
    8.4   Radar system considerations   278
        8.4.1   Carrier frequency and polarization   279
        8.4.2   Bandwidth   279
        8.4.3   Waveform and coherency   280
        8.4.4   Pulse repetition frequency   281
        8.4.5   Pencil-beam or ubiquitous radar   283
    8.5   Classification methods   285
    8.6   Conclusion   287
    Acronyms   288
    References   289

9   Hardware development and applications of portable FMCW radars   291
    Zhengyu Peng, Changzhi Li, Roberto Gómez-García and
    José-María Muñoz-Ferreras

    9.1   FMCW radar fundamentals   292
    9.2   Radar transceiver   296
          9.2.1   Transmitter   296
          9.2.2   Receiver   297
    9.3   Antenna and antenna array   298
          9.3.1   Beamforming   299
          9.3.2   Two-way pattern and MIMO   305
    9.4   Radar link budget analysis   308
    9.5   FMCW radar signal processing   309
          9.5.1   Range processing   310
          9.5.2   Range-Doppler processing   311
          9.5.3   Micro-Doppler   313
    9.6   Applications of micro-Doppler effects   314
          9.6.1   Gesture recognition   314
          9.6.2   Fall detection   316
          9.6.3   Human activity categorizing   319
    9.7   Summary   319
    References   319

10 Digital-IF CW Doppler radar and its contactless healthcare sensing   323
    Heng Zhao, Biao Xue, Li Zhang, Jiamin Yan,
    Hong Hong and Xiaohua Zhu

    10.1   Principles of digital-IF Doppler radar   323
    10.2   Overview of RF layer   326
    10.3   Implementation of digital-IF layer   327
           10.3.1   Direct IF sampling   328
           10.3.2   Digital quadrature demodulation   329
           10.3.3   Decimation   332
    10.4   DC offset calibration technique   335
    10.5   Applications to healthcare sensing   338
           10.5.1   Contactless beat-to-beat BP estimation using Doppler
                    radar   338
           10.5.2   Multi-sensor-based sleep-stage classification   342
    10.6   Summary   347
    References   347

11 L1-norm principal component and discriminant analyses of
    micro-Doppler signatures for indoor human activity recognition   351
    Fauzia Ahmad and Panos P. Markopoulos

    11.1   Introduction   351
    11.2   Radar signal model   353

11.3 L1-PCA-based classification   355
    11.3.1 L1-norm PCA   355
    11.3.2 L1-PCA through bit flipping   356
    11.3.3 Classifier   356
    11.3.4 Illustrative example   357
11.4 L1-LDA-based activity classification   359
    11.4.1 Problem formulation   359
    11.4.2 L1-LDA algorithm   360
    11.4.3 Classifier   360
    11.4.4 Illustrative example   361
11.5 Discussion   362
11.6 Conclusion   363
References   363

**12 Micro-Doppler signature extraction and analysis for automotive
   application**   **367**
René Petervari, Fabio Giovanneschi and María A. González-Huici

12.1 Introduction and state of the art   367
12.2 Micro-Doppler analysis in automotive radar   370
    12.2.1 Target detection techniques   371
    12.2.2 Tracking techniques   372
    12.2.3 Track-based spectrogram extraction   373
    12.2.4 Spectrogram processing   374
    12.2.5 Feature extraction and classification   375
12.3 Experimental validation   379
    12.3.1 Experimental radar system   379
    12.3.2 Experimental set-up   380
    12.3.3 Processing   381
12.4 Practicality discussion   387
12.5 Conclusion   388
References   389

Conclusion   **393**

Index   **395**

About the editors

Francesco Fioranelli is an assistant professor in the Microwave Sensing Signals and Systems (MS3) section, Department of Microelectronics, at the Delft University of Technology. He is a senior member of the IEEE, recipient of the 2017 IET RSN best paper award, Associate Editor for IET Electronic Letters and IET RSN. His research interests include distributed radar and radar-based classification.

Hugh Griffiths holds the THALES/Royal Academy Chair of RF Sensors in the Department of Electronic and Electrical Engineering at University College London, England. He is a fellow of the IET, fellow of the IEEE, and in 1997, he was elected to fellowship of the Royal Academy of Engineering. He is an IEEE AES Distinguished Lecturer. He has been a member of the IEEE AES Radar Systems Panel since 1989.

Matthew Ritchie is a lecturer at UCL in the Radar Sensing group. He currently serves as the chair of the IEEE Aerospace and System Society (AESS) for United Kingdom and Ireland, associate editor for the IET Electronics Letters journal and chair of the UK EMSIG group. He was awarded the 2017 IET RSN best paper award as well as the Bob Hill Award at the 2015 IEEE International Radar Conference.

Alessio Balleri is a reader in radar systems with the Centre of Electronic Warfare, Information and Cyber at Cranfield University. From June 2004 to December 2004, he was a visiting research scholar with the Department of Electrical and Computer Engineering at University of Illinois at Chicago. He was the Technical Program Committee Co-Chair for the IET International Radar Conference 2017 (Belfast, UK) and the Technical Co-Chair of the 2020 IEEE International Radar Conference (Washington, DC, USA). His research interests include radar signal processing and biologically inspired radar and sonar systems.

List of reviewers

Fabiola Colone, Università di Roma La Sapienza, Italy
Carmine Clemente, Strathclyde University, UK
Michail (Mike) Antoniou, University of Birmingham, UK
Changzhi Li, Texas Tech University, US
Ronny Harmanny, Lorenzo Cifola, Roeland Trommel, Thales NL, The Netherlands
Philip van Dorp, TNO, The Netherlands
Hong Hong, Nanjing University of Science and Technology, China
Børge Torvik, Daniel Gusland, FFI, Norway
Christopher Baker, University of Birmingham, UK
David Greig, Leonardo, UK
Julien Le Kernec, University of Glasgow, UK & UESTC, China

Preface

In recent years, we have witnessed a significant growth in interest in radar technology, beyond the more traditional and established scenarios in long-range surveillance for defence and security, to include novel applications such as short-range and indoor radar, biometric and biomedical radar, passive radar and automotive radar, amongst others. This growth has come at the same time as advances in computational power and progress in data processing and artificial intelligence, so that radar sensing systems will be ever more ubiquitous, integrated into everyday objects such as vehicles and mobile phones.

One of the many interesting features of radar sensing technology is the use of the Doppler effect to characterise the movements of targets or their components. Doppler radar has been used for a long time to distinguish target echoes from stationary clutter, but more recently, the concept of micro-Doppler has been proposed and widely studied not only to detect the presence of targets against the clutter background but also to classify targets [1,2]. Micro-Doppler signatures are generated by the additional echo modulation due to the small movements of the main body or target parts, such as rotor blades of helicopters or unmanned air vehicles (UAVs), or limb motion in the case of moving animals or humans. Characterising and analysing the modulation patterns and signatures due to micro-Doppler can provide a wealth of additional information about targets and enable their classification and identification.

This idea dates back to early classified work on jet engine modulation (JEM) in the 1960s for aircraft identification [3], which led to later and broader unclassified research in automatic and non-cooperative target recognition [4,5]. Today, we see micro-Doppler employed in a wide variety of applications: from distinguishing drones from birds and loaded drones from unloaded drones, to distinguishing humans carrying objects such as weapons from those who are not, from the characterisation of rotor modulations of different helicopters, to unobtrusive medical monitoring and the detection of vital signs from those buried after earthquakes.

The rationale and timing of this book are to provide readers with the most recent and significant contributions in this area of research as well as recent multifaceted applications. Each chapter is written by internationally leading researchers in the field from academia, industry and research organisations, to provide detailed demonstrations and explanations of the analysis of micro-Doppler signatures for classification.

Chapter 1 focuses on micro-Doppler signatures experimentally obtained using a multistatic radar system consisting of three separate nodes. The signatures can be used for classification of unarmed vs potentially armed personnel in outdoor scenarios, as well as identification of specific individuals. The benefit of the additional information from the multistatic nodes is clearly demonstrated.

Chapter 2 discusses the usage of passive radar exploiting Wi-Fi illuminators of opportunity, showing how the extracted micro-Doppler signatures can be used for indoor assisted living applications to classify human activities and gestures. This is an attractive short-range solution of increasing interest and application and avoids the privacy issues that may occur with video monitoring.

Chapter 3 describes signal processing methods based on sparsity and dictionary learning that can be applied to the analysis of human micro-Doppler signatures, with applications such as classification of hand gestures and the number of people moving in indoor environments.

Chapter 4 discusses deep learning techniques applied to the classification of radar micro-Doppler signatures, presenting architectures of interest for the neural networks and approaches to tackle the lack of a sufficient amount of data for training, for instance using generation of synthetic, simulated data.

Chapter 5 presents model-driven and data-driven techniques for classification based on micro-Doppler signatures, with a specific focus on personnel monitoring using ground-based radar systems for outdoor surveillance.

Chapter 6 discusses the use of radar micro-Doppler signatures combined together with other sensors, such as wearable or inertial sensors, in a multimodal information fusion framework. Results with different fusion schemes are presented in the context of ambient assisted living.

Chapter 7 focuses on the modelling and the experimental validation of radar micro-Doppler signatures of ballistic targets such as warheads and decoys, presenting a selection of spectrograms, features, and classification results.

Chapter 8 presents examples and analysis of micro-Doppler signatures of UAVs which are an increasingly important class of target, and their comparison with the signatures of large birds that may act as false targets in the detection/classification process.

Chapter 9 discusses hardware development and design for portable FMCW radar systems, with focus on human micro-Doppler signatures for applications such as gesture recognition, fall detection and classification of human activities.

Chapter 10 focuses on the use of CW radar and related Doppler signatures for contactless healthcare applications, especially the estimation and monitoring of vital signs such as beat-to-beat blood pressure and sleep stages.

Chapter 11 discusses two methods based on $L1$-norm to extract features and classify micro-Doppler signatures of indoor human activities via Principal Component Analysis and Discriminant Analysis; the interest in these methods with respect to their conventional $L2$-norm counterparts is the robustness to the presence of outliers in the training data, for example as a result of mislabelling while collecting data.

Chapter 12 presents an analysis and discussion on the use of radar micro-Doppler signatures in automotive applications, investigating the classification of various types of target of interest such as pedestrians and cyclists moving in different trajectories. Automotive radar is another application of increasing interest, with huge commercial potential.

References

[1] Chen, V.C., *The Micro-Doppler Effect in Radar*, Artech House, Dedham, MA, 2011 (second edition 2019).

[2] Chen, V.C., Tahmoush, D. and Miceli, W.J., *Radar Micro-Doppler Signatures: Processing and Applications*, The Institution of Engineering and Technology, Stevenage, 2014.

[3] Gardner, R.E., 'Doppler spectral characteristics of aircraft radar targets at S-band' Report 5656, Naval Research Laboratory, 3 August 1961.

[4] Bell, M.R. and Grubbs, R.A., 'JEM modelling and measurement for radar target identification', *IEEE Trans. Aerospace and Electronic Systems*, Vol. 29, No. 1, pp. 73–87, 1993.

[5] Blacknell, D. and Griffiths, H.D. (eds), *Radar Automatic Target Recognition and Non-Cooperative Target Recognition*, The Institution of Engineering and Technology, Stevenage, 2013.

Chapter 1
Multistatic radar micro-Doppler

*Matthew Ritchie[1], Francesco Fioranelli[2]
and Hugh Griffiths[1]*

As an introduction to this book, this chapter reviews how multistatic radar sensor networks can be utilised to sense a series of different moving targets to extract their micro-Doppler (μD) signatures and use these for classification purposes. The field of radar has advanced significantly since its inception in the early twentieth century. From these origins, typical radar systems have evolved from single transmit/ receive configurations into complex sensor networks characterised by a multiple nodal dynamic. The future of radar sensing is moving towards intelligent distributed sensing due to the many advantages this brings. There are still significant challenges that need to be addressed through innovative solutions, such as location and synchronisation even within a GNSS-denied environment, but promising solutions are already maturing in Technology Readiness Level (TRL).

The majority of the real experimental results presented within this chapter are derived from the NetRAD and NeXtRAD multistatic pulsed-Doppler radar sensor systems, which have been used to capture a diverse range of μD signatures of moving targets. These systems have been developed at University College London (UCL) in collaboration with the University of Cape Town (UCT). These radars are two of the few multistatic systems from which measurements have been openly published, and the data are therefore an important resource for the research community. Only through deploying real systems and capturing empirical measurements of targets, can a field of research advance to resolve real-world challenges. The measurements taken include those of individuals walking, carrying simulated rifle objects, as well as drone (i.e. UAV, unmanned aerial vehicle) and bird targets. Through capturing the movements of these targets in a variety of multistatic geometries, the advantages of a radar-sensing network have been quantified through the classification of the μD signatures observed.

This chapter begins with an introduction to the general area of multistatic radar sensing and the background theory behind this and then goes on to review a number

[1]Department of Electronic and Electrical Engineering, University College London, London, UK
[2]Department of Microelectronics, TU Delft, Delft, The Netherlands

of important outcomes from this multistatic radar system and the advantages and challenges associated with this type of measurement.

1.1 Introduction

Radar and radio frequency (RF) sensing, in general, is an area of engineering that has constantly innovated since its inception. One of the key trends currently within this area of engineering is the deployment of networks of sensors in order to better achieve their objectives, such as target detection, tracking and classification, communications, and electronic surveillance. Research into these types of configuration has been continuous throughout radar history, but with the advent of modern computing, communications, and positioning and timing solutions, their potential and practical implementation have become more relevant than ever.

There are multiple advantages RF sensor networks possess in comparison to single monostatic solutions, i.e. conventional systems where the transmitter and the receiver are co-located. A key one is that a network of sensors will be able to provide coverage over a wider area and to give simultaneous multiple observations of a target from different angles. This geometric diversity can be crucial when trying to sense challenging targets such as low observable (LO) platforms, as they are designed and shaped to scatter the incident radar signals in other directions away from the transmitter, thus reducing the intensity of the echoes backscattered towards the radar. Using a multistatic radar solution with different receiver nodes would allow the information in these radar echoes to be captured and exploited.

Other advantages include improved redundancy in case one or more nodes are malfunctioning or otherwise taken out, improved detection probability, wider coverage, increased accuracy for target classification thanks to multi-perspective view on the targets, and the possibility of more favourable clutter statistics for specific bistatic geometries. Furthermore, the use of passive receivers in conjunction with an active radar transmitter or broadcast illuminator of opportunity enables additional advantages in terms of reduced complexity of the hardware of each node, and improved covertness introduced from their passive operation.

A high-level taxonomy of radar system types can be seen in Figure 1.1. The classical radar system is a monostatic or quasi-monostatic sensor, which transmits and receives its signals from the same position. This chapter focuses on other types of radar sensors that are bistatic or multistatic in their configuration. These types of configuration use receiving and transmitting antennas that are located in different, physically separated positions. These sensors can either use a cooperative radar transmitter that is providing the RF energy in a coordinated way to illuminate the scene of interest, or use non-cooperative sources. The results shown here are formed using a bistatic/multistatic sensor network that has cooperative radar transmitters. Other types of radar also shown are those that use non-cooperative sources that may be another non-cooperative radar (often referred to as 'hitchhiking'), or simply broadcast signals, typically communication signals. When using broadcast signals, this type of radar sensor is typically called a passive bistatic radar (PBR), or just passive radar.

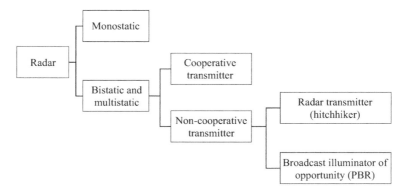

Figure 1.1 Taxonomy of types of radar systems

Some of the first radar sensors that were developed were bistatic in their configuration. This includes the radar used in the Daventry experiment of 1935 [1], the German *Klein Heidelberg* system that used the British Chain Home radar as illuminators [2], and early ionosphere measurements by Appleton and Barnett in 1925. In most of these cases, the reason for a bistatic solution was to exploit a powerful transmitter placed elsewhere, and to simplify the receiver architecture. These same reasons are valid today and support a resurgence of interest in PBR.

There are a few examples of multistatic radar networks available in the open literature, but they are often not a fully coherent multistatic system, but rather a series of independent monostatic radars that are networked. The American NEXRAD weather radar network is such an example, based on WSR-88D monostatic radar systems. This is a series of 159 radars sited across the USA with the objective of providing larger scale coverage of meteorological information particularly focused on extreme weather events [3]. Unlike the data analysed within this chapter, this type of radar network relies on singular monostatic radars fusing their data in a networked way. In comparison, a coherent multistatic network can transmit from one of the nodes with multiple nodes then receiving signals in a coherent synchronised manner. The advantage of this is the instantaneous measurement of the same target from multiple geometries providing N 'looks', which can then be fused to give the user more information.

There are also challenges when developing and deploying real-world multistatic radar sensors compared to monostatic devices. These challenges are mostly focused on the coordination of the sensor nodes and processing such that the fusion of the overall radar data is completed correctly. In order to coordinate multiple RF systems, synchronisation is vital, both at a command and control level and at the local oscillator (LO) level. The LOs at each of the radar node are required to be synchronous in order to be able to provide coherent sensing capabilities. This is especially important when the data extracted from the radar are used to characterise the Doppler and μD signatures of the targets. Without synchronised LOs, the Doppler data will be corrupted and hence, reliable μD analysis will be impossible. New solutions have

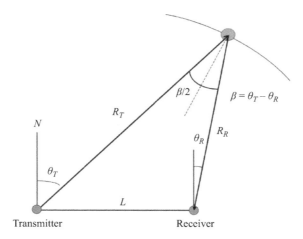

Figure 1.2   Bistatic range geometry. Derived from [5]

been developed in the field of stable oscillators, including those based on modern compact and relatively inexpensive Chip Scale Atomic Clock (CSAC) solutions and futuristic quantum-based clocks. These can be a powerful enabler for the development of multistatic radar solutions, as well as in many other position, navigation and timing solutions globally.

The rest of this chapter will review recent research on multistatic radar with particular focus on μD analysis, the applications that have motivated this research and the key trends in the quantitative analysis performed. The majority of this research is based on short-range classification of different targets or movements using real radar data from a multistatic radar system. Applications include human μD, with the focus on classification of potentially armed versus unarmed personnel, and the signatures of small UAVs.

1.2   Bistatic and multistatic radar properties

For a bistatic radar system, the range and Doppler values are modified from the typically derived monostatic equivalents. This is due to the influence of the geometry on the total range and on the perceived Doppler velocities of the targets. The bistatic range is typically noted as the total range from transmitter to target and target to receiver. Therefore, targets at the same bistatic range will lie on an iso-range ellipse with the transmitter and receiver locations as the two focal points [4]. This is shown in Figure 1.2 where R_T and R_R are the transmitter/receiver to target distances, respectively, L is the baseline separation between radar nodes, β is the bistatic angle, and θ_T and θ_R are the angles to the target from the transmitter and receiver positions.

The bistatic Doppler that is observed is also a function of geometry as seen in Figure 1.3. For the case that the transmitting and receiving radar nodes have zero

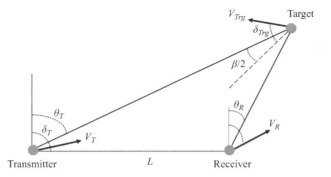

Figure 1.3   Bistatic Doppler diagram. Derived from [5]

velocity ($V_T = V_R = 0$), meaning that the radar nodes are stationary but the target has a non-zero velocity, then the Doppler frequency, f_D, as observed by the radar receiver is

$$f_D = \left(\frac{2V_{Trg}}{\lambda}\right)\cos\delta_{Trg}\cos\left(\frac{\beta}{2}\right) \tag{1.1}$$

There is a series of special cases for this resulting Doppler frequency [6] which are important to note. For example, when the bistatic angle is zero, the Doppler frequency reverts to the monostatic relationship as

$$f_D = \left(\frac{2V_{Trg}}{\lambda}\right)\cos\delta_{Trg} \tag{1.2}$$

When the target has a velocity vector that is in the radial direction, the Doppler frequency shift is

$$f_D = \left(\frac{2V_{Trg}}{\lambda}\right) \tag{1.3}$$

In the forward scatter geometry (target crossing the baseline connecting the transmitter and the receiver), which results in zero Doppler being observed by the radar receiver, also the range information is lost. However, in the near forward scatter geometry, experimental systems have shown that it is possible to extract some useful Doppler and range information for maritime targets crossing between nodes [7,8]. Zero Doppler is also observed when the velocity vector of the target is perpendicular to the bistatic bisector line due to the cosine relationship in (1.1). Hence, it is important to be able to provide a network of radar nodes to ensure that from one of the combinations of sensing transmitters/receiver pairs, an observable Doppler signature can be analysed. Another important aspect which is highlighted within the next section is that, due to this geometry relationship, the classification schemes generated to discriminate targets or actions will need inbuilt resilience to the changing of the perceived Doppler with respect to angle.

Table 1.1   NetRAD radar parameters list

Parameter	Value
Output power with HPA	+57.7 dBm
Output power without HPA	+23 dBm
Frequency	2.4–2.5 GHz
PRF	1 Hz–10 kHz
Pulse length	0.1–20 μs
Antenna gain	27 dBi
E-plane 3 dB beamwidth	11°
H-plane 3 dB beamwidth	8°

1.3   Description of the radars and experimental setups

The data presented in this chapter were all collected using the multistatic radar system NetRAD, developed over a number of years at UCL and the UCT [9–11]. NetRAD is a coherent pulsed radar consisting of three separate but identical transceiver nodes. It operates in the S-band frequency range (2.4 GHz), with programmable parameters in terms of bandwidth, pulse length, and pulse repetition frequency (PRF). For the majority of the data considered within this chapter, the system parameters were set as 45 MHz of bandwidth for the linear up-chirp modulation, 0.6 μs pulse duration, and 5 kHz PRF well sufficient to capture the signature of people walking at distances of several tens of metres and to include their whole Doppler and μD contributions without any aliasing. The radar had a transmitter power of approximately +23 dBm and used vertically polarised antennas providing approximately 27 dBi of gain and 10° beamwidth in both azimuth and elevation; the choice of vertical polarisation was made to match the dominant dimension of targets of interest, in the case of people who are taller than wider. A list of the wider parameters of the NetRAD radar system can be seen in Table 1.1. The system was initially developed to be a low power, coherent netted radar, which was capable of recording simultaneous monostatic and bistatic data by Derham and Doughty [11–13]. It was further adapted to a longer range wireless system by Sandenbergh and Al-Ashwal [14,15]. These adaptations focused on the addition of GPS disciplined oscillators (GPSDO), designed by UCT [14,16], which removed the requirement for cabled local oscillator (LO) synchronisation and allowed baselines of the order of kilometres and hence larger bistatic angles.

Although all three NetRAD nodes have transceiver capabilities, they were deployed with one node (node 1) as the monostatic transceiver, and the other two nodes as bistatic receivers. The nodes were deployed on a linear baseline, with 50 m separation between them, as shown in the photos of the system in Figure 1.4 and in the sketch representation of the experimental set-up in Figure 1.5. In the wireless synchronisation mode, the nodes have been utilised with baselines of up to 5 km, but for this short-range and low-power scenario, cabled configuration with approximately 50 m separation was sufficient to still achieve significantly wide bistatic angles in a

Figure 1.4   Photos of NetRAD system deployed for experimentation

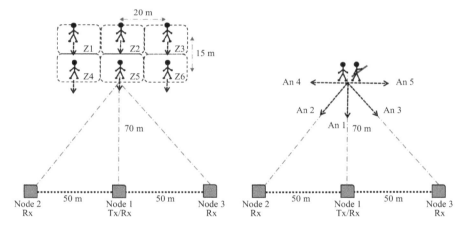

Figure 1.5   Sketch geometry of the experimental setup for the collection of data discussed in this chapter: first experiment on the left-hand side and second experiment on the right-hand side

smaller controlled scenario. The maximum separation was constrained by the requirement of a physical cable connecting the nodes to share the same clock signal for fully coherent operational mode. The experiments were performed in an open football field at the sports ground of UCL, at Shenley to the north of London. A large, open space, with distances to the target of the order of tens of metres was necessary to avoid the minimum range and 'blind zone' constraint of the pulse radar with the given pulse duration.

The two key experimental geometries described in this chapter are shown in Figure 1.5, where the node deployment is the same, but the target geometry is varied [17]. In the first experiment (sketched on the left-hand side of Figure 1.5), performed in July 2015, the targets were walking on a grid of six zones from the middle of each zone to the baseline of the radar. This was a set-up to simulate the radar surveillance of different zones of interest in a given area and to evaluate possible performance fluctuations in the classification accuracy. Note that the antennas were manually steered to focus on the middle of each zone before collecting the data. In each recording, there was only one subject walking, either empty-handed with arms free to swing while walking ('unarmed' case), or holding with both hands a metallic pole to simulate a rifle ('armed' case, with confined movements of arms). The pole was held with both hands in a manner similar to that in which a real rifle would be held and had size comparable to a real rifle. This two-case situation is an oversimplification compared to the many different circumstances that may be faced in the real world, but the objective was to better understand the principles using this simplified scenario. Parameters such as geometry, walking gait, and even gender will affect the μD signatures measured. The objectives of this work were more focused on understanding whether these two cases can be differentiated and if so what is the best way to fuse a classifier decision within a multistatic radar network. Its effect on the μD signatures has been shown in some of the previous work published by the authors [17–21]. In the first experiment, two subjects took part to the experiment, with the following body parameters: 1.87 m, 90 kg, average body type for person A, and 1.78 m, 66 kg, slim body type for person B, respectively. The total number of recordings was 360, considering three radar nodes, two people, six zones, and ten repetitions for each participant, of which half unarmed and half armed; the duration of each recording was 5 s to capture a few complete walking cycles.

In a later experiment, performed in February 2016, data with two people walking together at the same time were collected, as shown in Figure 1.5 on the right-hand side [19]. The people were walking along different aspect angles with respect to the radar baseline, either perpendicularly to it (aspect angle 1), or completely tangentially (aspect angles 4 and 5), or towards one of the bistatic radar nodes (aspect angles 2 and 3). Three different combinations of data were recorded, namely both people walking empty-handed, both people carrying metallic poles representing rifles, and only one person carrying a metallic pole (ensuring that this was carried by a different person in different recordings to create more variability in the μD signatures). The rationale of this experiment was to perform classification in the challenging case of two targets walking closely to each other in the same direction and same velocity, where one of them could be carrying the rifle. With the given radar resolution (approximately 3.3 m for 45 MHz bandwidth), the signature of the two targets was overlapped in the same range bins; also, as the targets were walking along the same trajectory and with similar speed, their μD signatures were also overlapped. For this experiment, a total number of 270 recordings were collected, considering three classes (both people armed, one person armed, and both people unarmed), three radar nodes, give aspect angles, and six repetitions for each class. Each recorded data was 5 s in duration, again to capture several periods of the walking cycle (approximately 0.6 s each).

The next-generation radar system following on from NetRAD is NeXtRAD [22,23]. This is an improvement on the NetRAD system in many ways and is the

Table 1.2 NeXtRAD radar parameters list

Parameter	Value
Output power X-band	400 W
Output power L-band	1.5 kW
X-band frequency range	8.5–9.2 GHz
L-band frequency range	1.2–1.4 GHz
Instantaneous bandwidth	50 MHz
X-band NF	2.5 dB
L-band NF	1.5 dB
PRF	1 Hz–10 kHz
Pulse length	Max 20 μs
Antenna gain	27 dBi

result of a joint development project between UCL and UCT. NeXtRAD is a higher power, dual frequency band system capable of operating in both X- and L-band frequencies. At X-band, the system is able to transmit one polarisation and receive simultaneous co- and cross-polarised data. The digital backend of the system is much more advanced compared to NetRAD and therefore allows for pre-distortion processing on waveforms or target-matched illumination thanks to waveform design and diversity. A list defining some of the parameters of the NeXtRAD system is shown in Table 1.2.

1.4 Analysis of multistatic radar data

In this section, examples of processing and results involving the use of multistatic radar data collected using NetRAD are reported. The results are mostly related to different applications of the analysis of human μD data in the context of ground surveillance, either to classify between unarmed and potentially armed personnel, or to identify specific individuals based on their walking gait. Additional results using UAVs as targets are also reported. Each subsection is organised around its individual research aspect or application and can be read as a self-contained part.

1.4.1 Micro-Doppler signatures

The data recorded by NetRAD are saved as a series of pulses (range profiles) that can be organised into a Range-Time-Intensity (RTI) 2D matrix. Moving target indication filtering can be applied to highlight only the signatures of targets of interest. Then the objective is to identify the range bins containing the target signatures, typically by applying a CFAR (constant false alarm rate) algorithm. If the target traverses a number of range bins, it is possible to correct this range migration by re-aligning the echoes into a single range bin or simply coherently summing over these range bins. Following this step, a time-frequency transform is used to characterise the Doppler and μD

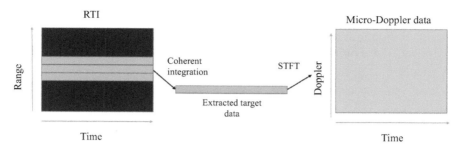

Figure 1.6   Processing diagram of RTI to micro-Doppler

signatures of the targets that are present. This process is shown diagrammatically in Figure 1.6. For the majority of the results discussed in this chapter, a short-time Fourier transform (STFT) was applied, with 0.3 s Hamming windows overlapped by 95%; these parameters were chosen to produce a smooth final representation of the μD signature of the walking subjects. It is noted that these parameters can be tuned to the dynamics of the target that is being observed and improved classification results may be obtained by varying them to each specific dataset.

In practical multistatic radar systems, one of the key challenges is maintaining coherency to nanosecond accuracy over the period of recording. As mentioned in Section 1.3, the NetRAD system uses GPSDO clocks to achieve this. These do not necessarily provide a perfect synchronisation and an element of drift may be observed between the relative phase of the two measurements. In order to correct for this residual drift, two signals of opportunity can be used to offset and calibrate the phase of the bistatic data. This can be either the direct transmission from the sidelobe of the transmitter which appears within the bistatic data at the range bin equivalent to the bistatic separation, or alternatively a bright static scatterer/target within the scene that both monostatic and bistatic nodes observe. By extracting the phase angle for this signal and offsetting all phase angles in the rest of the bistatic data, the apparent drift is corrected for. This correction is fundamental for Doppler processing, as the residual drift will otherwise produce a significant offset in the recorded targets' Doppler shifts.

The choice of STFT is common in the literature for its simplicity, as it consists of a sequence of fast Fourier transforms (FFTs) operations, although it suffers from the trade-off between time and Doppler resolution. This can be addressed by resorting to alternative time-frequency transforms or complementary techniques such as empirical mode decomposition, Wigner–Ville distributions, and wavelet transforms, but at the price of increased complexity. Another common approach for the characterisation of μD signatures is the cadence velocity diagram (CVD) analysis, which applies a further FFT across the time dimension of the result of the STFT (typically on the spectrogram). CVD was successfully applied to analyse the tumbling motions of ballistics targets in [24] and to analysing the movement of people and horses within [25]. In both cases, the CVD was found to be effective and help enhance the overall classification of the targets observed. The disadvantage is that it is an additional processing step as the CVD is derived from the spectrogram; therefore, if

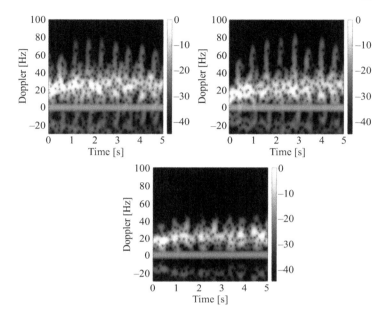

Figure 1.7 *Example of spectrograms for human micro-Doppler signatures: two people unarmed (top left), one person armed and one unarmed (top right), and both people armed (bottom)*

processing overhead is a strict requirement for a given system, it may not be suitable to use this.

The square absolute value of the STFT is generally referred to as a spectrogram, which can be plotted as an image showing the Doppler/velocity patterns of the different body parts as a function of time. Figure 1.7 shows examples of spectrograms generated from the NetRAD data, namely for the case of two people walking together described in the previous section. The three spectrograms refer to the case of both people unarmed, one armed and one unarmed, and both armed, respectively. All the three cases present the typical human μD signature, with the main, more intense contribution (in bright white colour) due to the torso of the subject, and the additional pattern of peaks (in darker grey colour) due to the swinging of limbs while walking. The case of both people unarmed is relatively easy to distinguish from the others, as these peaks are significantly reduced because of the confined movements of the arms while holding the metallic poles, as described in the previous sections. On the other hand, distinguishing between the situation of one person armed and both persons unarmed is challenging, and the two spectrograms appear very similar as the μD signature of the two subjects are overlapped.

The following subsections will present results related to specific applications, with details on the feature extraction and classification approach implemented, and a discussion on the performance that was achieved.

1.4.2   Feature diversity for multistatic radar systems

In the area of μD signature classification, there are many papers that present results of classification success based on selected features [19,26]. These features are often chosen based on experimental experience or the empirical knowledge of the investigator after working within this field. This then results in bespoke feature selection that is often completely different between publications/researchers, making the direct comparison of performance challenging. These empirical features may be based on the parameters of the physical movement that the target is exhibiting, often tailored to the specific dataset and nuances of the radar or experimental set-up used. For example, in the context of human μD, these empirical features may be the period of the gait of a person's walk or the overall extent of the signature along the Doppler axis. These values are useful as they can be related to the tangible parameters of a targets motion and its physical movements but require more manual processing to extract them, setting a series of parameters and thresholds to define where peaks of energy in the signatures are located [20].

Other features consider the μD signature as a bulk matrix from which features can be extracted in an automatic way, and more recently deep learning and neural networks have been used to perform such extraction [27]. While this approach is effective as the deep layers of the networks are capable of abstracting high-level underlying information in the signatures, this process can be challenging to reverse-engineer to explain the decisions taken by the network. For features generated by treating the μD as a full matrix, simple values such as the variance, standard deviation, entropy, or total power within a certain Doppler range can be considered. Alternatively, the application of transformation such as Singular Value Decomposition (SVD) or Principal Component Analysis (PCA) has been considered. SVD is a bulk matrix transform, based on eigenvalue decomposition, which has been successfully applied to a number of μD feature extraction problems and in each case has produced effective results [28,29]. The SVD is a matrix decomposition technique that enables an initial matrix to be represented by the product of two matrices (U and V) and a diagonal matrix S. The SVD of matrix A is shown in the following equation:

$$A = USV^{T} \tag{1.4}$$

U and V are the left- and right-hand singular value matrices derived from A. The S matrix contains the singular values decomposed from the original matrix A and is a diagonal matrix made up of non-negative real numbers. This is shown diagrammatically within Figure 1.8. This process of extracting the key values from the original matrix aims to transform the signals into a space prioritising the key contributions. It is then hypothesised that extracting features from these matrices will provide improved classification performance.

The choice of features that are extracted from the radar signatures is one of the upmost important aspects of radar signal processing applied to discrimination of targets. It has been shown in multiple publications that the outcome of a successful classifier is mostly dependent on the selection of features that yield good separation of different classes of targets in feature space [17,26]. If poor features are selected, then the separation of samples within the feature space is minimal and even the most

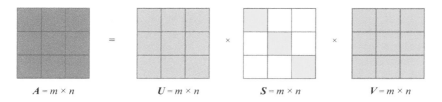

Figure 1.8   SVD decomposition of original matrix A into the left (U) and right (V) hand matrices as well as the singular values matrix S

elaborate classification processing techniques will not be able to make an accurate decision.

Earlier work in [26] proposed to go beyond feature selection performed once and for all, to adopt adaptive selection of features when classifying μD signatures in a monostatic case. In this publication, it was described that there are two key methods of optimally selecting subsets of features for classification, the first being a filter-based method and the second a wrapper method. For the filter method, the effectiveness of the features available is quantified using metrics such as class separability, distance measures such as Euclidean or Bhattacharya, and mutual information. Wrapper methods look to search the feature space in order to find feature combinations that create the best classification accuracy. This avoids the use of generic metrics for the individual features and directly computes the performance with a specific classifier. After applying these selection methods, those features that have been found to have the best performance are then prioritised when selecting a subset of the initial available set. Then, a simple validation was provided on how the selection of the subset of optimal features would vary as a function of the aspect angle from the radar to the target, demonstrating the concept of adapting the selection of features to the scenario under test and the target behaviour.

The concept of adaptive features was then applied to μD data generated by NetRAD in order to explore how it can be extended to a multistatic network and quantify the benefit in classification performance [19]. The main research hypothesis of this work was to explore the effect on the final classification accuracy of using different combinations of features at different radar nodes of the multistatic radar network. This approach was named 'feature diversity', whereby each node not only perceived the μD signature slightly differently due to the geometric diversity in aspect and bistatic angle, but also the features extracted at each node could be different to maximise the classification performance.

In this work [19], the following features were considered for each radar node:

1.   mean bandwidth of the Doppler
2.   mean of the Doppler centroid
3.   standard deviation of the Doppler bandwidth
4.   standard deviation of the Doppler centroid
5.   standard deviation of the first right singular vector
6.   mean of the first right singular vector

7. standard deviation of the first left singular vector
8. mean of the first left singular vector
9. standard deviation of the diagonal of the U matrix
10. mean of the diagonal of the U matrix
11. sum of U matrix
12. sum of V matrix

These sets of features can be divided into two categories: 1–4 are derived directly from the μD spectrogram, whereas 5–12 are evaluated from the SVD analysis of the μD signatures mentioned before. The Doppler centroid can be considered as an evaluation of the centre of gravity of the Doppler signatures, and the bandwidth the evaluation of the width of these signatures either side of this point. The centroid (f_c) and bandwidth (B_c) [30] were evaluated as in the following equations:

$$f_c(j) = \frac{\sum_i f(i) F(i,j)}{\sum_i F(i,j)} \qquad (1.5)$$

$$B_c(j) = \sqrt{\frac{\sum_i \left(f(i) - f_c(j)\right)^2 F(i,j)}{\sum_i F(i,j)}} \qquad (1.6)$$

where the $F(i,j)$ represents the power of the spectrogram in the ith Doppler bin and jth time sample and $f(i)$ is the Doppler frequency in Hz of the ith Doppler bin.

Using the aforementioned features, three classifiers were tested as part of this analysis of feature diversity, namely the Naïve Bayesian (NB) classifier, the diagonal-linear discriminant analysis and the K-nearest neighbour. These were selected for their simplicity as the focus of the work was to evaluate trends in monostatic and bistatic radar measurements and the effectiveness of the features selected. The input features to these classifiers were extracted from 180 recordings from 3 radar nodes, 5 aspect angles, 2 classes (armed/un-armed) with 6 repeats per class.

When evaluating the effect of the selection of different features, a brute force calculation of all possible combinations was performed. Considering the above-mentioned 12 features and assuming to use only 1 feature per node, 1,728 (12^3) combinations were tested for each classifier. Figure 1.9 shows examples of how the classification accuracy changes as a function of the combinations of features used at each multistatic radar node, with the constraint of using a single feature per node. The results from the NB classifier were used in this case, and the dashed line denotes the average classification. Given a certain dwell time and aspect angle, the accuracy can change significantly, more than 20%, depending on the combination of features used at multiple radar nodes. This shows how the optimal selection of features at different nodes, 'feature diversity', adds an extra level of complexity in multistatic systems but can deliver improved performance if knowledge of the most suitable combination of features can be inferred or obtained for a certain scenario.

1.4.3 Multiple personnel for unarmed/armed classification

In this section, we focus on a different challenge in outdoor human μD classification [17], namely the separation of the signatures of cases where two people are walking

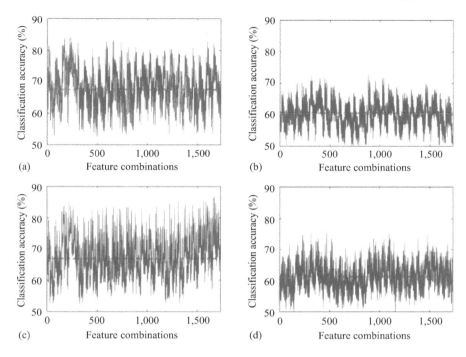

Figure 1.9   *Variation of classification accuracy depending on a combination of the
12 available features that are used with the constraint of a single feature
per node. (a) Angle 1 and dwell time 1 s, (b) angle 4 and dwell time 1 s,
(c) angle 1 and dwell time 2.5 s, and (d) angle 4 and dwell time 2.5 s
(from [19]). Note that the aspect angles are indicated in Figure 1.5*

together at close distance, and one of them may or may not be armed, with the
spectrograms being very similar in these two cases as shown in Figure 1.7. This
challenge is close to a real situation where perhaps multiple individuals are within
the radar beam at the same time and differentiating these signals would be a priority
of a functional deployed system. If the two, or more, people are walking in opposite
directions, obviously their signatures would be separated by the opposite frequency
central Doppler component. If the case is that they are walking in the same direction,
the challenge is more significant as the radar would observe a coherent sum of all μD
contributions that will overlap if the targets are present in the same range bin. The
experimental set-up is shown in the right-hand side of Figure 1.5. In this challenging
case, simple features based on SVD right and left vectors and their moments are not
suitable for the discrimination of samples of the different classes. An example of
these samples extracted from the monostatic node data for both people walking along
aspect angle 1 is shown in Figure 1.10(a), with dots and diamond markers used for
the two classes. Note that aspect angle 1 refers to the subjects walking in a straight
trajectory towards the baseline of the multistatic radar (Figure 1.5).

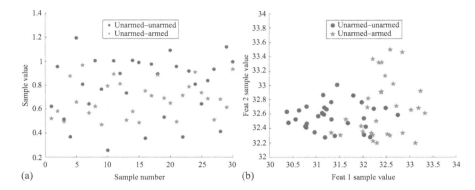

Figure 1.10   Feature samples for SVD-based feature (a) and centroid-based features (b) in the case of discrimination between multiple armed–unarmed personnel (from [17])

The samples related to the two situations of interest appear to be mixed, without a clear separation between the two classes; hence, it is expected that a classifier will perform poorly on these samples. On the contrary, features based on the mean of the centroid and bandwidth of the μD signature appear to perform well or at least provide enhanced inter-class separation, as shown, for example, in Figure 1.10(b) for aspect angle 1 data. It is believed that the SVD-based feature is simply related to the overall range of Doppler frequencies covered by the μD signature, whereas the centroid-based features can take into account how the signature energy is spread within that range of frequencies, whether this is more or less spread from the main Doppler component. The centroid features can therefore help discriminate between two μD signatures with similar overall bandwidth as those in Figure 1.7(a) and (b), whereas the SVD feature cannot.

Figure 1.11 shows the classification accuracy for the multiple personnel data aiming at distinguishing the cases when both people are unarmed and only one person is armed. An NB classifier trained with 40% of the available data was used in this example, with results shown as a function of the different data duration used to extract features and for each considered aspect angle. Note that multistatic data have been combined at decision level with a binary voting approach between the three nodes. The classification accuracy is higher, consistently above 80%, for aspect angles 1–3, where the subjects were walking towards one of the nodes in the baseline, whereas it decreases for aspect angles 4 and 5, where the subjects were walking tangentially to the baseline; hence, their Doppler signatures were much attenuated. At a given aspect angle, the data duration used for feature extraction has an impact on the accuracy with differences up to 10%. It is also interesting to observe that the optimal duration (i.e. that providing the best classification accuracy) changes with different aspect angles suggesting that for an actual system operating in real time in the field the optimal feature extraction approach will depend on the target trajectory. This confirms the observation made in the previous section and the results in Figure 1.9,

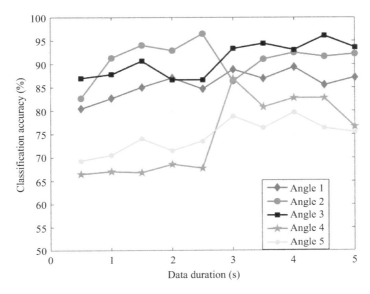

Figure 1.11   Classification accuracy per aspect angle using centroid-based features as a function of data duration for feature extraction (multiple personnel data, from [17])

whereby diversity of features at each multistatic radar node can improve the overall classification accuracy.

1.4.4   Simple neural networks for multistatic radar data

Deep learning has attracted significant interest in recent years in almost all fields of engineering and radar is not excluded from this, in particular for the automatic classification of targets of interest. This section evaluates the results produced from applying neural network classification methods to multistatic radar μD in order to improve the classification accuracy. The main attractiveness of deep learning methods is that feature extraction and selection can be performed directly by and within the neural network, with minimal or no inputs from the radar operator. This is a significant step forward compared with the handcrafted features or the effective but still limited features based on SVD, centroid and bandwidth, and CVD representations mentioned in previous sections [27]. While a dedicated chapter of this book deals with recent developments of deep learning methods for radar data, in this section a simple example of neural network developed specifically for the analysis of multistatic NetRAD data is reported and discussed [21]. The classification problem refers to the armed versus unarmed discrimination as in the previous two subsections. The experimental set-up was shown in Figure 1.5 on the left-hand side, with six different zones to monitor for the presence of unarmed personnel in front of the radar. The input to the neural network was the radar μD spectrogram treated as a matrix of pixels.

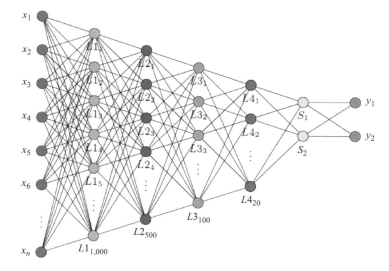

Figure 1.12   Sketch of the auto-encoders-based neural network implemented for multistatic radar data (from [21])

In order to reach an adequate compromise between capabilities of the network and feasibility of its implementation, auto-encoders (AEs) were chosen as these had been used extensively in several image-classification problems. The architecture consisted of four stacked AEs with a sigmoid activation function, followed by a softmax regression layer, as shown in Figure 1.12. The initial input layer is followed by multiple hidden layers, including 1,000, 500, 100, and 20 neurons, respectively, to compress the size of the data while extracting the relevant features; the final softmax layer consists of two neurons to match the binary classification problem under test, i.e. armed versus unarmed personnel.

Three different neural networks with the same structure were implemented for each multistatic radar node and trained with 60% of the available data. As the number of samples was insufficient for training even after segmenting the original spectrograms of length 5 s into shorter images, data augmentation was applied by adding Gaussian noise, thus taking the number of training samples for each node from 200 to 2,000 samples of duration 0.5 s. The results of the three different networks were then combined in an ensemble approach by using decision weights that are calculated assessing the individual classification performances of each network with an additional 20% of validation data; note that these decision weights are not related to the internal weights of the networks, which are fine-tuned at the training stage. Finally, the remaining 20% of data were used to test the overall classification algorithm with the three networks and their fused final decision.

A much simpler two-layer feed-forward network (FFN) was also implemented for comparison and used in the same approach as an alternative to the more complex AEs-based network. Figure 1.13 reports the classification accuracy in tabular forms

	Node/zone	1		2		3		4		5		6	
		S	W	S	W	S	W	S	W	S	W	S	W
	1	96.5	230.2	85.4	60.6	68.6	27.3	92.2	102.4	81.1	43.8	93.3	135.9
	2	84.9	52.5	65.9	24.5	88.3	59.7	36.6	11.1	88.9	67.0	74.1	33.2
	3	70.0	25.3	93.0	144.4	72.9	26.9	96.0	209.6	87.6	62.3	89.6	81.9
(a)	Combined decision	99.6		99.3		98.4		99.3		99.7		99.7	

	Node/zone	1		2		3		4		5		6	
		S	W	S	W	S	W	S	W	S	W	S	W
	1	100.0	3,000	98.8	270.0	98.8	159.3	99.8	497.5	99.8	492.5	100.0	3,000
	2	99.0	159.3	99.0	184.7	99.5	3,000	98.5	184.6	99.5	3,000	89.5	2.53
	3	99.8	885.0	98.3	219.4	99.7	497.5	100.0	3,000	99.3	350.3	100.0	3,000
(b)	Combined decision	100		98.5		99.5		100		100		100	

Figure 1.13   Classification accuracy of the feed-forward neural network (a) and auto-encoders-based network (b) for different multistatic radar nodes and combined decision (from [21])

for the simpler FFN (top table) and the deep AEs network (bottom table); the accuracy is reported for each individual node and for the final fused result in each of the six surveillance zones depicted in Figure 1.5. The decision weights are also reported for completeness. The AEs network outperforms significantly the FFN approach at the level of individual radar nodes, but the results become comparable when the combined decision fusion approach is considered. One could therefore argue that, at least for this relatively simple binary classification, there is no need of implementing the more complex deep learning solution. More recent research shows, however, that deep learning algorithms perform much better than simpler counterparts when classifying more challenging multiple class problems. What the best approach to implement such deep learning classification approaches into a multistatic radar network is that remains an outstanding and fascinating research problem.

1.4.5   Personnel recognition using multistatic radar data

In the previous sections, it was shown how μD radar information can be used to discriminate actions performed by different subjects, and it is well known how this can also be used to discriminate different classes of targets, for example vehicles or different models of aircraft and helicopters. Other studies such as [31] have shown that monostatic coherent radar sensors are capable of recognising different animals, for example human and cattle or humans and dogs. This section looks to evaluate the results produced in identifying a specific individual from their μD data using a multistatic radar network.

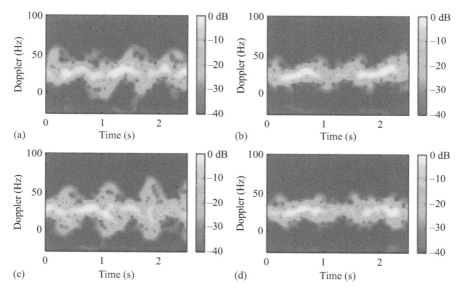

Figure 1.14   Example of spectrograms for human micro-Doppler signatures from two people who were either armed or unarmed. (a) Person A, unarmed, (b) person A, armed (c) person B unarmed, (d) person B armed (from [32])

All the aforementioned classification problems aim to discriminate among rather different sub-classes, but μD radar has been also used for the identification of individuals based on their walking gait and specific movements. This is a harder problem as the differences between the signatures will be more subtle and more challenging to capture in an automatic classification framework. This issue was investigated within [32] where the NetRAD multistatic radar system was deployed to gather human μD of people walking; the experiment was performed in July 2015, with the same geometry shown in Figure 1.5. The μD signals from two individuals who were walking armed or unarmed can be seen within Figure 1.14. In the unarmed case, higher Doppler contributions are observed within the signals present, because the person had freely swinging arms. In the armed case, the Doppler bandwidth is suppressed due to the restriction of the object they are holding. In addition, some small differences can be observed between signals of the two people by comparing sub-figures (a) to (c) and (b) to (d). Person A when unarmed has a slightly more asymmetric gait compared to person B. The difference when both individuals are 'armed' is more subtle, but further analysis is applied to evaluate if they can still be differentiated.

This research focused on the application of neural networks, specifically a deep convolutional neural network (DCNN) and treated the input spectrogram data as red, green and blue (RGB) imagery. This has been shown to be an effective means of classifying radar signals, although the authors acknowledge that this approach has been driven by image processing domain expertise. Radar information can be

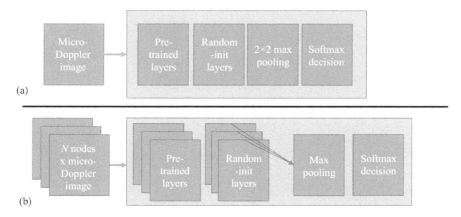

Figure 1.15 *Neural network architecture for a micro-Doppler classification*
process for multistatic radar inputs that uses (a) one node per network
and (b) three nodes as a joint input

portrayed as an image, but there is, for example additional information (such as the phase resulting from the application of the STFT) that is not considered in this way. Furthermore, a spectrogram has inherently physical and kinematic meaning on the targets' movements associated to its pixels and position in the image, whereas this is not necessarily true for an optical image. For example, in an optical image, the same object translated into different areas of the image can still be classified correctly as belonging to a certain class. In a spectrogram, the fact that the signature is positive or negative has strong implications on the trajectory of the target (towards or away from the radar) and on its velocity. This to say that some radar domain expertise is still necessary when processing data with DCNNs or other image-processing inspired methods.

A key aspect of leveraging the result from a multistatic radar network is how best to fuse the output results. In this specific example, it was shown to be effective to fuse the multistatic DCNN result rather than taking individual DCNNs from each radar node and then try to fuse the overall decision. A comparison between a singular neural network classification process versus a fused multistatic decision was shown in [32] and a high-level diagram for this can be seen in Figure 1.15.

For the results in [32], a neural network used 200 samples of Doppler spectrograms data as training data. The μD spectrograms were configured to be 128 (frequency bins) by 125 (time samples) in size. They were preprocessed in order to create grey-scale images between 0 and 255 in value when converting from the dB scale original data, which had a scale of 0 to -40 dB (see Figure 1.14). In order to relate the Doppler signature data to image processing techniques that use RGB scale images, these three channels of data were replaced with three versions of the Doppler spectra data. These three versions were generated with differing temporal window sizes within the STFT, which enabled to capture different details of the underlying movements thanks to the different resolutions of the three windows. The DCNN architecture of the neural network shown in Figure 1.15 used 3 pre-training layers, these were inserted from a pre-trained neural network named VGG-f that focuses on optical image classification [33].

The results from this personnel recognition analysis showed that the neural network was able to classify using data from individual radar nodes with an average accuracy ranging 97%–99.9%, depending on the amount of training data used. It was also shown that using the Multistatic-DCNN shown in Figure 1.15(b) enabled to increase the minimum, worst-case result in a series of 15 cross-validation experiments in comparison to using independent networks at each node followed by decision fusion through voting (*VMo-DCNN*); this made the result more robust across different train/test partitions of the available data.

1.4.6   Multistatic drone micro-Doppler classification

The NetRAD and its successor the NeXtRAD radar systems have both been successfully deployed to measure the μD signatures of drone targets, in particular small quadcopters and hexacopters. Further details on μD of drone targets can be seen in a different chapter of this book, but the focus of this short section is to introduce the multistatic sensing aspect of these drone measurements. The interest in small drones and UAVs is related to their potential misuses, accidentally or on purpose, for malicious activities such as illegal filming of private areas or key national infrastructures, collisions hazard with other aircraft or assets on the ground, smuggling of illicit substances such as drugs into prisons, and possible weaponised drones. Radar is a promising sensing technology for drones, due to its long-range and weather- or light-independent sensing capability, although drones are difficult targets because of their low radar cross-section (RCS) and agile manoeuvrability that makes them hard to track [34].

For experiments involving drone signatures, the NetRAD sensor was deployed in the same field environment as for the tests measuring people walking described in the previous sections. A comparison of the μD of a person walking and a quadcopter (DJI Phantom 2) captured by both a monostatic and bistatic NetRAD node is shown in Figure 1.16, along with a photograph of one of the NetRAD antennas deployed in the field pointing at the drone. The spectrograms show the typical oscillatory μD response from the person walking towards the radar in positive Doppler, and a contribution from the drone body in the negative Doppler, as the drone was flying away. The measurement was taken in vertical polarisation, hence, the stronger response from the vertically orientated limbs of a person, and no observable μD contributions from the rotor blades. The simultaneous measurement from a bistatic node can be seen on the right-hand side of Figure 1.16; this shows a very similar plot to the monostatic result, but there are increased Doppler sidelobes showing images around the target due to imperfections in the distributed local oscillator, a key challenge to multistatic radar as discussed in the early part of this chapter.

The sensor was then used to capture a number of measurements of the quadcopter target with a payload attached, the results of which were published in [29,35]. The

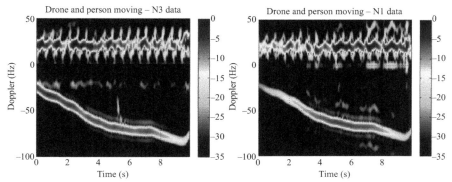

Figure 1.16 *(Top) NetRAD deployed in the field for the measurement of drones.*
Micro-Doppler signatures of a person walking and a quadcopter
drone flying in opposite directions from (bottom right) monostatic and
(bottom left) bistatic radars

objective of these measurements was to identify whether the μD signatures contained sufficient information to be able to identify if the drone was carrying different weights. The tests were completed for both hovering and moving scenarios with the weight of 0, 200, 300, 400, and 500 g attached. These measurements were taken with the expectation that this type of decision would be a future challenge for radar sensors. It is often predicted that swarms of UAVs will be implemented in both civil and military applications in the near future. DARPA has a current challenge, named *OFFensive Swarm-Enabled Tactics*, focusing on pushing innovation in this area where tactical drone swarms of up to 250 UAVs could be controlled at once. From a radar-sensing perspective, this may overload a radar sensor and force the device to try and prioritise key targets that are deemed the most significant threat. These targets may be the UAVs with the heaviest payload and hence, a means of discrimination based on this is desirable.

As part of the analysis performed, a comparison was made between features that were extracted from the range-time domain versus those that were extracted from the Doppler-time domain. This is important as it quantifies how much additional information is provided when completing automatic target recognition in the Doppler

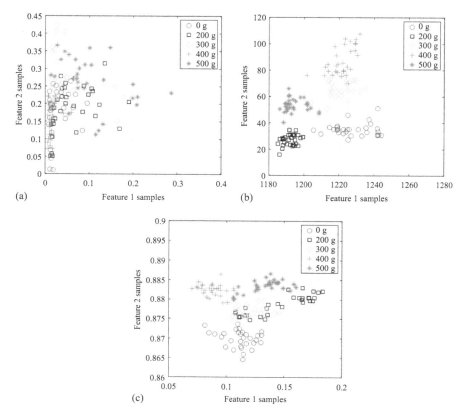

Figure 1.17   Feature space plots for features extracted from radar data for drone hovering with different payloads: (a) time domain features of mean and standard deviation of the target signature; (b) centroid-based features from the micro-Doppler signatures; (c) SVD-based features from the micro-Doppler signatures (from [29])

domain over just the time domain. The data captured were initially taken with a hovering drone at a range of only 60 m, which provided a high SNR for the target body and rotor blades. This was an ideal situation compared to a real-world challenge where the target may be much further away, very agile in its movement, and in a cluttered area, but it was important to establish feasibility of this technique in a simple scenario first. Measurements of 30 s were taken and then broken down into samples of 1 s each for feature extraction and classification. In the time domain, the mean and standard deviation of the complex reflected signature were used for classification, whereas in the Doppler domain the two different feature sets previously mentioned were extracted, namely centroid-based and SVD-based. The centroid-based features were the weighted centre of gravity of the Doppler spectrogram and the bandwidth relative to this. These were evaluated for each of the 1 s slices of data, although further testing did investigate the effect this window size had on the classification

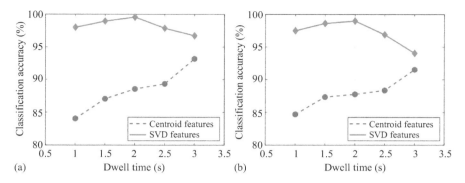

Figure 1.18   Classification accuracy as a function of the spectrogram duration for feature extraction (dwell time) when distinguishing different payloads on a flying drone; (a) diagonal-linear classifier and (b) Naïve Bayes classifier (from [29])

result. The SVD-based features of mean and standard deviation of the U matrix were selected. Figure 1.17 shows 2D feature spaces or different features considered, where the different markers' shapes indicate different payloads; the data were extracted from the same radar node (node 1 in this case). The centroid-based features provided the best separation between the different classes of payloads, and the SVD-based features also provided some separation. On the contrary, time-domain features were in this case not suitable as the separation was rather poor.

The results from the analysis showed that it was possible to classify if the drone had a payload using the μD data with accuracy higher than 90%–95%. It was also investigated if changing the window size of the samples that were provided for training and testing would affect the classifiers' results, which is directly related to the dwell time of the radar on the target, i.e. how long the section of available spectrogram for feature extraction was. Figure 1.18 shows the classification accuracy for dwell times from 1 to 3 s in the case of a diagonal-linear classifier and NB classifier, for both centroid-based and SVD-based features; note that the drone in this case was flying back and forth in front of the radar baseline and not just hovering. Longer dwell time appears to affect positively the classification accuracy when centroid-based features are used, whereas the effect for SVD features is less prominent.

Another significant challenge for radar-based detection of drones is the differentiation between drones and non-drones, whereby these potential false alarms can be cars or trees swaying in the wind detected by the side lobes of the radar, but mostly birds. Birds can generate a radar return comparable in RCS to that of small drones given the similarity in size of the two targets with the risk of 'flooding' the radar detector with many echoes that can saturate the processing unit to perform tracking and surveillance of all these targets. Reliable classification of drone and non-drone targets is therefore a hot topic.

The NetRAD sensor was used to gather data on bird targets and drone targets in a controlled experiment, where the system could be set-up to capture the movements

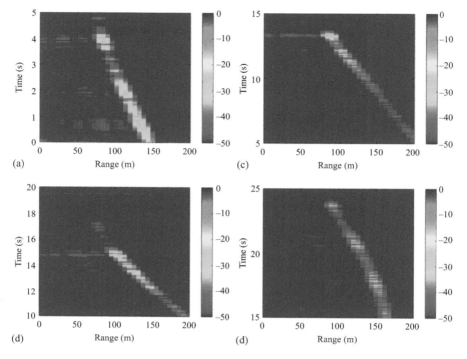

Figure 1.19   Range-time-intensity plots for hooded vulture (a), eagle owl (b), barn owl (c), and DJI phantom drone (d) collected by monostatic radar node (from [36])

of the two classes over the exact same ranges to make a direct comparison. The results from the analysis of this data were published within [36]. The geometry of the radar deployment was a straight baseline between two of the radar nodes, with an additional receive only node at the monostatic site. The baseline between the monostatic and bistatic nodes was limited to approximately 28 m due to the geometry of the site. In this configuration, it was possible to measure simultaneous HH- and HV-polarised data at the monostatic site with two nodes, and bistatic HH data at the third node. The polarimetry data were collected with the assumption that this information can be used to discriminate between different target classes [37]. The flapping motion of a wing beat should induce a pattern in the ratio of horizontal to vertical signatures, whereas a rotor blade is likely not to have the same pattern.

The targets used were a quadcopter DJI Phantom UAV and three different species of birds (barn owl, eagle owl, and hooded vulture). Figures 1.19 and 1.20 show examples of the RTI plots and μD for these targets. While the traces for the birds and drone are fairly similar in the range-time domain, the μD signatures are clearly different, as in the drone case the radar can observe the μD signatures of the rotor blades, which are only present when the drone flies. Note that Figure 1.20(d) has a much wider Doppler frequency axis in order to show the contributions from the rotor

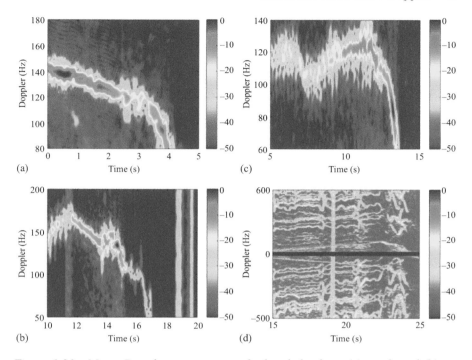

*Figure 1.20   Micro-Doppler spectrograms for hooded vulture (a), eagle owl (b),
barn owl (c), and DJI phantom drone (d) collected by monostatic
radar node (from [36])*

blades. The bird targets have a flapping signature followed by a period of gliding when
no μD patterns are seen only bulk velocity from the translational movement of the
bird. It is interesting to observe how the wing-flapping signature is more evident for
the smaller bird, the barn owl, in Figure 1.20(c), whereas the largest bird, the vulture,
did not need to flap its wing much to fly for the short distance of this experiment;
hence, not so much periodic μD modulations are seen in Figure 1.20(a).

While this initial experimental study showed how features in the μD signatures could support the discrimination of drone versus non-drone flying targets, birds
mostly, the applicability of this approach remains an outstanding research question.
Fixed-wing models of UAVs may have much less visible blade modulations than the
quadcopter used in this experiment or other larger models. Furthermore, the RCS of
the blades is typically much lower than the RCS of the drone and therefore a clear
signature of blades in the μD domain may not always be obtainable, especially in
the conditions of low SNR. Finally, extracting the μD signature requires the drone
to be well placed in the radar beam to be detected and tracked, and performing these
operations for all possible targets before being able to classify them as drones or
birds may be too computationally onerous or require too long latency for practical
usage.

Figure 1.21 NeXtRAD system deployment geometry for drone measurements

The NeXtRAD radar sensor was used as part of a drone measurements campaign in December 2018 in collaboration between UCL, UCT, and the South African Institute of Maritime Technology (IMT). This radar is capable of measuring in both X- and L-bands and has three radar nodes that can be wirelessly synchronised to provide distributed sensing. The results shown in this section are gathered when the radar system was using two nodes to sense the drone target in a monostatic – bistatic pair. The data presented here are from a short-range baseline test where the radar nodes were separated by only 150 m, while the target was at a range of 475 m. Further data were captured at wider bistatic angles, but only this simple initial subset of data is evaluated here. The resulting bistatic angle for this geometry was approximately 10° (see Figure 1.21), which is fairly narrow but would still provide a slightly different perspective on the target. This type of geometry is the typical length of a destroyer class ship and therefore is representative of a deployment of a bistatic pair of sensors at the bow and stern of such a vessel. The targets measured in this case included a DJI Matrice Hexacopter drone and bird targets of opportunity. The hexacopter is larger than the DJI Phantom that was measured by the NetRAD sensor but the ranges from the radar to the targets are much longer in this geometry.

μD signatures are shown for both X-band (Figure 1.22) and L-band (Figure 1.23) from measurements taken in horizontal polarisation. The experiments analysed included monostatic and bistatic captures from the same geometry on repeated flights of the same drone. The spectrograms shown are normalised to a peak of 0 dB showing a dynamic range of 50 dB and are not calibrated. The drone target was clearly observed

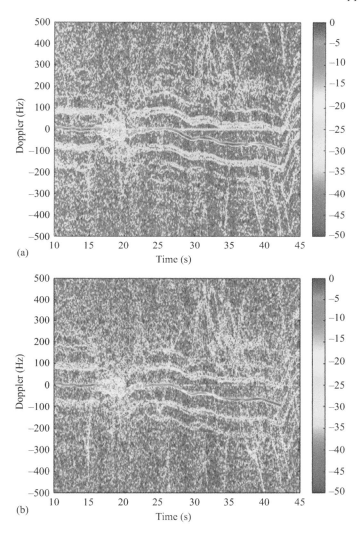

*Figure 1.22   NeXtRAD spectrogram of a hexacopter drone micro-Doppler
signatures from HH polarised X-band channel (a) monostatic and
(b) bistatic*

by both the radar nodes and therefore could be detected and classified in a fused
signal-processing framework. The properties of the drone signatures are quite different between frequency bands but are comparable between monostatic and bistatic
observations. The rotor blade contributions, often referred to as the horizontal HERM
lines, are seen to be stronger in L-band compared to X-band and are strongest in the
bistatic L-band dataset.

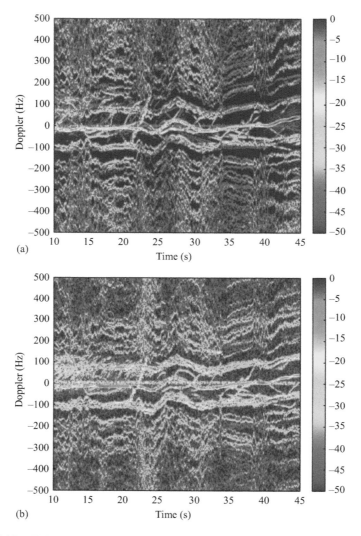

*Figure 1.23   NeXtRAD spectrogram of a hexacopter drone micro-Doppler
signatures from HH polarised L-band channel (a) monostatic and
(b) bistatic*

The signature of the drone body versus the peak rotor blade contributions was
analysed and is reported in Table 1.3. The body SNR was found to be approximately
21–24 dB in all measurements while the drone was flying at a range of approximately
475 m. The rotor blade contributions were found to be higher in the monostatic data
by 3.6 and 5.1 dB, for X- and L-band, respectively. The results presented are only

Table 1.3   SNR of body and rotor blade contributions from NeXtRAD measurements

Feature	SNR (dB)			
	X-band mono	X-band bistatic	L-band mono	L-band bistatic
Body	21.6	22.23	24.34	21.2
Rotor blades	11.64	8.08	15.5	10.4

for a single geometry and further investigation is required to fully understand the potential suitable deployment geometries that could be utilised to maximise SNR, and therefore detection and classification probability. Clearly, the μD signatures produced by quadcopter/hexacopter drones can be used to identify these targets and initiate results have shown that monostatic captures produce stronger contribution from the rotor blade but further data and analysis are needed to confirm this initial result.

1.5   Concluding remarks

This chapter has reviewed the recent research outputs in the area of multistatic μD. The advantages of sensing in a multistatic way have been defined quantitatively for a number of challenging scenarios using experimental data. In each of the cases discussed, the multistatic nature of the measurements enabled an improvement on the outcome, often the classification result, for a given experiment. Clearly, the increased amount of data and their spatial diversity provided by a network of radar sensors has improved the probability of extracting the information required for detection or classification of targets of interest.

Multistatic radar-sensing does come with additional costs and challenges alongside the proven benefits. In order to operate coherently the biggest challenge for the radars is synchronisation. In order to achieve this, GPSDOs have been successfully employed to produce clear coherent data and multiple nodes. For experimental static radar systems, this configuration has sufficed, but in any military scenario you are unable to rely on such a vulnerable timing solution. Some alternatives to GPS do exist, including Chip Scale Atomic Clock solutions, but these are limited in how long they can provide synchronisation for and would require an initial inter-sensor synchronisation to start. Currently, a great deal of research is being invested into alternative navigation solutions for this very reason, including new innovative concepts such as gravity sensors and quantum clocks. As these new concepts are developed, radar system engineers will need to leverage them as timing sources for multistatic RF sensor networks.

The research presented here used a ground-based static system that was fixed in the parameters it utilised during a capture. As RF sensors are increasing in their complexity and capabilities, it is envisioned that in the near future sensors will be much more adaptive in order to work within congested and contested environments, as well as optimise themselves for the tasks they are looking to achieve. The parameter space for optimisation for a single monostatic radar is a complex one that has led to a large area of work on cognition. This parameter space increases exponentially for an intelligent multistatic network that could adaptively change them, creating a fascinating yet rather challenging research problem.

Acknowledgements

The authors acknowledge the support of several funders that have enabled research into multistatic radar at UCL, including the IET for the A. F. Harvey Prize, EPSRC, the Royal Academy of Engineering, FFI Norway, and the US ONR(G), as well as the Institute of Maritime Technology (IMT) in South Africa for their support in the NeXtRAD drone measurement campaign.

The authors are grateful to the early developers of the NetRAD system: Chris Baker, Karl Woodbridge, Shaun Doughty, Tom Derham, Waddah Al-Ashwal, and Stephan Sandenbergh; and also to recent trials team members Amin Amiri, Jarez Patel, Sevgi Gürbüz, Saad Alhuwaimel, and all who helped generate the experimental results described in this chapter.

References

[1] H. D. Griffiths, "Early history of bistatic radar," *EuRAD Conference 2016*, London, 6/7 October 2016, pp. 253–257.

[2] H. D. Griffiths and N. Willis, "Klein Heidelberg—the first modern bistatic radar system," *IEEE Trans. Aerosp. Electron. Syst.*, vol. 46, no. 4, pp. 1571–1588, 2010.

[3] National Research Council, *Weather Radar Technology Beyond NEXRAD*. Washington, DC: The National Academies Press, 2002.

[4] N. J. Willis and H. D. Griffiths, *Advances in Bistatic Radar*. London: SciTech Publishing, 2007.

[5] M. C. Jackson, "The geometry of bistatic radar systems," *IEE Proc., F*, vol. 133, no. 7, pp. 604–612, 1986.

[6] N. J. Willis, "Bistatic radars and their third resurgence: passive coherent location," *IEEE Radar Conference*, 2002.

[7] A. De Luca, L. Daniel, M. Gashinova, and M. Cherniakov, "Target parameter estimation in moving transmitter moving receiver forward scatter radar," *Proc. Int. Radar Symp.*, 2017, pp. 1–7.

[8] M. Gashinova, K. Kabakchiev, L. Daniel, E. Hoare, V. Sizov, and M. Cherniakov, "Measured forward-scatter sea clutter at near-zero grazing angle: analysis of spectral and statistical properties," *IET Radar Sonar Navig.*, vol. 8, no. 2, pp. 132–141, 2014.

[9]   M. Ritchie, A. Stove, K. Woodbridge, and H. D. Griffiths, "NetRAD: mono-static and bistatic sea clutter texture and Doppler spectra characterization at S-band," *IEEE Trans. Geosci. Remote Sens.*, vol. 54, no. 9, pp. 5533–5543, 2016.

[10]   S. Doughty, K. Woodbridge, and C. Baker, "Characterisation of a multistatic radar system," *2006 European Radar Conference, Manchester*, September 2006, pp. 5–8.

[11]   T. Derham, S. Doughty, K. Woodbridge, and C. J. Baker, "Realisation and eval-uation of a low cost netted radar system," *Proc. CIE International Conference on Radar*, 2006.

[12]   S. R. Doughty, "Development and performance evaluation of a multistatic radar system," PhD thesis, University College London, 2008.

[13]   T. E. Derham, S. Doughty, K. Woodbridge, and C. J. Baker, "Design and evaluation of a low-cost multistatic netted radar system," *IET Radar Sonar Navig.*, vol. 1, no. 5, pp. 362–368, 2007.

[14]   J. S. Sandenbergh and M. R. Inggs, "A common view GPSDO to synchronize netted radar," *IET International Conference RADAR 2007*, Edinburgh, pp. 1–5.

[15]   W. A. Al-Ashwal, "Measurement and modelling of bistatic sea clutter," PhD thesis, University College London, 2011.

[16]   J. S. Sandenbergh, M. R. Inggs, and W. A. Al-Ashwal, "Evaluation of coherent netted radar carrier stability while synchronised with GPS-disciplined oscil-lators," *2011 IEEE Radar Conference*, Kansas City, MO, 23–17 May 2011, pp. 1100–1105.

[17]   F. Fioranelli, M. Ritchie, and H. D. Griffiths, "Centroid features for clas-sification of armed/unarmed multiple personnel using multistatic human micro-Doppler," *IET Radar Sonar Navig.*, vol. 10, no. 9, pp. 1702–1710, 2016.

[18]   F. Fioranelli, M. Ritchie, and H. D. Griffiths, "Classification of unarmed/armed personnel using the NetRAD multistatic radar for micro-Doppler and singular value decomposition features," *IEEE Geosci. Remote Sens. Lett.*, vol. 12, no. 9, pp. 1933–1937, 2015.

[19]   F. Fioranelli, M. Ritchie, S. Z. Gürbüz, and H. D. Griffiths, "Feature diversity for optimized human micro-Doppler classification using multistatic radar," *IEEE Trans. Aerosp. Electron. Syst.*, vol. 53, no. 2, pp. 640–654, 2017.

[20]   F. Fioranelli, M. Ritchie, and H. D. Griffiths, "Multistatic human micro-Doppler classification of armed/unarmed personnel," *IET Radar Sonar Navig.*, vol. 9, no. 7. pp. 857–865, 2015.

[21]   J. S. Patel, F. Fioranelli, M. Ritchie, and H. D. Griffiths, "Multistatic radar classification of armed vs unarmed personnel using neural networks," *Evol. Syst.*, vol. 9, pp. 135–144, 2018.

[22]   S. Alhuwaimel, S. Coetzee, P. Cheng, *et al.*, "First measurements with NeXtRAD, a polarimetric X/L Band radar network," *IEEE Radar Conference*, Seattle, WA, 8–12 May 2017.

[23]   M. R. Inggs, S. Lewis, R. Palamà, M. A. Ritchie, and H. D. Griffiths, "Report on the 2018 trials of the multistatic NeXtRAD dual band polarimetric radar," *IEEE Radar Conference*, Boston, MA, 22–26 April 2019, pp. 1–6.

[24]   A. R. Persico, C. Clemente, L. Pallotta, A. De Maio, and J. Soraghan, "Micro-Doppler classification of ballistic threats using Krawtchouk moments," *IEEE Radar Conference*, Philadelphia, PA, 2–6 May 2016.

[25]   A. W. Miller, C. Clemente, A. Robinson, D. Greig, A. M. Kinghorn, and J. J. Soraghan, "Micro-Doppler based target classification using multi-feature integration," *IET Conference on Intelligent Signal Processing*, 2–3 December 2013.

[26]   S. Z. Gürbüz, B. Erol, B. Çaglıyan, and B. Tekeli, Operational assessment and adaptive selection of micro-Doppler features," *IET Radar Sonar Navig.*, vol. 9, no. 9, pp. 1196–1204, 2015.

[27]   S. Z. Gürbüz and M. G. Amin, "Radar-based human-motion recognition with deep learning: promising applications for indoor monitoring," *IEEE Signal Process. Mag.*, vol. 36, no. 4, pp. 16–28, 2019.

[28]   M. Ritchie and A. Jones, "Micro-Doppler gesture recognition using Doppler, time and range based features," *IEEE Radar Conference*, Boston, MA, 22–26 April 2019.

[29]   M. Ritchie, F. Fioranelli, H. Borrion, and H. D. Griffiths, "Multistatic micro-Doppler radar feature extraction for classification of unloaded/loaded micro-drones," *IET Radar Sonar Navig.*, vol. 11, no. 1, pp. 116–124, 2017.

[30]   M. Greco, F. Gini, and M. Rangaswamy, "Non-stationarity analysis of real X-Band clutter data at different resolutions," *IEEE Radar Conference*, Verona, NY, 24–27 April 2006.

[31]   D. Tahmoush and J. Silvious, "Remote detection of humans and animals," *Proc. Appl. Imag. Pattern Recognit. Work.*, 2009, pp. 1–8.

[32]   Z. Chen, G. Li, F. Fioranelli, and H. D. Griffiths, "Personnel recognition and gait classification based on multistatic micro-Doppler signatures using deep convolutional neural networks," *IEEE Geosci. Remote Sens. Lett.*, vol. 15, no. 5, pp. 669–673, 2018.

[33]   K. Chatfield, K. Simonyan, A. Vedaldi, and A. Zisserman, "Return of the devil in the details: Delving deep into convolutional nets," *BMVC 2014 – Proc. Br. Mach. Vis. Conf. 2014*, 2014, pp. 1–11.

[34]   J. S. Patel, F. Fioranelli, and D. Anderson, "Review of radar classification and RCS characterisation techniques for small UAVs or drones," *IET Radar Sonar Navig.*, vol. 12, no. 9, pp. 911–919, 2018.

[35]   F. Fioranelli, M. Ritchie, H. D. Griffiths, and H. Borrion, "Classification of loaded/unloaded microdrones using multistatic radar," *Electron. Lett.*, vol. 51, no. 22, pp. 1813–1815, 2015.

[36]   M. Ritchie, F. Fioranelli, H. D. Griffiths, and B. Torvik, "Monostatic and bistatic radar measurements of birds and micro-drone," *IEEE Radar Conference*, Philadelphia, PA, 2–6 May 2016.

[37]   B. Torvik, K. E. Olsen, and H. D. Griffiths, "Classification of birds and UAVs based on radar polarimetry," *IEEE Geosci. Remote Sens. Lett.*, vol. 13, no. 9, pp. 1305–1309, 2016.

Chapter 2

Passive radar approaches for healthcare

Karl Woodbridge[1], Kevin Chetty[2],
Qingchao Chen[2] and Bo Tan[3]

The increase in life expectancy over the last few decades has resulted in a dramatic rise in the number of elderly people in most societies, with 25% of the UK population projected to be over 65 by 2051. Reluctance to rely on voluntary carers or potentially expensive institutional care provision has resulted in a growing interest in technologies to enable older people to live independently. These assistive technologies are now often integrated into the framework known as ambient assisted living (AAL). This framework includes a variety of sensor technologies to monitor a wide range of regular domestic metrics ranging from power usage right through to individuals' pattern of life and specific human activities. Micro-Doppler (μD) information from radio-frequency (RF)-based AAL sensors can potentially provide a wide range of health-related data ranging from long-term activity monitoring to remote alarm triggers in emergency situations like falls. Privacy issues, however, mean that many of the sensors being investigated are unlikely to be acceptable in many domestic situations. Radar-based sensors are however generally relatively unobtrusive, have potentially very good μD capability and have been shown to be able to identify a wide range of movement patterns. A range of radar systems is therefore being investigated for healthcare applications. Many of the radar sensors being examined for AAL are active, wide bandwidth systems which are typically of higher complexity, expensive, difficult to install and a potential source of interference to other systems and sensors. As a result, there is now significant interest in the use of passive radar, which is generally of low cost, low impact and adaptability to a wide range of expectation-maximisation (EM) environments. Use of passive radar sensors however depends on the availability of suitable transmitters of opportunity and requires complex processing to extract the useful Doppler data. Passive radar has widely been studied over recent years mainly for long-distance surveillance of targets such as aircraft, ships and more recently drones, using transmitters such as FM, DAB and communications or GNSS satellites. In the AAL context, however, these transmitters are generally not suitable

[1]Department of Electronic and Electrical Engineering, University College London, London, UK
[2]Department of Security and Crime Science, University College London, London, UK
[3]Faculty of Information Technology and Communication Sciences, Tampere University, Tampere, Finland

for indoor human activity monitoring due to low signal strength and the relatively low Doppler velocities involved. The rapid roll-out of wireless communications and resulting widespread availability of Wi-Fi in domestic and commercial environments has provided a very suitable transmitter of opportunity for a wireless-based passive radar for indoor human activity monitoring within an AAL context.

In this chapter, we initially review some of the radar technologies that have been investigated for healthcare applications. We then focus on the use of wireless communications networks (Wi-Fi) as the most widely available illuminator of opportunity for passive radars and describe the basic characteristics of these systems. The advantages and drawbacks of using a Wi-Fi passive radar in AAL environments are then discussed with emphasis on forms of μD-based human activity monitoring, most relevant to healthcare. Following this review of the literature, detailed investigations for the use of a Wi-Fi passive radar for human activity monitoring are described, including μD extraction methods and classification techniques. The use of Wi-Fi channel state information (CSI) rather than RF carrier data to obtain finer Doppler resolution is then discussed, and further experimental results using CSI in a healthcare setting are presented. Advanced activity recognition methods using machine-learning (ML) approaches are also becoming of significant research interest and the use of some of these methods on real data Wi-Fi passive radar data is also described. The chapter concludes by summarising the future challenges for deployment of passive radar in healthcare, including both social issues such as privacy and operational issues such as reliability and availability and expected requirements for alert and monitoring systems. Finally, the future landscape for passive radar in healthcare is discussed, including the potential performance improvements possible using new wireless network standards and emerging illuminators of opportunity.

2.1   Introduction

2.1.1   Ambient assisted living context

Life expectancy has increased dramatically over recent decades. Recent research shows that in the United Kingdom over 25% of the population is expected to be more than 65 years of age by 2051. With a downfall at health level in later years, this trend projects a potentially unsustainable increase in requirement for state or institutional care. In tandem with a growing resistance amongst the elderly to becoming institutionalised or relying on cares there is now a major push towards developing ways of maintaining independent living for elderly people in their home environment by increasing their confidence and autonomy. Many new technologies are being leveraged to achieve this goal with a key requirement being unobtrusive activity monitoring in home environments. Early alert to activity indicators such as lack of movement or indication of a fall could be critical to life. Longer term pattern of life monitoring is also a useful indicator of developing conditions and health issues which would benefit from early medical intervention. It is these drivers that have resulted in the evolution of the AAL concept. As previously mentioned, one of the key aims of AAL is the use of ICT sensors and communications technologies to support the preservation of health and functional capabilities of the elderly by

- promoting a better and healthier lifestyle for individuals at risk;
- enhancing security and preventing social isolation;
- supporting cares, families and care organisations.

This framework is now becoming widely used to focus on the integration of a range of sensors and e-healthcare technologies to monitor elderly and disabled people. AAL is also becoming increasingly embedded in the Smart Home and IoT concepts with an increasing number of research papers investigating the integration of healthcare monitoring within these frameworks.

2.1.2 Sensor technologies in healthcare

The main requirements of AAL technology are the ability to recognise human motion and activities and specifically the development of reliable classifiers to identify key movements of interest (e.g. falls, wheelchair upsets). In addition, the collection of longer term tracking, positioning and activity data and integration of data with that from other sensors would also provide valuable extra capability and compatibility. In this chapter we concentrate on the application of radar sensor technology to this application but a wide range of other sensors have been investigated for human activity recognition within the AAL framework. This includes systems based on passive infrared (PIR) [1–4] and vision-based sensing [5–11] as well as an increasing number of smart phone-based self-monitoring technologies [12–14]. Many wearable devices have also been investigated, frequently using accelerometers and gyroscopes [15–17] or RF identification (RFID) tags [18–20], to provide direct physical movement information about the subjects. There is also a significant amount of published work on wireless body area networks for personal health monitoring [21–24]. A number of shortcomings in the use of many of these sensor technologies in home environments have been identified. Wearable sensors suffer from low movement update rates of typically less than 5 Hz. In addition, people may forget to wear or drop their on-body sensors due to physical discomfort. PIR sensors are able to only provide the coarse-grained room level existence [25] while RFID-based devices employ complex transmitters and receivers and require pre-planning in order to optimally site the positions of the nodes [26]. Similar to other on-body sensors, RFID tags or transmitters can also be easily damaged, lost or forgotten [27]. Ultra wide band (UWB) radar activity recognition systems need heavy pre-deployment set-up and UWB components are more expensive than other technologies. Smartphone technologies have data accuracy and reliability limitations and require a relatively high level of user interaction. Video systems such as MS Kinect and Intel RealSense have been investigated in a number of healthcare projects [28]. However, in general, video camera systems require optimal lighting conditions and the deployment of video cameras in home environments raises many acceptability and privacy issues.

Unobtrusive low cost and passive sensor technology for activity monitoring with impacts is therefore becoming an increasingly attractive option for AAL. Passive radar-based systems can potentially provide this capability provided there is a suitable ambient e-m environment. The rapid roll-out of wireless communications (Wi-Fi) in public and private areas has provided a very suitable and widely available illuminator of opportunity for healthcare scenarios. The high Doppler resolution potentially available with such systems can be used to provide advantages in human motion and activity recognition using μD. Therefore there is now a growing body of research in this area including the development of advanced activity recognition and classification methods to facilitate low false alarm, real-time monitoring.

This chapter provides an overview of the current status of passive radar technologies that have been investigated for healthcare applications, focusing on μD-based activity monitoring using a Wi-Fi passive radar. Specific examples of activity recognition using μD data arising from typical human motions are presented and a range of classification methods are discussed. Data obtained using Wi-Fi CSI rather than RF carriers in order to obtain finer Doppler resolution are also presented and advanced activity recognition methods using artificial intelligence and ML approaches are also described. Finally, the current state-of-the-art in this technology is summarised along with the benefits and challenges of deploying this type of system in real-world healthcare scenarios.

2.2   Passive radar technology for human activity monitoring

2.2.1   Current radar technology research with application to healthcare

Radar sensors have been investigated for some years as important tools in human activity monitoring. Much of this activity was originally focused on defence and security applications, but in recent years, interest has grown in the Smart Home and healthcare areas. The majority of radar systems deployed in these scenarios have been active systems generally using μD signatures to characterise human motions and activities. There is by contrast much more limited research in the application of passive radar to healthcare which is the focus of this chapter. The use of an active radar in healthcare is addressed in another chapter of this publication, so only a few representative examples of this research are summarised here to set the passive radar technology in contrast.

Human activity monitoring using active radar systems has achieved a successful detection of a range of healthcare-related movement patterns of interest. This includes gait analysis [29,30], fall detection [31–33] and vital signs monitoring [34–38]. As a follow-on to these measurements, there has also been a considerable amount of research on μD feature extraction and classification methods to accurately assess the motions of interest and reduce false alarm. Recently, this has been increasingly focused on application of ML-based methods [39–42]. An active radar deployed in these scenarios has been mainly based on UWB or continuous wave (CW) radar technology. A CW radar can give high resolution Doppler data but very limited range information, whereas a UWB radar can give both Doppler and high-resolution range information. UWB radars can also potentially resolve closely spaced targets using impulse radio or stepped frequency CW technology [43].

There are clearly a number of advantages of using an active radar over passive systems. This includes the availability of well-matched suitable waveforms giving high bandwidth and high range resolution and enabling the use of standard radar signal processing. Higher power in relation to passive systems is normally also possible improving the signal-to-noise ratio which can be important in cluttered environments or when detecting weak signals such as those from vital signs. Generally, these systems can be designed to obtain high-resolution μD data from a range of human motions, and activities relevant to the healthcare scenario and active system are a useful tool for research in this area. However, when considering real-world widescale deployment of such sensors in the AAL context, a number of disadvantages are immediately apparent.

Cost and complexity
Active radar systems need specialised transmission and reception equipment and may also incur licensing cost. Some systems reported also require multi-node geometry or complex array antennas. They are therefore generally of high cost compared to those of passive systems and unlikely to be suitable for widespread use in private care homes or domestic scenarios.

Transmission issues
By definition, active system transmits potentially high-power RF which introduces spectrum congestion issues and may cause interference with other systems in nearby frequency bands and in areas outside the desired surveillance zone.

Installation and maintenance
Long-term activity monitoring requires semi-permanent installation of the detection equipment. This can result in high installation and maintenance costs for active radar systems.

User interactions
Active radar systems are not generally designed for use by untrained users, because it may cause difficulties in user operation and acceptability.

In view of the above issues, attention has increasingly turned in recent years to the potential use of passive radar systems in the AAL context. The key requirement is the availability of some suitable transmissions of opportunity in the appropriate AAL scenarios.

2.2.2 Passive radar technologies for activity monitoring

There has been a considerable amount of research carried out over many years on the use of μD to characterise human activities. As mentioned previously and discussed in other parts of this chapter, this has primarily involved the use of active radar. Activity monitoring using a passive radar has received very less attention. This has been partly due to the applications scenarios which were originally largely focused on the security and defence fields. Many of these applications are in outdoor, low interference scenarios where performance is paramount as opposed to issues of cost or complexity making active radars the obvious system of choice. The receive-only nature of passive radar systems means that they are generally of low cost and can achieve high Doppler resolution because of the increased integration times possible The traditional illuminators of opportunity for passive monitoring such as FM, DVB or GPS are not however, in general, suitable for human activity monitoring in typical AAL indoor environments due to the signal characteristics and low indoor signal strength. A passive radar is of a specific form of a bistatic radar which dates back from the earliest operational radar systems. For more details on the early development and basic principles of passive radars, the reader is referred to the papers by Farina and Kuschel [44] and Griffiths and Baker [45] as well as the references therein.

New potential illuminators of opportunity connected with communication networks have appeared in recent years which have provided the resurgence of interest in the use of passive radars for a range of civil applications. The rapid development of the mobile phone network over the last 20 years resulted in a surge in research into the use of GSM signals for passive radar monitoring. This has, however, been largely concentrated on the detection of vehicles such as aircraft [46], trains [47] and maritime targets [48]. GSM is however also not very suitable for indoor activity monitoring due to the low bandwidth and unpredictable signal strength and availability.

In more recent years, the widespread roll-out of wireless communication networks, especially Wi-Fi, a technology for wireless local area networking based on the IEEE 802.11 standards, has provided a relatively wide bandwidth freely available source of transmissions for indoor human activity monitoring. A more detailed discussion on the waveform properties and other characteristics of Wi-Fi transmissions are given in Section 2.3.1 of this chapter. Wi-Fi radar came to general attention during the development of the 802.11 communication networks and subsequently there has been a considerable amount of research published in the general area of human movement detection. Early research utilised both the beacon [49,50] and orthogonal frequency division multiplexing (OFDM)-based data [51] signals from Wi-Fi access points (APs) to detect the Doppler return from moving targets. Subsequently, a large amount of work was reported on the improvement of the processing schemes, see for example [52]. There has also been considerable interest in through-wall detection using Wi-Fi radars [53,54]. With the current increasing interest and investment in IoT, Smart Home and AAL, generally, there is now a rapidly growing area of research into the use of Wi-Fi passive sensing for detecting and characterising both large and small human motions. This research has shown a general trend of progression from using basic RF-received signal strength (RSS) echo signal to the use of Wi-Fi CSI for finer grained movement resolution and, more recently, the application of ML methods to improve activity classification.

Much of the initial research in this area used traditional RSS methods to achieve localisation of human subjects [55,56] using cross-correlation processing of the RF signals. A number of papers subsequently reported on specific activity characterisation [57–61] in indoor AAL type environments. In these studies μD data from the subjects were generally gathered followed by off-line characterisation into a number of typical movement classes. Vital signs detection including respiration has also been investigated by a number of research groups [62–64] and some very fine movement detection has been achieved. These studies and related research demonstrated the capability of using Wi-Fi for a range of human activity monitorings, applicable especially to domestic healthcare environments.

Using the basic Wi-Fi RSS signal has some disadvantages including the necessity of intensive classifier training and resolution limitations for finer movements. Therefore there has been a growing interest also in using Wi-Fi CSI to improve accuracy and to increase the range of movement detection possibilities and therefore to increase the application possibilities in the healthcare area. In [65] a subspace-based method is reported utilising the angle and time of arrival measurements for fine-grained device-free localisation while in [66] phase information and RSS profiles are used to identify people falling for applications in tele-healthcare. Doppler information has also been exploited in Wi-Fi CSI for activity recognition and requirements for interference filtering and de-noising [67] have been discussed in the literature. Wang *et al.* [68] have reported measurements of reflections from different body parts using the short-time Fourier transform and discrete wavelet transform to profile the energy of each frequency component. Recently there have been reports of significant databases of human movements being collected using CSI models. Wang *et al.* [69] proposed a CSI-based human activity recognition and monitoring system (CARM) and have collected a database of over 1400 samples. Guo *et al.* [70] have constructed a Wi-Fi activity data set with a classification accuracy greater than 93% using a support vector machine (SVM) method. Wi-Fi CSI information has also recently been applied to the detection of vital signs. CSI data from a range of activities specific to residential healthcare have been reported by Tan *et al.* [71] including through wall detection of respiration. Detection of respiration in a sleeping person is of interest for medical monitoring and Wi-Fi CSI data were used by Liu *et al.* [72] to track a person's breathing during sleeping. Zhang [73] exploits Fresnel zones in Wi-Fi signal propagation to examine the effect of the location and orientation in respiration detection.

In some cases, especially relating to smaller movements and gestures it becomes increasingly difficult to classify the motions accurately. Therefore there has been considerable recent interest in the application of ML methods to Wi-Fi data to improve classification performance. Al-qaness [74] recently reported the use of CSI phase and amplitude in tandem with an ML algorithm to improve performance. Wang *et al.* [75] applied deep learning methods to Wi-Fi CSI data for fingerprinting, activity recognition and vital sign monitoring in an IoT framework.

A deep learning approach to performance optimisation using channel hopping has also recently been reported [76]. Yousefi *et al.* [77] propose deep learning methods such as long–short-term memory recurrent neural networking to improve performance. Deep learning using a convolutional neural network (NN) has also been reported to give good results when used with Wi-Fi CSI data for respiration monitoring [78].

From the above review, it is clear that a Wi-Fi-based passive radar is becoming a key tool for low-cost unobtrusive healthcare monitoring. There is a focus on the detection of emergency situations such as falls, but longer term monitoring of daily activities and vital signs is also an important research goal. A major challenge is the unambiguous interpretation of μD data to identify specific motions and reduce false alarms to acceptable levels. Much work is therefore underway on advanced classification techniques using ML-based approaches. It is also important to note that, as mentioned, Wi-Fi is based on the IEE 802.11 standards which are continually evolving. Next and later generation Wi-Fi transmission standards may give improved capability and have to be considered in assessing the performance of future passive Wi-Fi healthcare monitoring systems. In the next section of this chapter, we consider these issues and describe some of basic design issues and characteristics of Wi-Fi-based passive radars. Some of the main techniques for human activity recognition using μD data from a Wi-Fi passive radar area are then discussed and illustrated using both simulated and real experimental data.

2.3   Human activity recognition using Wi-Fi passive radars

2.3.1   *Fundamentals of Wi-Fi passive radars*

Wi-Fi is a globally adopted technology with more than 50% of the world population now connected to the internet [79]. In the United Kingdom, the Office for National Statistics reported that in 2017, 90% of households in United Kingdom had internet access, an increase from 89% in 2016 and 57% in 2006 [80]. In developed countries especially, the use of the existing ubiquitous Wi-Fi communications infrastructure to identify physical activities and monitor long-term behavioural patterns therefore offers the potential to provide a low-cost and pervasive means of remote patient monitoring within AAL frameworks.

The IEEE 802.11 protocol (Wi-Fi) [81] defines the standards for wireless local area networks. The most recently deployed version is 802.11ac (Wi-Fi5), which was released in 2014 and operates only in the 5 GHz band. Wi-Fi5 is a faster and more scalable implementation of its predecessor 802.11n (Wi-Fi4) which is still commonly used and operates in both 2.4 and 5 GHz bands.

Both Wi-Fi4 and 5 employ OFDM, but the key differentiators to enable the faster speeds in Wi-Fi5 are (i) the use of denser 256 quadrature amplitude modulation (QAM) rather than 64 QAM employed in Wi-Fi4; and (ii) the ability to support eight spatial streams in its multiple inputs, multiple output (MIMO) functionality, as opposed to the previous four streams. In addition, beamforming technology has been built into the 802.11ac standards to allow increased signal strengths to be directed towards devices which the Wi-Fi AP is communicating. In terms of capacity, the maximum theoretical data-transfer speed supported by Wi-Fi5 is 1,300 Mbps which is around three times faster than that achievable by Wi-Fi4 (450 Mbps) and corresponds to a transmission bandwidth of 160 MHz – a significant increase from the maximum 40 MHz bandwidth supported by Wi-Fi4. However, the first wave of products conforming to Wi-Fi5 standards has been built around 80 MHz (433 Mbps), and in reality factors such as the number of antennas on the Wi-Fi APs and clients act to significantly reduce these quoted theoretical data throughput rates.

Physical activity recognition is a vital ingredient for AAL applications; and yet accurate, low power and user acceptable solutions remain elusive. Video depth cameras and wearable technologies for example can provide high levels of accuracy and represent the state-of-the-art in indoor sensing. However, they are often criticised for being too invasive. Moreover, both approaches require expensive equipment and dense deployments. Video monitoring is also limited by ambient light conditions and requires purpose built and expensive infrastructure. Conversely, wearable devices tend to have limited battery life and have high risk of non-compliance, especially amongst the elderly.

Wi-Fi-based passive radar technology addresses many of the important barriers which are obstructing the large-scale adoption of sensing technologies in home environments for AAL and e-healthcare applications. Compared to the other state-of-the-art modalities, the RF nature of Wi-Fi radar alleviates the difficulties associated with optimal lighting conditions and contrast experienced by optical systems. It also has a wide coverage area, dependent on the characteristics of the receiver antenna, and its through-the-wall sensing capability [53] mitigates issues of blind spots from physical barriers suffered by camera systems.

The use of already-existing Wi-Fi signal transmissions means that the set-up requires only the distribution of low-cost receiver nodes within the surveillance area. Moreover, the uncooperative nature of passive Wi-Fi radars (PWR) eliminates the requirement for patients to carry or wear any kind of cooperative active devices such an RFID tag/transponder or a wearable smart watch, making it an overall unobtrusive technology.

Compared to active radar systems used for AAL [38–43], the receive-only nature of PWR permits long integration times and therefore enables high sensitivity to small motions, which can aid various μD classification tasks. Furthermore, for technology manufacturers, the high-cost and regulatory issues around spectrum licensing associated with transmission-based systems are non-existent.

A significant advantage that a Wi-Fi passive radar has over other passive radar systems is that the Wi-Fi AP constantly transmits a *beacon signal* [81] even when the Wi-Fi AP is idle and is not an active user on the network. The bandwidth of the beacon signal is lower for those associated with Wi-Fi data transmissions and with typical interval times of 100 ms, the duty cycle is significantly lower. Though detection performance is suboptimal, a number of research studies have shown that the Wi-Fi beacon signal is a viable opportunistic signal in passive radars [55,82–84]. A further issue is that the OFDM-modulated, QAM-encoded communication signals are not ideally suited for radar sensing compared to tailored waveforms such as chirps, typically used in radars. Finally, another key challenge in indoor Wi-Fi passive radars for healthcare is related to the computationally intensive cross-correlation processing involved in extracting a target range and Doppler information from reflected signals; consideration must be given to how processing is employed within distributed or centralised network topologies to manage data throughputs.

In the following sections, we outline a body of work that has been carried out at University College London (UCL) to demonstrate a number of activity recognition capabilities for healthcare applications using Wi-Fi passive radars. The studies are based on both μD and CSI measurements and involve bespoke signal processing and ML approaches.

2.3.2   Micro-Doppler activity recognition

In this section, we describe a typical pipeline for μD activity recognition which includes constructing the PWR signal model, μD feature extraction and activity classification. We also illustrate the effectiveness of this approach using experimental data in a typical AAL-type environment.

2.3.2.1   Signal model

PWR utilises the existing Wi-Fi AP as transmitters of opportunity. The reference signal $ref(t)$ can be described as a linear combination of the 'clean' transmitted Wi-Fi signal $x_{source}(t)$ and the reflections from static objects, characterised by pth path delay τ_p and the corresponding complex magnitude and static phase A_p^{ref}:

$$ref(t) = \sum_p A_p^{ref} x_{source}(t - \tau_p) \tag{2.1}$$

Similarly, the signal in the surveillance channel $sur(t)$ is composed of echoes from all moving targets in the illuminated area of interest, which can be characterised as the pth path delay τ_p, its Doppler shift $f_{d,p}$ and relevant magnitudes and phase A_p^{sur}:

$$sur(t) = \sum_p A_p^{sur} x_{source}(t - \tau_p) \exp^{j2\pi f_{d,p}t} \tag{2.2}$$

It is assumed that the reference and surveillance signals are separated via the spatially directional antennas. In general, a target can be identified by cross-correlating the reference and surveillance signals and using the fast Fourier transform (FFT) to find the exact delay τ and frequency shift f of the reflected signal. This can be represented by the following cross-ambiguity function (CAF) [85,86]:

$$CAF(\tau, f) = \int_{-\infty}^{+\infty} \exp^{-j2\pi ft} ref^*(t - \tau) \times sur(t) dt \tag{2.3}$$

2.3.2.2   Micro-Doppler feature extraction

After extracting the range-Doppler map which is the matrix defined by (2.3), CFAR can be applied to extract the frequency components at the target's range bin. The post-processing method to extract μD illustrated in Figure 2.1 can be divided into two stages: data alignment and data pre-processing. In the *data alignment stage*, we first extract the active time bins of targets' motion and interpolate these time–frequency matrices into a fixed size of 2.5 s. In the *data pre-processing stage*, we perform conventional data whitening on training and testing data first and principal component analysis (PCA) on these whitened data. The output of PCA can be regarded

Figure 2.1   *Pipeline for micro-Doppler feature extraction method*

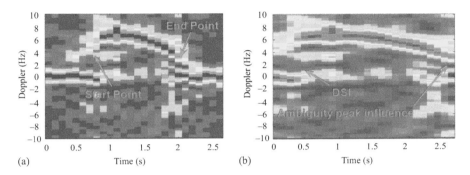

Figure 2.2   *(a) Start and end-point illustration; (b) the useful part of the μDS is adjusted and interpolated into the same size of temporal bins, with the DSI and the ambiguity peak examples*

as data features after dimension reduction. More details of each stage are illustrated in the following sections.

Data alignment stage

To set up a data set, the most important requirement is to keep the data samples of the same size. It is straightforward to maintain the same number of frequency bins through the CAF operation by dividing signal samples within the certain time period into a fixed number of batches. However, due to the time variations between differing motion classes, it is difficult to keep the fixed number of time bins (2.5 s) for each μD sample. To sum up, we wish all μD samples with the same and fixed number of time bins for classification method. To tackle this, we design the following two procedures:

- Automatic start- and end-point detection
- Adjust the data sample size by interpolation

To illustrate the two steps clearly, an example is shown in Figure 2.2(a) and (b). In Figure 2.2(a), the start- and end points of an activity are identified by the red arrows and we denote the μD between them as the active μD. Next in Figure 2.2(b), the active μD is extracted and is interpolated in temporal domain to a fixed size using bi-cubic interpolation [87].

In this example, we use a method for μD alignment using the standard deviation of frequency vector x_t at a specific time stamp t. Conventional CFAR-based detection may not be suitable in the indoor environment (Figure 2.2(b)) for the following reasons:

- The ambiguity peaks in μD could mislead the CFAR to provide false detections, as shown in Figure 2.2(b) as they are similar to the Doppler bins induced by the target movements.
- The direct signal interference (DSI) effect will generate strong peaks on the zero Doppler line, which again may mislead the CFAR detector, as shown in Figure 2.2(b). In addition, as DSI will be an important feature to distinguish μD, we have not performed any DSI elimination method till this point.

Without ambiguity elimination and DSI, an intuitive way to identify the active μDS periods is to check whether the non-zero frequency bins have large powers; from another perspective, the variance of the Doppler spectrum is a good feature to distinguish active or inactive μD. To eliminate the effects of zero Doppler clutters when calculating the distribution variances, we adopt the weighted standard deviation in (2.4) and (2.5), where I is the weights of the frequency bins, which are larger on higher frequency bins. We choose the indicator function as $I[l] = l^2$, where $WMean(\bullet)$ calculates the vector average and $abs(\bullet)$ calculates the absolute value:

$$WMean\,(\mathbf{x}_t) = \frac{\sum_{l=1}^{N} I[l] \times x_t[l]}{N} \tag{2.4}$$

$$Wstd\,(\mathbf{x}_t) = \frac{\sum_{l=1}^{N} I[l] \times abs(x_t[l] - Mean(x_t))^2}{N} \tag{2.5}$$

In general, we first select the active μD region and then choose the smallest and largest time bin as the start- and end point. The active time bins are selected once the weighted standard deviations of continuous three-time bins are larger than the threshold $Thres = \gamma \times \min(Wstd(x_t))$, where parameter γ indicates the scaling factor and $\min(Wstd(x_t))$ calculates the minimum of the Doppler energy time history. As the active μD may have a different number of time bins, we interpolate them in the time domain to a fixed number using the conventional bi-cubic interpolation methods [88]. In this way, the transformed data can be denoted as $X_{i,Fix} \in \mathbb{R}^{N_f \times N_t}$, where N_f is the number of Doppler bins and N_t is the interpolated number of time bins. Next, we transform the data sample-by-simple concatenating $X_{i,Fix}$ to a vector $d_i \in \mathbb{R}^{N_{total}}$, where $N_{total} = N_f \times N_t$. Next, we again concatenate d_i to matrix form $D_{PWR} \in \mathbb{R}^{N_{total} \times N_{PWR}}$, where N_{PWR} is the total number of data samples.

Data pre-processing stage
We adopt the standard data whitening and PCA [89] based on the implementation of singular value decomposition to extract the features with reduced dimensions as $D \in R^{N_r \times N_{PWR}}$, where N_r is the total number of reduced feature dimension. For evaluation, we denote the subset $D_{Red,T}$ *and* $D_{Red,S}$ as the training and test data set, respectively, where $D_{Red,T} \in R^{N_r \times N_{PWR,T}}$, $D_{Red,S} \in R^{N_r \times N_{PWR,S}}$ and $N_{PWR,S}$ $N_{PWR,T}$ represent the number of the test and training set samples, respectively.

2.3.2.3 Classification methods

Classification is important for real-time healthcare activity monitoring and there is an abundance of classifiers from which to choose. In this section, we apply the sparse representation classifier (SRC) on the PCA features because it provides a more compact representation under the sparse coding scheme and is (i) robust to noise and (ii) intrinsically discriminative [90]. In addition, SRC does not require a large number of parameters to learn, which are suitable for the relatively small-scale μD data set.

The core idea of the SRC is that we do not use a conventional dictionary, such as the Fourier basis, but utilise the labelled training μDS samples as the dictionary $D_{Red,T}$. Suppose that the dictionary is within C classes, denoted as $D_{Red,T} = [D_1, D_2, \ldots, D_C]$. SRC first represents the test μD data, denoted as $d_{test} \in R^{N_r \times 1}$ from $D_{Red,S}$ as the linear combination of the training sample dictionary $D_{Red,T}$ constrained by moderate sparsity level. Second, the test sample is assigned to the label c if and only if the residual error r_c using all atoms from the class c is the smallest. More specifically, the SRC can be interpreted as two steps, including sparse coding and classification.

Sparse coding: The sparse representation of d_{test} shown in (2.6) can be obtained using the greedy algorithm-based solvers.

Classification: After the optimal s is estimated, the classification step assigns the class c to the test sample d_{test} by selecting the best class c^* to reconstruct the d_{test} with the minimum residual, as shown in (2.7). $D_{Red,T,c}$ indicates the cth column of the dictionary and s_c indicates the cth scalar of the vector $s \in \mathbb{R}^{N_{PWR,T} \times 1}$.

$$s = \text{argmin}_s \|d_{test} - D_{Red,T} \, s\|_2 + \alpha \|s\|_0 \tag{2.6}$$

$$c^* = \text{argmin}_c \|d_{test} - D_{Red,T,c} \, s_c\|_2 \tag{2.7}$$

Several methods have been utilised to solve this optimisation problem, for example the $l1$-solver [91] or the orthogonal matching pursuit (OMP)-based greedy methods [92]. The $l1$-solver is computationally expensive, and the OMP is noted for its low convergence rate and inaccuracy [93]. Therefore, we choose subspace pursuit [93] as it has a faster convergence rate than the $l1$-solver without loss of accuracy.

2.3.2.4 Experiments and results

Implementation

The passive radar system used in our experiments is built around a software-defined radio (SDR) architecture. In this system, two Universal Software Radio Peripheral N210s synchronised with a MIMO cable are used to down-convert the RF Wi-Fi signals centred at 2.462 GHz. An FPGA (Xilinx Spartan 3A-DSP 3400) and a 100 Mbps, 14-bit analogue-to-digital converter are used to digitise the signal. The data are recorded and transferred into a laptop via a gigabit Ethernet port for real-time processing. The experimental scenario is shown in Figure 2.3, where a data set of six activity motions is collected in the test area of the figure. The details of the data set are shown in Table 2.1.

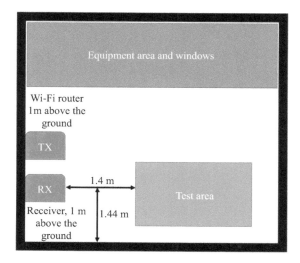

Figure 2.3   Experimental scenario

Table 2.1   Data set description of the six activities in the experiments

Activity and index	Description
M1(20)	Subject picks up from the ground and stands up
M2(20)	Subject sits down on a chair
M3(20)	Subject stands up from a chair
M4(57)	Subject falls down
M5(62)	Subject stands up after falling
M6(10)	Subject gets up from a lying down position

M1(20) indicates the first motion including 20 micro-Doppler samples.

Analysis of micro-Doppler signatures

We show the corresponding μD in Figure 2.4. The six μD signatures exhibit different patterns which can be described using the following visual discriminative characteristics:

- the maximum Doppler shift,
- time duration of the μD,
- whether Doppler frequency traverses gradually from negative to positive or just shows clear negative or positive values,
- the strength of the zero Doppler line caused by DSI or multipath.

In general, the maximum Doppler frequencies of these six μDs range from 2 to 4.5 Hz and movement patterns with different maximum Doppler frequency shifts will help distinguish different signatures. The second discriminative feature relates

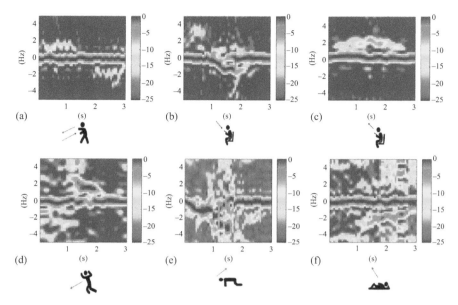

Figure 2.4 *Six micro-Doppler signatures related to the six activities in Table 2.1: (a) M1: subject picks up from the ground and stands up, (b) M2: subject sits down on a chair, (c) M3: subject stands up from a chair, (d) M4: subject falls down, (e) M5: subject stands up after falling, (f) M6: subject gets up from a lying down position*

to the relative direction of motion, indicated by the sign (positive or negative) of the Doppler shift frequency: some motions induce Doppler frequencies that transverse from positive to negative (e.g. M1 and M2), while others induce only positive or negative Doppler frequencies (M3, M4, M5 and M6). Although M1 and M2 both have the similar patterns (from positive to negative), the time duration of each signature segment increases the discrimination, such as the shorter duration of positive Doppler frequency in M2 than the positive Doppler frequency in M1. The final distinguishable feature is the presence of the zero Doppler line during the motion. A clear example is the comparison between the M5 and M6, where the Doppler signature patterns are similar, but the latter has a strong zero Doppler line while former does not. The reason why M5 exhibits no zero Doppler line is when the target gets out of the floor, the bulk motion blocks the direct signal to the receiver. Although these selected empirical features agree closely with the intuitive visual interpretation, obtaining these features accurately requires complex feature selection methods such as detecting accurate Doppler patterns and estimating the mentioned empirical features. In addition, these methods are prone to be erroneous and have a big influence on the classification outcome. In practice, it will always be difficult to represent a high-dimensional data set using just four to six empirical features.

Table 2.2   Activity recognition results (%)

Feature+classifier	Trained on 20%	Trained on 40%	Trained on 60%
PCA+SVM	32.9	57.0	60.0
PCA+SRC	82.0	81.0	88.0

Table 2.3   Confusion matrix of SVM method, trained on 20% data

	M1	M2	M3	M4	M5	M6
M1	0.0	0.0	0.0	0.0	100.0	0.0
M2	0.0	0.0	0.0	0.0	100.0	0.0
M3	0.0	0.0	0.0	0.0	100.0	0.0
M4	0.0	0.0	0.0	0.0	100.0	0.0
M5	0.0	0.0	0.0	0.0	100.0	0.0
M6	0.0	0.0	0.0	0.0	100.0	0.0

Table 2.4   Confusion matrix of SRC method, trained on 20% data

	M1	M2	M3	M4	M5	M6
M1	100.0	0.0	0.0	0.0	0.0	0.0
M2	18.8	75.0	6.2	0.0	0.0	0.0
M3	50.0	6.3	43.8	0.0	0.0	0.0
M4	6.5	0.0	4.3	89.1	0.0	0.0
M5	0.0	0.0	0.0	2.0	96.0	2.0
M6	12.5	0.0	0.0	0.0	75.0	12.5

Activity recognition results

In Table 2.2, we can see that, compared with the widely utilised SVM method, SRC with PCA features outperforms SVM on average 28% in the recognition results.

We also provide confusion matrix related to these methods in Table 2.3, and due to the space limitations, we only report the confusion matrix using 20% training data. In Table 2.3, SVM with PCA feature achieves the worst results where all test data are assigned to class 6, which gives rise to the lowest recognition results. Through observations of the confusion matrices from Table 2.4, SRC is mostly confused by μD among [M1, M2, M3] and between [M5, M6]. These can be explained by their visual similarities in Figure 2.4(a)–(c) and between (e) and (f) of Figure 2.4. Specifically, μD in both M1 and M2 transverses from positive to negative and with a very similar maximum Doppler frequency. In addition, μDs in M5 and M6 are similar to their Doppler periods, both of which are 1 s, and the maximum frequencies are around 3 Hz.

2.3.2.5 Summary

This section introduces the pipeline for PWR-based μD signature classification, including the signal model and processing, μD feature extraction and classification design. Through experimental data validation, it has been shown that the SRC classifier with the PCA feature extraction achieves better results compared with the conventional distance-based classifiers for activity monitoring. The laboratory experiments show the potential for this approach in a healthcare environment. The next section extends the conventional μD classification to be applied to the data obtained using Wi-Fi CSI.

2.3.3 Using Wi-Fi CSI informatics to monitor human activities

In the previous section, PWR and μD applications of PWR were introduced. The passive Wi-Fi μD data capture, feature analysis and classification mechanisms were discussed in detail with the support of experimental data in residential healthcare and laboratory environments. Previous studies have shown that conventional passive Doppler radars can be a candidate for monitoring daily activity, emergent events and vital signs [94] in residential or clinical healthcare scenarios. In this section, we extend the boundary of classic passive radars to a more generalised level by learning from the channel estimation (CE) and CSI concepts in wireless communication research. The techniques of CE and CSI, representation, data capture methods and utilisation of wireless CSI for activity monitoring are described in this section.

2.3.3.1 From micro-Doppler (μD) to channel state information (CSI)

Classic passive radar

A classic passive radar relies on reference and surveillance signals which represent reference and surveillance channel respectively. Usually, we assume that the reference channel is the direct path between the wireless signal source and reference antenna while the surveillance channel focuses on the path which contains the targets. By applying the CAF, (2.3) in Section 2.3.2, the time and frequency differences can be easily visualised on a range-Doppler plot. In addition, the signal bandwidth determines the time (range) resolution and the integration time determines the frequency (Doppler) resolution. In order to detect the target and describe the motion state of the target as accurately as possible in the classic passive context, we expect that the reference channel is an ideal direct path between signal source and receiving antenna without reflections from target and other moving or static objects, while the surveillance channel contains as much target reflection as possible without DSI components and other object reflections which may mask the target returns.

Emerging applications like human activity monitoring, gaming and indoor target tracking often take place in indoor environments. In these scenarios, ideal reference and surveillance signals are difficult to be achieved due to the probable high density of objects in the surveillance area, interference from devices in the same spectrum and the high dynamic range of the target motion. From the application aspect, the system is often required to have very high range (sub-metre level) and Doppler resolution (1–3 Hz) to accurately represent the body gesture, limb movement and displacement.

However, wideband signals are rarely present in the common residential situations. PWR using, for example 802.11 signals often works with a 20 or 40 MHz bandwidth (corresponding to 7.5 and 3.75 m range resolution) which is difficult to be used for fine-grained movement detection or accurate indoor location tracking. The system also needs 0.5–1 s integration time which is equivalent to 20 or 40 million samples for cross-correlation and FFT calculation ((2.1) in Section 2.3.2) to achieve the high Doppler resolution required for gesture recognition. The performance and application scenarios of standard Wi-Fi passive radars are therefore limited for the detection and classification of small movement such as gestures or signs of life. Therefore there has been growing interest in utilising CSI for this type of application.

Channel state information
In contrast to standard Wi-Fi passive radars that use the directly reflected RF signal data, CSI-based approaches try to tackle the problem by interpreting how the moving human target impacts the RF signal propagation environments. In wireless communications, CSI refers to the individual channel properties which describe the propagation environment from transmitter to receiver [95]. Usually, the CSI effect can be modelled by the following equation:

$$Y = \mathbf{H} \circ X + N \tag{2.8}$$

where X with size $1 \times N$ and Y with size $M \times N$ are transmitting and receiving signals in frequency domain, N with size $M \times N$ denotes the noise vector which is often considered as additive white Gaussian noise, while $\mathbf{H}$ with size $M \times N$ is the CSI matrix. Here operation 'o' denotes element-wise multiplication. In the context of this section, N denotes the number of the subcarriers, and M denotes the number of receiving antennas. The CSI data symbolise scattering, fading and power decay effects on the wireless signal in a certain propagation environments. The parameters used to describe the channel properties are normally the number of paths and attenuation factor, time delay, frequency shift and angle of arrival of each path. The method to obtain the estimated version $\widehat{\mathbf{H}}$ is called CE. To estimate $\widehat{\mathbf{H}}$, researchers have developed blind estimation and pilot-based estimation methods [96]. Pilot-based CSI estimation is widely used in 802.11 [97] and 3GPP standards [98] based mobile system like 4G, LTE/LTE-A and 5G. Pilot-based CE uses a common method in which the transmitter sends a predefined pilot sequence which is known by both transmitter and receiver. The pilot is often located in the front of the data frame or distributed in the data symbols. The receiver estimates the CSI according to the post-propagation pilot sequence after sample level synchronisation. The detailed estimation methods will vary substantially due to the high diversity of wireless communications signals [96]. We will introduce OFDM signal-based CSI in detail in the next section. The format, accuracy and embodiment of $\widehat{\mathbf{H}}$ vary according to signal format, propagation media and requirement of applications. In wireless communications systems, the estimated CSI is often used for equalising the adverse channel effect on the wireless signal. CSI is the key information for high-quality wireless communications physical layer processing such as IQ data reception, MIMO detection, demodulation and channel decoding in the fading wireless channel. Usually the CSI parameters are only used for bottom layer signal processing and are not open to the end user. The most common CSI parameters that can be observed from the application level are the RSS indicator, an indirect measure of signal attenuation. In the next section, we will discuss methods of obtaining the CSI based on the Wi-Fi signal.

It is not difficult to see that CSI data will involve a group of time-varying parameters if there are moving objects in the wireless propagation environment. So, if there is a moving human in the area of interesting, the human activity will be embodied in the time-varying parameters of CSI. Thus, human activities can be recognised by interpreting the time-varying characteristics of estimated CSI parameters. CSI is usually monitored to combat adverse channel effects in wireless communications. In the Wi-Fi passive radar scenario it can be used for understanding the characteristics of both static and dynamic objects in the surveillance area. The use of CSI data is therefore potentially an important extra method of activity monitoring in the healthcare area.

Classic passive radar requires both a reference and surveillance antenna. CSI-based human activity detection/sensing system can just use the antenna or antennas which are already on the Wi-Fi receiver to monitor all the possible wireless signals, rather than trying to differentiate between reference and surveillance channel signals. A CSI-based passive detection/sensing system therefore does not require either a two-channel receiver or dedicated IQ signal processing. It can also largely use the existing Wi-Fi chipsets with corresponding divers. In practice, the known pilot sequence in the receiver effectively acts as the 'reference signal' in a CSI-based human activity detection/sensing system.

2.3.3.2 Wi-Fi CSI signal capture and parameter extraction

Signal capture

In theory, any wireless signal such as Bluetooth, ZigBee, mobile as well as Wi-Fi could be captured and analysed for CSI data. In this chapter, we use the Wi-Fi 802.11 signal as an example to illustrate wireless CSI since this is one of the most widely deployed and heavily used indoor wireless communication systems. In addition, the long duration bursts, multiple carrier frequencies and multiple transmit antennas normally in range make it possible to break through the conventional bandwidth limited range resolution problem. It also enables the extraction of angle information using Wi-Fi CSI data.

Considering the multiple carriers and multiple antennas available, $\mathbf{H}$ in (2.8) can be rewritten as the following format:

$$\mathbf{H} = [\boldsymbol{H}_1; \quad \boldsymbol{H}_2; \quad \ldots; \quad \boldsymbol{H}_m] \tag{2.9}$$

where, for convenience, we use the frequency domain form, assume that $\boldsymbol{H}_m$ is the CSI vector on the mth receiving antenna and that it can be extended in the following way:

$$\boldsymbol{H}_m = \begin{bmatrix} h_{1,m} & h_{2,m} & \cdots & h_{n,m} \end{bmatrix} \tag{2.10}$$

where $h_{n,m}$ is the value on the mth receiving antenna and nth subcarrier, which is complex and the result of multipath effect.

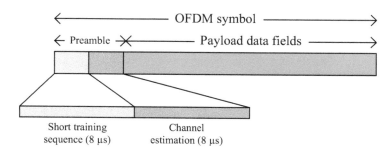

Figure 2.5   Preamble sequence in 802.11 standard signal

Due to multipath propagation paths in the indoor environment, each CSI value $h_{n,m}$ can be considered as the superimposed result of the signal from different paths:

$$h_{n,m} = \sum_{k=1}^{K} \alpha_k \exp\left(-j2\pi(f_n + \Delta f_k)\right)\exp\left(-j2\pi(f_n \cdot \tau_k)\right) \tag{2.11}$$

where α_k, Δf_k and τ_k are the attenuation factor, frequency shift and time delay of kth path. f_n is the frequency point of the nth subcarrier. The term $\exp\left(-j2\pi(f_n \cdot \tau_k)\right)$ is the phase shift on nth subcarrier caused by path delay τ_k. From (2.8) and (2.11), we can see that four factors – attenuation, frequency shift, time delay and direction of the signal – determine the CSI matrix **H**. In order to estimate the CSI matrix **H**, the 802.11 standards use a pilot sequence (labelled as a preamble sequence in Figure 2.5) to implement CE. CE uses the second half of the preamble sequence which has to be implemented after sample level synchronisation. This is aided by the first half of the preamble which is defined as a training sequence. We will not go into frame synchronisation in detail as it is out of scope of this chapter. It should also be noted that the preamble sequence length may vary according to the specific 802.11 version being utilised. In this chapter, we only introduce the main principles and do not discuss these detailed differences.

Here, we use S and R to denote the original and receive CE sequences in the preamble. If we assume that the CE is taking place in the frequency domain, the estimation of each CSI value on the mth antenna H_m can be simply written as

$$\widehat{H}_m = \frac{\mathcal{F}(R)}{\mathcal{F}(S)} \tag{2.12}$$

where $\mathcal{F}(\cdot)$ is the Fourier transformation. Estimation methods for CSI **H** are a well-established research area in wireless communications, but this is not the focus of this chapter. CE methods for 802.11 systems are comprehensively described using the metric of minimum mean square error in [97]. In practice, researchers can also use dedicated devices or commodity tools like SDR for high accuracy CE. As CE is the basis of much physical layer processing, most commodity 802.11 chipsets carry this out at the ASIC level; however, **H** is often considered as an intermediate processing result and is not open to end users. However, some researchers have recently developed special drivers for Qualcomm Atheros QCA9558 [99] and Intel 5300 80211n MIMO [100] chipsets, respectively, to export estimated CSI to end users for further applications. In [99], $P \times M \times 56$ CSI is the output for a 20 MHz system and $P \times M \times 114$ is the output for a 40 MHz system, where P is the number of transmitting antennas, and M is the number of the receiving antennas. The numbers of subcarriers are 56 and 114. Reference [100] exports 28 CSI frequency groups for bandwidth, which is about 1 group for every 2 subcarriers at 20 MHz or 1 in 4 at 40 MHz. For a multiple-antenna receiver, the estimated CSI can be denoted as

$$
\mathbf{H} = \begin{bmatrix} h_{1,1} & h_{1,2} & \cdots & h_{1,M} \\ h_{2,1} & h_{2,2} & \cdots & h_{2,M} \\ \vdots & \vdots & \ddots & \vdots \\ h_{N,1} & h_{N,2} & \cdots & h_{N,M} \end{bmatrix}
\tag{2.13}
$$

Each entry $h_{n,m}$ is a complex number representing the channel effect described in (2.11). These CSI tools make the use of software-based processing possible.

Parameter extraction

From Section 2.3.3.1, we know that the CSI '**H**' is the representation of the wireless signal propagation properties which can be described by attenuation, delay, frequency offset and direction. Among these parameters, attenuation can be easily read from most wireless network interface card (NIC) drivers. In this section, we introduce how the other three parameters are extracted from **H**.

The multiple-antenna configuration of the recent Wi-Fi devices provides an opportunity to determine the direction of the reflecting object by using the multiple-signal classification (MUSIC) algorithm [101]. If we take any row from **H** and make as $\mathbf{H}_{:,n}$, define the auto-correlation matrix as

$$
R_{H_n H_n} = \mathbb{E}[\mathbf{H}_{:,n} \mathbf{H}_{:,n}^*]
\tag{2.14}
$$

where $*$ denotes the conjugated transpose, $\mathbb{E}[\cdot]$ denotes the statistical expectation. Then, if we take the eigenvalue decomposition of $R_{H_n H_n}$ and we sort the eigenvalue according to descending order $[\lambda_1, \lambda_2, \ldots, \lambda_M]$, then, the corresponding eigenvectors can be written as $[e_1; e_2; \ldots; e_M]$. We define the top D eigenvectors as the signal subspace $\mathbf{E}_{nS} = [e_1; e_2; \ldots; e_D]$, the rest of the eigenvectors are defined as noise subspace $\mathbf{E}_{nN} = [e_{D+1}; e_{D+2}; \ldots; e_M]$. We can further define the receiving steering vector as $\mathbf{a}(\theta) = [a_1(\theta), a_2(\theta), \ldots, a_M(\theta)]^T$, where $a_M(\theta)$ is determined by the geometry of the receiving array. For example, $a_M(\theta) = \exp(j2\pi(d/\lambda) \cdot \sin(\theta))$ in a uniform linear array (ULA) and $a_M(\theta) = \exp(-j(2\pi r/\lambda)\cos(\theta - (2\pi m/M)))$ in a uniform circular array (UCA), where d is the interval between ULA antennas, r is the radius of the UCA, and λ is the wavelength of the signal. The combined directions of the objects are the potentials θs that make the target function larger:

$$
P(\theta) = \frac{\mathbf{a}^*(\theta)\mathbf{a}(\theta)}{\mathbf{a}^*(\theta)\mathbf{E}_{nN}\mathbf{E}_{nN}^*\mathbf{a}(\theta)}
\tag{2.15}
$$

Similar subspace-based MUSIC methods can also be applied to the frequency domain signal for delay estimation. If we take any column from $\mathbf{H}$ and call this $\mathbf{H}_{m,:}$, then the frequency auto-correlation matrix $R_{H_m H_m}$ can be defined similarly to (2.14). By taking the eigenvalue decomposition on the $R_{H_m H_m}$, signal subspace and noise subspace based on the frequency domain signal can be found and denoted by E_{mS} and E_{mN}. The frequency domain steering vector (also called the manifold vector) is defined as $g_n(\tau) = [1, e^{-j2\pi f_1 \tau}, \ldots, e^{-j2\pi f_N \tau}]$, where τ is the delay of the signal, and f_N is the frequency point of Nth subcarrier. The time delay of signal propagation can be identified by looking for the τ value which makes target function larger:

$$P(\tau) = \frac{g^*(\tau)g(\tau)}{g^*(\tau)E_{mN}E_{mN}^*g(\tau)} \tag{2.16}$$

The subspace method can also be extended to a 2D search for joint angle and delay estimation (JADE) [102]. In this case, the steering vector is redefined as $u(\theta, \tau) = a(\theta) \otimes g(\tau)$, where $\otimes$ represents the Kronecker product. Correspondingly, we can apply the Kronecker transformation of $\mathbf{H}$. Following similar steps, a 2D target function can be obtained for JADE:

$$P(\theta, \tau) = \frac{u^*(\theta, \tau)u(\theta, \tau)}{u^*(\theta, \tau)E_N E_N^* u(\theta, \tau)} \tag{2.17}$$

The subspace-based method from (2.14) to (2.17) is not the only method that can be used for estimating the delay and direction of wireless signal. Another important estimation category is maximum a posteriori estimation. The most cited algorithm in this category is space-alternating generalised EM [103]. The EM-based methods have been shown to provide a high-joint resolution of the complex amplitude, delay, Doppler shift and direction of the impinging signals. Besides [103,104] introduced discrete Fourier transform of Wi-Fi CSI for Doppler information extraction, [105] patented a conjugation multiplication of Wi-Fi CSI from two antennas for Doppler shift estimation and [106] proposed CSI-speed model to reveal the relation between a moving target and Wi-Fi CSI.

Due to the moving nature of the human targets and their impact on the signal propagation environment, the estimated CSI and secondary parameters will present time-varying features. Establishing the correlation relationships between the time-varying features and the monitored human activities through a machine provides an avenue for CSI techniques to be a promising tool for activity recognition.

2.3.3.3   Wi-Fi CSI-enabled healthcare applications

With the continuing rapid roll-out of Wi-Fi-enabled devices and the release of CSI tools [99,100], researchers have started intensively investigating the applications of Wi-Fi CSI beyond the traditional communications environment. From 2014, a large range of CSI-based human activity/gesture research has been conducted and this has now become a well-established research topic. Researchers have demonstrated a wide range of motion detection and recognition capability via a variety of approaches which can be found in some recent review and survey papers [107–109]. Human motions detected range from visible whole-body movement to inconspicuous movements such as respiration or heartbeat. This opens up a range of promising applications in residential or clinical healthcare.

Gesture and activity recognition using Wi-Fi CSI

As described in Section 2.3.3.2, it is now possible to acquire high-quality CSI data from commodity NICs like the Intel 5300 or Qualcomm Atheros QCA9558, as well as secondary parameters such as delay, direction, Doppler and attenuation profiles. The work in [110] and QGesture in [111] carefully investigated the details of the phase-rotating pattern along antennas and subcarriers and traced these back to the causes of this phase rotation in the spatial and frequency domains. This was related to the direction and distance of the reflecting objects. This then enabled the development of direction and distance-based motion detection and gesture recognition system using traditional discrimination and classification methods. CARM in [106,112] and [113] discovered periodical phase change caused by dynamic human movement. Similar to μD signature analysis in radars, CARM also leverages imaging-processing techniques to recognise time-Doppler patterns. With the increasing amount of research in this field, researchers [114] have also tried to use model-based methods to replace pure pattern-based techniques for a more theoretical and tractable approach. These studies proved that secondary parameters-based methods work well. However, the complications of the indoor propagation environment including clutter, interference, the uncontrollable geometry of human motion passive sensing and the high dynamic range of human activity introduce many uncertainties that impact the recognition performance. In these situations it is very difficult or not possible to use traditional feature-based classification methods for activity recognition. To address this issue, there has been a shift towards the application of ML techniques such as NNs and deep learning in [115–117]. Particularly, the multi-link data fusion method in [118] extracted environment-independent features and removed subject-specific information dependence from the recognition processing.

Gait awareness and fall detection

Most of the research discussed above is focused on resolving the recognition problem for general activities in a universal environment. This is a challenging task even using CSI data because of the large amount of uncertainty in an uncontrollable environment. However, if we scale down to concentrate on specific features of Wi-Fi CSI and/or a limited range of human activities, there is more likelihood of achieving reliable performance. Two typical specific healthcare-related activities on this track are gait awareness and fall detection. In [119], Doppler information extracted from Wi-Fi CSI was used to analyse the correlation between the periodical Doppler and motion parameters such as torso speed, footstep size and leg speed. Using fine-grained features, the authors showed comparatively robust human identification performance by using an SVM. This method shows promise for monitoring a number of important healthcare-related movements such as freeze of gait associated with Parkinson's disease. Specific activities or events may indicate an immediate emergency situation, like falling. Fall detection has therefore become an important topic and has become a unique genre of CSI-based healthcare applications. In order to differentiate falling from other activities some researchers [120,121] have used phase as the base parameter for segmentation, assisted by the time–frequency declining pattern to discriminate fall and fall-like activities. In [122], a one-class SVM followed by a random forest is applied to a group of selected CSI features, such as normalised STD, RSS offset, motion period and median absolute deviation for very high accuracy fall detection.

Inconspicuous movement detection and applications
Apart from whole-body movements, which are obviously visible, some inconspicuous movements like chest expansion and contraction and even movements inside the body such as heartbeat can also be observable indirectly from the Wi-Fi CSI data. In order to visualise the small CSI value variance caused by these very small movements, researchers need to choose the proper secondary parameters to use which are most sensitive to the movement being monitored. The other key requirement is to mitigate the effect of returns from other movements. In both BreathTrack [123] and TinySense [124], the periodical change of phase is used as the main parameter to identify the observed respiration. In these studies the authors also report leveraging time delay and angle information to eliminate multipath effects. Using a similar approach, other studies have shown heart rate detection in addition to respiration [125,126]. It is also significant that other papers [124,126–128] discovered some beneficial effects from monitoring in the Fresnel zone of the radio signal propagation. A strategy to increase the density of the Fresnel zone to improve the sensitivity of the detection has been proposed in [124].

Use of Wi-Fi CSI-based device-free passive respiration and heartbeat detection in daily life therefore currently seems promising; however, various challenges still need to be overcome for the technique to move into real-world applications. For example, limitations are associated with the high dynamic range of any torso movements and body rotations. Though researchers in [129] proposed Hempel filtering to eliminate this effect, the reliability of the results in these high dynamic motion situations is still uncertain. Monitoring vital signs in non-dynamic applications will obviously have less problems with these issues. Sleep monitoring is one such application which has obvious advantages compared to the traditional polysomnography approach. A recent study [130] proposed an energy feature-based Gaussian mixture model for monitoring foreground sleeping motion like rollover using Wi-Fi CSI. Simultaneous respiration and posture monitoring is claimed in [127,131] even with the presence of multiple people. There have also been investigations into an extension of sleep monitoring into human emotion evaluation based on detecting physical expressions of emotion. EmoSene in [128] takes the advantage of Fresnel zone and similar features to those discussed in [122] to identify emotions like happiness, sadness, anger and fear, even gender-related features.

Future trends for CSI-based monitoring
Recent studies on Wi-Fi CSI-based passive sensing have shown that a wide range of human motions and activities can be detected and recognised from a commodity Wi-Fi

device using fine-grained CSI information. Many of these activities are important in healthcare applications. However, this is still an emerging research area and there is still plenty of scope for further development to move this research on to practical and innovative healthcare applications. There are a number of likely future trends with some of the key ones expected to be as follows:

- *Data fusion:* Fusion of Wi-Fi CSI data with other types of sensor data such as video, accelerometers and inertial [132,133].
- *Life vitality*: Wi-Fi CSI can also be used as a high-level health condition indicator by monitoring activity level, life style and quality [134–137].
- *Wearable RF devices:* There is increasing work on the design of wearable Wi-Fi devices for RF sensing of activities [138,139].
- *Multiple sources:* There is interest in investigating the benefits and drawbacks of using multiple signal sources in a typical indoor complex wireless environment.
- *Multi-person sensing:* Some studies are now being carried out on expanding Wi-Fi CSI sensing to include the capability of monitoring multiple people in the surveillance area [140].

Apart from the above trends, it is important to note that most of the current Wi-Fi CSI research is based on data collected in comparatively small-scale data gathering in controlled environments. Extension of these methods to large scale of data collection in diverse environments may be important for deploying this technology widely in healthcare applications. This poses a significant challenge if a wide range of practical healthcare applications are to be developed. The use of NNs and deep learning-based methods may be crucial in enabling the activity detection performance to be sufficiently robust for the future healthcare market.

2.4 The future for passive radar in healthcare

The increasing demand for AAL technologies which are now emerging in e-healthcare, coupled with significant changes in the urban wireless landscape, presents a number of opportunities for passive radar technology to evolve as the healthcare sensor of choice. However, there are also a number of inhibitors which should be mitigated in the system design and deployment if passive radar is to generate any commercial traction as a medical technology.

As described previously, the IoT is emerging as the next step-change in the evolution of the internet having the potential to connect as many as 20 billion smart devices by 2020 [141]. An IoT vertical seeing significant growth within the connected landscape is Smart Homes, and the widespread uptake of wireless technologies in home ranging from smart TVs and voice assistants to connected lights and locks has resulted in a high density of RF signals which could be used as the basis for passive sensor networks. Wi-Fi would be the current frontrunner for such a passive sensor network given its ubiquitous and pervasive nature, though short-range protocols such as Zig-Bee and Bluetooth would also offer feasible solution within the passive radar context. Moreover, the emergence of the latest generation of both Wi-Fi5 and 5G technologies to support the demand for higher data rates with broader coverage and lower costs and latency offers possibilities for passive radar systems to operate with higher spatial resolutions of the order of image quality, and with improved contextual awareness, which could lead to more accurate prognosis and earlier diagnosis of diseases. Furthermore, a passive radar is seen as attractive in these types of scenarios compared to their counterpart active radar systems as the 'green' nature of the technology means that it does not contribute to the already-congested RF spectrum, nor would it be subject to the high costs and in-depth due diligence associated with obtaining a spectrum license to transmit signals.

A number of AAL services such as CanaryCare and Hive Link in the United Kingdom [142] which accumulate and analyse the 'big' data sets generated by home IoT devices and camera systems are now appearing as commercial offerings to provide pattern-of-life insights and anomaly alerts for healthcare professionals. Though these AAL services are already demonstrating to be popular, this level of fine-grain monitoring does raise a number of user privacy and ethical issues [143,144] so any possible future integration of personalised information into such services must be carefully considered as research has shown the importance of public acceptance to ensure uptake of new and potentially controversial technologies [145]. Additionally, as new technological capabilities begin to generate new types of data sets within homes that identify people and make inferences about their intimate personal habits, the actual real-world deployments of such systems will face barriers from increasing government regulation focusing on data privacy, such as GDPR. As described previously however, passive radar systems are significantly less exposed to these types risks and could therefore significantly enhance the capabilities of existing AAL services by providing location and activity information, without generating any type of identifiable information such as direct images (which are produced by optical camera systems) or through metadata, for example where a cooperative device that is being tracked has embedded details of the user. In addition, the cost of integration would be minimal in terms of the minimal hardware requirements necessary to deploy a system and the reduced level of regulatory approvals associated with using already-existing signal transmissions.

2.5   Conclusions

Driven by rising life expectancies across both developed and developing countries, the world population is currently at its highest level in human history [146], and the increasing average age of the population is bringing with it significant implications globally; from financial impacts on interest rates and land prices, to education provision. In healthcare especially, an ageing population can put significant strains on medical and care services such as the increased costs of having to treat conditions which typically occur in later life, and the high number of regular touch-points required to manage elderly care by professionals. Technological solutions are however seen as a way to circumvent many of the issues currently experienced within a nations healthcare system, and passive radar is one such technology that has the potential to offer benefits in e-Healthcare and AAL with a global reach. In this chapter, the authors have provided insights into the various approaches, techniques and performance metrics which are emerging in research and are paving the way for passive radars to potentially become a medical device for wide scale and unobtrusive remote monitoring of patients.

The growth of communication infrastructures within the urban landscape has resulted in a plethora of illuminators of opportunity available for passive radar (GPS, FM, DAB, TV, etc.). However, the pervasive and ubiquitous nature of Wi-Fi communications in all walks of life, especially within homes, makes Wi-Fi APs the most appropriate and widely used transmitter for exploitation and is expected to continue to be as the standard evolves to accommodate higher data rate communications (e.g. high-definition video streaming), IoT drivers and the increasing number of mobile devices such as smart phones. Moreover, Wi-Fi signal characteristics, such as its high bandwidth OFDM modulation and constant beacon signal transmissions, make it an attractive solution for in-home remote monitoring.

Activity recognition can provide valuable information for disease diagnosis and prognosis, and in Section 2.3 we demonstrate time–frequency signal-processing techniques for extracting μD components associated with small body movements which make up larger actions. Furthermore, we show how ML methodologies can be employed to make sense of full μD signatures corresponding to various activities ranging from picking up an object and falling over, to sitting and standing from a chair. Possible applications here to generate emergency alarms, for example, after a patient trips over and also to provide early indications of failing health from anomalies in life patterns or changes in physiology (i.e. a patients begins to limp) would be invaluable in a healthcare setting. Moreover, the ability of passive radar to create this level of context around a recognisable action, whilst also maintaining user anonymity gives it a unique selling point compared to competitor technologies such as video. Although other technologies such as wearables could also offer these capabilities without identifying patients, they do suffer from low levels of compliance, especially amongst the elderly. The application of ML for classifying human μD signatures is a noteworthy innovation in the field of radar. However, unlike in the field of machine vision which has access to large databases of labelled training images, RF radar μD data sets are few and far-between. The lack of large volumes of training data – known as the cold start problem – needs to be addressed to allow significant advancement of this emerging area of radar research.

Finally, this chapter describes advancements in exploiting the CSI which is a measure employed in the field of communications that provides information about the channel properties which describe the propagation environment between a radio transmitter and receiver. It describes how the analysis and extraction of information from CSI models can further enhance the flow of information into a classifier to determine activities, including vital signs such as breathing, and even gesture recognition which offers possibilities for more immersive 2-way interactions between patients and their assistive technologies, which could ensure higher user uptake. Additionally, CSI measurements from more developed passive radar hardware such as receiver antenna arrays enables subspace-based direction finding approaches that allow joint estimate of both angular bearing and range, again adding context through location and providing new feature sets for ML techniques. The increasing availability of off-the-shelf radar hardware, particularly though software-defined systems where radio functions are now performed in software, is facilitating the evolution of passive radar technology from lab-based systems into real-world commercial products.

References

[1]   O'Brien, A., McDaid, K., Loane, J., Doyle, J., and O'Mullane, B., 'Visualisation of movement of older adults within their homes based on PIR sensor data', *6th International Conference on Pervasive Computing Technologies for Healthcare (PervasiveHealth) and Workshops*, San Diego, USA, May 2012. pp. 252–259.

[2]   Popescu, M., Hotrabhavananda, B., Moore, M., and Skubic, M., 'VAMPIR – An automatic fall detection system using a vertical PIR sensor array', *6th International Conference on Pervasive Computing Technologies for Healthcare (PervasiveHealth) and Workshops*, San Diego, USA, May 2012. pp. 163–166.

[3]   Wang, Y., Shuang, C., and Yu, H., 'A noncontact-sensor surveillance system towards assisting independent living for older people', *23rd International Conference on Automation and Computing (ICAC)*, Columbus, USA, July 2017, pp. 1–5.

[4]   Yazar, A., and Cetin, E., 'Ambient assisted smart home design using vibration and PIR sensors', *21st Signal Processing and Communications Applications Conference (SIU)*, Haspolat, Turkey, 24–26 April 2013, pp. 1–4.

[5]   Guan, Q., Yin, X., Guo, X., and Wang, G., 'A Novel Infrared Motion Sensing System for Compressive Classification of Physical Activity', *IEEE Sensors Journal*, 2016;**16**(8):2251–2259.

[6]   Yao, B., Hagras, H., Alghazzawi, D., and Alhaddad, M.J., 'A Big Bang–Big Crunch Type-2 Fuzzy Logic System for Machine-Vision-Based Event Detection and Summarization in Real-World Ambient-Assisted Living', *IEEE Transactions on Fuzzy Systems*, 2016;**24**(6):1307–1319.

[7]   Richter, J., Findeisen, M., and Hirtz, G., 'Assessment and care system based on people detection for elderly suffering from dementia', *2015 IEEE 5th International Conference on Consumer Electronics – Berlin (ICCE-Berlin)*, Berlin, Germany, 2015, pp. 59–63.

[8]   Liciotti, D., Massi, G., Frontoni, E., Mancini, A., and Zingaretti, P., 'Human activity analysis for in-home fall risk assessment', *2015 IEEE International Conference on Communication Workshop (ICCW)*, London, UK, 2015, pp. 284–289.

[9]   Amine Elforaici, M.E., Chaaraoui, I., Bouachir, W., Ouakrim, Y., and Mezghani, N., 'Posture recognition using an RGB-D camera: Exploring 3D

body modeling and deep learning approaches', *2018 IEEE Life Sciences Conference (LSC)*, Montreal, Canada, 2018, pp. 69–72.

[10] Solbach, M.D., and Tsotsos, J.K., 'Vision-based fallen person detection for the elderly', *2017 IEEE International Conference on Computer Vision Workshops (ICCVW)*, Venice, Italy, 2017, pp. 1433–1442.

[11] Ezatzadeh, S., and Keyvanpour, M.R., 'Fall detection for elderly in assisted environments: Video surveillance systems and challenges', *2017 9th International Conference on Information and Knowledge Technology (IKT)*, Tehran, Iran, 2017, pp. 93–98.

[12] Mirjalol, S., and Whangbo, T.K., 'An authentication protocol for smartphone integrated Ambient Assisted Living system', *2018 International Conference on Information and Communication Technology Convergence (ICTC)*, Jeju, 2018, pp. 424–428.

[13] Ballı, S., Sagbaş, E.A., and Korukoglu, S., 'Design of smartwatch-assisted fall detection system via smartphone', *26th Signal Processing and Communications Applications Conference (SIU)*, Izmir, Turkey, 2018, pp. 1–4.

[14] Menhour, I., Abidine, M.B., and Fergani, B., 'A new framework using PCA, LDA and KNN-SVM to activity recognition based smartphone's sensors', *6th International Conference on Multimedia Computing and Systems (ICMCS)*, Rabat, Israel, 2018, pp. 1–5.

[15] Zainudin, M.N.S., Sulaiman, M.N., Mustapha, N., and Perumal, T., 'Monitoring daily fitness activity using accelerometer sensor fusion', *2017 IEEE International Symposium on Consumer Electronics (ISCE)*, Kuala Lumpur, Malaysia, 2017, pp. 35–36.

[16] Boateng, G., Batsis, J.A., Proctor, P., Halter, R., and Kotz, D., 'GeriActive: Wearable app for monitoring and encouraging physical activity among older adults', *2018 IEEE 15th International Conference on Wearable and Implantable Body Sensor Networks (BSN)*, Las Vegas, USA, 2018, pp. 46–49.

[17] Saadeh, W., Altaf, M.A.B., and Altaf, M.S.B., 'A high accuracy and low latency patient-specific wearable fall detection system', *2017 IEEE EMBS International Conference on Biomedical & Health Informatics (BHI)*, Orlando, USA, 2017, pp. 441–444.

[18] Yao, W., Chu, C., and Li, Z., 'The use of RFID in healthcare: Benefits and barriers', *2010 IEEE International Conference on RFID-Technology and Applications*, Guangzhou, China, 2010, pp. 128–134.

[19] Shinmoto Torres, R.L., Ranasinghe, D.C., Shi, Q., and Sample, A.P., 'Sensor enabled wearable RFID technology for mitigating the risk of falls near beds', *2013 IEEE International Conference on RFID (RFID)*, Penang, 2013, pp. 191–198.

[20] Wang, P., Chen, C., and Chuan, C., 'Location-aware fall detection system for dementia care on nursing service in Evergreen Inn of Jianan Hospital', *2016 IEEE 16th International Conference on Bioinformatics and Bioengineering (BIBE)*, Taichung, 2016, pp. 309–315.

[21]   Al Rasyid, M.U.H., Uzzin Nadhori, I., Sudarsono, A., and Luberski, R., 'Analysis of slotted and unslotted CSMA/CA Wireless Sensor Network for E-healthcare system', *2014 International Conference on Computer, Control, Informatics and Its Applications (IC3INA)*, Bandung, Indonesia, 2014, pp. 53–57.

[22]   Büsching, F., Bottazzi, M., Pöttner, W., and Wolf, L., 'DT-WBAN: Disruption tolerant wireless body area networks in healthcare applications', *2013 IEEE 9th International Conference on Wireless and Mobile Computing, Networking and Communications (WiMob)*, Lyon, France, 2013, pp. 196–203.

[23]   Sarkar, S., and Misra, S., 'From Micro to Nano: The Evolution of Wireless Sensor-Based Health Care', *IEEE Pulse*, 2016;**7**(1):21–25.

[24]   Vora, J., Tanwar, S., Tyagi, S., Kumar, N., and Rodrigues, J.J.P.C., 'FAAL: Fog computing-based patient monitoring system for ambient assisted living', *2017 IEEE 19th International Conference on e-Health Networking, Applications and Services (Healthcom)*, Dalian, China, 2017, pp. 1–6.

[25]   Yamazaki, T., 'The Ubiquitous Home', *International Journal of Smart Home*, 2007;**1**:17–22.

[26]   Chan, M., Estève, D., Escriba, C., and Campo, E., 'A Review of Smart Homes—Present State and Future Challenges', *Computer Methods and Programs in Biomedicine*, 2008;**91**:55–81.

[27]   Kalimeri, K., Matic, A., and Cappelletti, A., 'RFID: Recognizing failures in dressing activity', *2010 4th International Conference on Pervasive Computing Technologies for Healthcare*, Munich, Germany, 2010, pp. 1–4.

[28]   Chen, V.C., *The micro-Doppler effect in radar*, Boston: Artech House;2011.

[29]   Seifert, A., Amin, M.G., and Zoubir, A.M., 'New analysis of radar micro-Doppler gait signatures for rehabilitation and assisted living', *2017 IEEE International Conference on Acoustics, Speech and Signal Processing (ICASSP)*, New Orleans, USA, 2017, pp. 4004–4008.

[30]   Seyfioğlu, M.S., Serinöz, A., Özbayoğlu, A.M., and Gürbüz, S.Z., 'Feature diverse hierarchical classification of human gait with CW radar for assisted living', *International Conference on Radar Systems (Radar 2017)*, Belfast, UK, 2017, pp. 1–5.

[31]   Cippitelli, E., Fioranelli, F., Gambi, E., and Spinsante, S., 'Radar and RGB-Depth Sensors for Fall Detection: A Review', *IEEE Sensors Journal*, 2017;**17**(12):3585–3604.

[32]   Amin, M.G., Zhang, Y.D., Ahmad, F., and Ho, K.C.D., 'Radar Signal Processing for Elderly Fall Detection: The Future for In-Home Monitoring', *IEEE Signal Processing Magazine*, 2016;**33**(2):71–80.

[33]   Diraco, G., Leone, A., and Siciliano, P., 'Detecting falls and vital signs via radar sensing', *2017 IEEE SENSORS*, Glasgow, 2017, pp. 1–3.

[34]   Shyu, K.-K., Chiu, L.-J., Lee, P.-L., Tung, T.-H., and Yang, S.-H., 'Detection of Breathing and Heart Rates in UWB Radar Sensor Data Using FVPIEF-Based Two-Layer EEMD', *IEEE Sensors Journal*, 2019;**19**(2):774–784.

[35] Hasebe, S., Sasakawa, D., Kishimoto, K., and Honma N., 'Simultaneous detection of multiple targets' vital signs using MIMO radar', *2018 International Symposium on Antennas and Propagation (ISAP)*, Busan, South Korea, 2018, pp. 1–2.

[36] Prat, A., Blanch, S., Aguasca, A., Romeu, J., and Broquetas, A., 'Collimated Beam FMCW Radar for Vital Sign Patient Monitoring', *IEEE Transactions on Antennas and Propagation*, 2018;**67**(8):5072–5080.

[37] Wang, K., Zeng, Z., and Sun, J., 'Through-Wall Detection of the Moving Paths and Vital Signs of Human Beings', *IEEE Geoscience and Remote Sensing Letters*, 2019;**16**(5):717–721.

[38] Cho, H.-S., and Park, Y.-J., 'Detection of Heart Rate Through a Wall Using UWB Impulse Radar', *Journal of Healthcare Engineering*, 2018;**2018**:7 pages, Article ID 4832605.

[39] Du, H., He, Y., and Jin, T., 'Transfer learning for human activities classification using micro-Doppler spectrograms', *2018 IEEE International Conference on Computational Electromagnetics (ICCEM)*, Chengdu, China, 2018, pp. 1–3.

[40] He, Y., Yang, Y., Lang, Y., Huang, D., Jing, X., and Hou, C., 'Deep learning based human activity classification in radar micro-Doppler image, *2018 15th European Radar Conference (EuRAD)*, Madrid, Spain, 2018, pp. 230–233.

[41] Chen, Q., Ritchie, M., Liu, Y., Chetty, K., and Woodbridge, K., 'Joint fall and aspect angle recognition using fine-grained micro-Doppler classification', *2017 IEEE Radar Conference (RadarCon)*, Seattle, USA, 2017, pp. 0912–0916.

[42] Rana, S.P., Dey, M., Brown, R., Siddiqui, H.U., and Dudley, S., 'Remote vital sign recognition through machine learning augmented UWB', *12th European Conference on Antennas and Propagation (EuCAP 2018)*, London, UK, 2018, pp. 1–5.

[43] Schreurs, D., Mercuri, M., Soh, P.J., and Vandenbosch, G., 'Radar-based health monitoring', *2013 IEEE MTT-S International Microwave Workshop Series on RF and Wireless Technologies for Biomedical and Healthcare Applications (IMWS-BIO)*, Singapore, 2013, pp. 1–3.

[44] Farina, A., and Kuschel, H., 'Guest Editorial Special Issue on Passive Radar (Part I)', *IEEE Aerospace and Electronic Systems Magazine*, 2012;**27**(10):5.

[45] Griffiths, H.D., and Baker, C.J., 'Passive Coherent Location Radar Systems. Part 1: Performance Prediction', *IEE Proceedings – Radar, Sonar and Navigation*, 2005;**152**(3):153–159.

[46] Krysik, P., Samczynski, P., Malanowski, M., Maslikowski, L., and Kulpa, K., 'Detection of fast maneuvering air targets using GSM based passive radar', *2012 13th International Radar Symposium*, 2012, Warsaw, Poland, pp. 69–72.

[47] Chetty, K., Chen, Q., and Woodbridge, K., 'Train monitoring using GSM-R based passive radar', *2016 IEEE Radar Conference (RadarCon)*, Philadelphia, USA, 2016, pp. 1–4.

[48] Zemmari, R., Daun, M., Feldmann, M., and Nickel, U., 'Maritime surveillance with GSM passive radar: Detection and tracking of small agile targets', *2013 14th International Radar Symposium (IRS)*, Dresden, Germany, 2013, pp. 245–251.

[49] Li, W., Tan, B., and Piechocki, R., 'Opportunistic Doppler-only indoor localization via passive radar', *2018 IEEE 16th Intl Conf on Dependable, Autonomic and Secure Computing, 16th Intl Conf on Pervasive Intelligence and Computing, 4th Intl Conf on Big Data Intelligence and Computing and Cyber Science and Technology Congress (DASC/PiCom/DataCom/Cyber SciTech)*, Athens, 2018, pp. 467–473.

[50] Guo, H., Woodbridge, K., and Baker, C.J., 'Evaluation of WiFi beacon transmissions for wireless based passive radar', *2008 IEEE Radar Conference, Rome*, 2008, pp. 1–6.

[51] Falcone, P., Colone, F., Bongioanni, C., and Lombardo, P., 'Experimental results for OFDM WiFi-based passive bistatic radar', *2010 IEEE Radar Conference*, Washington, DC, 2010, pp. 516–521.

[52] Colone, F., Falcone, P., Bongioanni, C., and Lombardo, P., 'WiFi-Based Passive Bistatic Radar: Data Processing Schemes and Experimental Results', *IEEE Transactions on Aerospace and Electronic Systems*, 2012;**48**(2):1061–1079.

[53] Chetty, K., Smith, G.E., and Woodbridge, K., 'Through-the-Wall Sensing of Personnel Using Passive Bistatic WiFi Radar at Standoff Distances', *IEEE Transactions on Geoscience and Remote Sensing*, 2012;**50**(4):1218–1226.

[54] Domenico, S.D., Sanctis, M.D., Cianca, E., and Ruggieri, M., 'WiFi-Based Through-the-Wall Presence Detection of Stationary and Moving Humans Analyzing the Doppler Spectrum', *IEEE Aerospace and Electronic Systems Magazine*, 2018;**33**(5–6):14–19.

[55] Milani, I., Colone, F., Bongioanni, C., and Lombardo, P., 'WiFi emission-based vs passive radar localization of human targets', *2018 IEEE Radar Conference (RadarConf18)*, Oklahoma City, USA, 2018, pp. 1311–1316.

[56] Tan, B., Woodbridge, K., and Chetty, K., 'A real-time high resolution passive WiFi Doppler-radar and its applications', *2014 International Radar Conference*, Lille, 2014, pp. 1–6.

[57] Tan, B., Burrows, A., Piechocki, R., *et al.*, 'Wi-Fi based passive human motion sensing for in-home healthcare applications', *2015 IEEE 2nd World Forum on Internet of Things (WF-IoT)*, Milan, 2015, pp. 609–614.

[58] Chen, Q., Tan, B., Woodbridge, K., and Chetty, K., 'Indoor target tracking using high Doppler resolution passive Wi-Fi radar', *2015 IEEE International Conference on Acoustics, Speech and Signal Processing (ICASSP)*, Brisbane, QLD, 2015, pp. 5565–5569.

[59] Li, W., Tan, B., and Piechocki, R. 'Passive Radar for Opportunistic Monitoring in E-Health Applications', *IEEE Journal of Translational Engineering in Health and Medicine*, 2018;**6**:1–10.

[60] Mahmood Khan, U., Kabir, Z., and Hassan, S.A., 'Wireless health monitoring using passive WiFi sensing', *2017 13th International Wireless*

Communications and Mobile Computing Conference (IWCMC), Valencia, 2017, pp. 1771–1776.

[61] Chen, Q., Tan, B., Chetty, K., and Woodbridge, K., 'Activity recognition based on micro-Doppler signature with in-home Wi-Fi', *2016 IEEE 18th International Conference on e-Health Networking, Applications and Services (Healthcom)*, Munich, 2016, pp. 1–6.

[62] Li, W., Tan, B., and Piechocki, R. 'Non-contact breathing detection using passive radar', *2016 IEEE International Conference on Communications (ICC)*, Kuala Lumpur, 2016, pp. 1–6.

[63] Chen, Q., Chetty, K., Woodbridge, K., and Tan, B., 'Signs of life detection using wireless passive radar', *2016 IEEE Radar Conference (RadarConf)*, Philadelphia, USA, 2016, pp. 1–5.

[64] Tang, M., Wang, F., and Horng, T., 'Vital-sign detection based on a passive WiFi radar', *2015 IEEE MTT-S 2015 International Microwave Workshop Series on RF and Wireless Technologies for Biomedical and Healthcare Applications (IMWS-BIO)*, Taipei, 2015, pp. 74–75.

[65] Li, X., Li, S., Zhang, D., *et al.*, "Dynamic-music: Accurate device-free indoor localization', *2016 ACM UbiComp*. New York, NY, USA: ACM, 2016, pp. 196–207.

[66] Wang, H., Zhang, D., Wang, Y., *et al.*, 'RT-Fall: A Real-Time and Contactless Fall Detection System with Commodity WiFi Devices', *IEEE Transactions on Mobile Computing*, 2017;**16**(2):511–526.

[67] Wang, W., Liu, A.X., and Shahzad, M., 'Gait recognition using WiFi signals', *2016 ACM UbiComp*. New York, NY, USA: ACM, 2016, pp. 363–373.

[68] Wang, W., Liu, A.X., Shahzad, M., *et al.*, 'Understanding and modeling of WiFi signal based human activity recognition', *21st ACM MobiCom*. New York, NY, USA: ACM, 2015, pp. 65–76.

[69] Wang, W., Liu, A.X., Shahzad, M.K., Ling, K., and Lu, S., 'Device-Free Human Activity Recognition Using Commercial WiFi Devices', *IEEE Journal on Selected Areas in Communications*, 2017;**35**(5):1118–1131.

[70] Guo, L., Wang, L., Liu, J., *et al.*, 'A novel benchmark on human activity recognition using WiFi signals', *2017 IEEE 19th International Conference on e-Health Networking, Applications and Services (Healthcom)*, Dalian, 2017, pp. 1–6.

[71] Tan, B., Chen, Q., Chetty, K., Woodbridge, K., Li, W., and Piechocki, R., 'Exploiting WiFi Channel State Information for Residential Healthcare Informatics', *IEEE Communications Magazine*, 2018;**56**(5):130–137.

[72] Liu, X., Cao, J., Tang, S., Wen, J., and Guo, P., 'Contactless Respiration Monitoring via Off-the-Shelf WiFi Devices', *IEEE Transactions on Mobile Computing*, 2017;**15**(10):2466–2479.

[73] Zhang, D., Wang, H., and Wu, D., 'Toward Centimeter-Scale Human Activity Sensing with Wi-Fi Signals', *Computer*, 2017;**50**(1):48–57.

[74] Al-qaness, M.A.A., 'Indoor micro-activity recognition method using ubiquitous WiFi devices', *2018 Ubiquitous Positioning, Indoor Navigation and Location-Based Services (UPINLBS)*, Wuhan, 2018, pp. 1–7.

[75] Wang, X., Wang, X., and Mao, S., 'RF Sensing in the Internet of Things: A General Deep Learning Framework', *IEEE Communications Magazine*, 2018;**56**(9):62–67.

[76] Wang, F., Gong, W., Liu, J., and Wu, K., 'Channel Selective Activity Recognition with WiFi: A Deep Learning Approach Exploring Wideband Information', *IEEE Transactions on Network Science and Engineering*, 2020;**7**(1):181–192.

[77] Yousefi, S., Narui, H., Dayal, S., Ermon, S., and Valaee, S., 'A Survey on Behavior Recognition Using WiFi Channel State Information', *IEEE Communications Magazine*, 2017;**55**(10):98–104.

[78] Khan, U.M., Kabir, Z., Hassan, S.A., and Ahmed, S.H., 'A deep learning framework using passive WiFi sensing for respiration monitoring', *GLOBE-COM 2017–2017 IEEE Global Communications Conference*, Singapore, 2017, pp. 1–6.

[79] Baller, S., Dutta, S., and Lanvin, B. 'The Global Information Technology Report 2016. Innovating the Digital Economy'. Insight Report by the World Economic Forum, 2016. ISBN: 978-1-944835-03-3.

[80] Prescott, C. 'Internet Access – Households and Individuals, Great Britain: 2017'. Report by the UK Office for National Statistics. Release Date: August 2017.

[81] IEEE Std 802.11ac-2013, 'Part 11: Wireless LAN Medium Access Control (MAC) and Physical Layer (PHY) Specifications'. IEEE Computer Society, 2013.

[82] Li, W., Tan, B., Piechocki, R., and Craddock, I., 'Opportunistic physical activity monitoring via passive WiFi radar', *IEEE International Conference on e-Health Networking, Applications and Services (Healthcom)*, 2016, pp. 1–6.

[83] Milani, I., Colone, F., Bongioanni, C., and Lombardo, P., 'Impact of beacon interval on the performance of WiFi-based passive radar against human targets', *IEEE International Microwave and Radar Conference (MIKON)*, May 2018, pp. 190–193.

[84] Shi, F., Chetty, K., and Julier, S., 'Passive activity classification using just WiFi probe response signals', *2019 IEEE Radar Conference (RadarCon)*, Boston, 2019, pp. 1–6.

[85] Palmer, J.E., Harms, H.A., Searle, S.J., and Davis, L., 'DVB-T Passive Radar Signal Processing', *IEEE Transactions on Signal Processing*, 2013;**61**(8): 2116–2126.

[86] Woodward, P.M., *Probability and Information Theory with Applications to Radar*. International Series of Monographs on Electronics and Instrumentation (Vol. 3). Elsevier, 2014.

[87] de Boor, C., *A Practical Guide to Splines* (Vol. 27, p. 325). New York: Springer-Verlag, 1978.

[88] Fritsch, F.N., and Carlson, R.E., 'Monotone Piecewise Cubic Interpolation', *SIAM Journal on Numerical Analysis*, 1980;**17**(2):238–246.

[89] Gerbrands, J.J., 'On the Relationships Between SVD, KLT and PCA', *Pattern Recognition*, 1981;**14**(1–6):375–381.

[90] Wright, J., Ma, Y., Mairal, J., Sapiro, G., Huang, T.S., and Yan, S., 'Sparse Representation for Computer Vision and Pattern Recognition', *Proceedings of the IEEE*, 2010;**98**(6):1031–1044.

[91] Donoho, D.L., and Elad, M., 'Optimally Sparse Representation in General (Non-Orthogonal) Dictionaries via $\ell 1$ Minimization', *Proceedings of the National Academy of Sciences*, 2003;**100**(5):2197–2202.

[92] Tropp, J.A., and Gilbert, A.C., 'Signal Recovery from Random Measurements via Orthogonal Matching Pursuit', *IEEE Transactions on Information Theory*, 2007;**53**(12):4655–4666.

[93] Dai, W., and Milenkovic, O., 'Subspace Pursuit for Compressive Sensing Signal Reconstruction', *IEEE Transactions on Information Theory*, 2009;**55**(5):2230–2249.

[94] Li, W., Tan, B., and Piechocki, R.J., 'Non-contact breathing detection using passive radar', *2016 IEEE International Conference on Communications (ICC)*, Kuala Lumpur, 2016, pp. 1–6.

[95] Ma, Y., Zhou, G., and Wang, S. 'WiFi Sensing with Channel State Information: A Survey', *ACM Computing Surveys*, 2019;**52**(3):1–36.

[96] Liu, Y., Tan, Z., Hu, H., Cimini, L.J., and Li, G.Y., 'Channel Estimation for OFDM', *IEEE Communications Surveys & Tutorials*, Fourthquarter 2014;**16**(4):1891–1908.

[97] Yuan, H., Ling, Y., Sun, H., and Chen, W., 'Research on channel estimation for OFDM receiver based on IEEE 802.11a', *6th IEEE International Conference on Industrial Informatics*, Daejeon, 2008, pp. 35–39.

[98] Pratschner, S., Zöchmann, E., and Rupp, M., 'Low Complexity Estimation of Frequency Selective Channels for the LTE-A Uplink', *IEEE Wireless Communications Letters*, 2015;**4**(6):673–676.

[99] Xie, Y., Li, Z., and Li, M., 'Precise power delay profiling with commodity WiFi', *Proceedings of the 21st Annual International Conference on Mobile Computing and Networking (MobiCom '15)*, ACM, New York, USA, pp. 53–64.

[100] Halperin, D., Hu, W., Sheth, A., and Wetherall, D. 'Linux 802.11n CSI Tool'. 2011. Available from: https://dhalperi.github.io/linux-80211n-csitool/ (accessed 01 May 2020).

[101] Schmidt, R.O., 'Multiple Emitter Location and Signal Parameter Estimation', *IEEE Transactions on Antennas and Propagation*, 1986;**AP-34**:276–280.

[102] van der Veen, A.J., Vanderveen, M.C., and Paulraj, A., 'Joint Angle and Delay Estimation Using Shift-Invariance Techniques', *IEEE Transactions on Signal Processing*, 1998;**46**(2):405–418.

[103] Fleury, B.H., Tschudin, M., Heddergott, R., Dahlhaus, D., and Ingeman Pedersen, K., 'Channel Parameter Estimation in Mobile Radio Environments Using the SAGE Algorithm', *IEEE Journal on Selected Areas in Communications*, 1999;**17**(3):434–450.

[104] Di Domenico, S., De Sanctis, M., Cianca, E., and Ruggieri, M., 'WiFi-Based Through-the-Wall Presence Detection of Stationary and Moving Humans Analyzing the Doppler Spectrum', *IEEE Aerospace and Electronic Systems Magazine*, 2018;**33**(5–6):14–19.

[105] Zhang, D., and Li, X., *Method for Determining a Doppler Frequency Shift of a Wireless Signal Directly Reflected by a Moving Object*. US Patent Application US20190020425A1, 2017.

[106] Wang, W., Liu, A.X., Shahzad, M., Ling, K., and Lu, S., 'Understanding and modeling of WiFi signal based human activity recognition', *MobiCom'15*. New York, NY, USA: ACM, 2015, pp. 65–76.

[107] Kim, S., 'Device-free activity recognition using CSI & big data analysis: A survey', *2017 Ninth International Conference on Ubiquitous and Future Networks (ICUFN)*, Milan, 2017, pp. 539–541.

[108] Yousefi, S., Narui, H., Dayal, S., Ermon, S., and Valaee, S., 'A Survey on Behaviour Recognition Using WiFi Channel State Information', *IEEE Communications Magazine*, 2017;**55**(10):98–104.

[109] Wang, C., Chen, S., Yang, Y., Hu, F., Liu, F., and Wu, J., 'Literature Review on Wireless Sensing—Wi-Fi Signal-Based Recognition of Human Activities', *Tsinghua Science and Technology*, 2018;**23**(2):203–222.

[110] Qian, K., Wu, C., Yang, Z., Liu, Y., He, F., and Xing, T., 'Enabling Contactless Detection of Moving Humans with Dynamic Speeds Using CSI', *ACM Transactions on Embedded Computing Systems*, 2018;**17**(2), Article 52.

[111] Yu, N., Wang, W., Liu, A.X., and Kong, L., 'QGesture: Quantifying Gesture Distance and Direction with WiFi Signals', *Proceedings of the ACM on Interactive, Mobile, Wearable and Ubiquitous Technologies*, 2018;**2**(1):23 pages, Article 51.

[112] Wang, W., Liu, A.X., Shazad, M., Ling, K., and Lu, S., 'Device-Free Human Activity Recognition Using Commercial WiFi Devices', *IEEE Journal on Selected Areas in Communications*, 2017;**35**(5):1118–1131.

[113] Guo, X., Liu, B., Shi, C., Liu, H., Chen, Y., and Chuah, M.C., 'WiFi-enabled smart human dynamics monitoring', *Proceedings of the 15th ACM Conference on Embedded Network Sensor Systems (SenSys'17)*, Rasit Eskicioglu (Ed.). New York, NY, USA: ACM, 2017, Article 16.

[114] Wu, D., Zhang, D., Xu, C., Wang, H., and Li, X., 'Device-Free WiFi Human Sensing: From Pattern-Based to Model-Based Approaches', *IEEE Communications Magazine*, 2017;**55**(10):91–97.

[115] Huang, H., and Lin, S., 'WiDet: Wi-Fi based device-free passive person detection with deep convolutional neural networks', *Proceedings of the 21st ACM International Conference on Modeling, Analysis and Simulation of Wireless and Mobile Systems (MSWIM'18)*. New York, NY, USA: ACM, 2018, pp. 53–60.

[116] Lee, C., and Huang, X., 'Human activity detection via WiFi signals using deep neural networks', *2018 IEEE/ACM International Conference on Utility and Cloud Computing Companion (UCC Companion)*, Zurich, 2018, pp. 3–4.

[117] Wang, F., Gong, W., and Liu, J., 'On Spatial Diversity in WiFi-Based Human Activity Recognition: A Deep Learning Based Approach', *IEEE Internet of Things Journal*, 2019;**6**(2):2035–2047.

[118] Zhu, H., Xiao, F., Sun, L., Wang, R., and Yang, P., 'R-TTWD: Robust Device-Free Through-the-Wall Detection of Moving Human with WiFi', *IEEE Journal on Selected Areas in Communications*, 2017;**35**(5): 1090–1103.

[119] Wang, W., Liu, A.X., and Shahzad, M., 'Gait recognition using WiFi signals', *Proceedings of the 2016 ACM International Joint Conference on Pervasive and Ubiquitous Computing (UbiComp'16)*. New York, NY, USA: ACM, 2016, pp. 363–373.

[120] Wang, H., Zhang, D., Wang, Y., Ma, J., Wang, Y., and Li, S., 'RT-Fall: A Real-Time and Contactless Fall Detection System with Commodity WiFi Devices', *IEEE Transactions on Mobile Computing*, 2017;**16**(2):511–526.

[121] Wang, Y., Wu, K., and Ni, L.M., 'WiFall: Device-Free Fall Detection by Wireless Networks', *IEEE Transactions on Mobile Computing*, 2017;**16**(2): 581–594.

[122] Khamis, A., Chou, C.T., Kusy, B., and Hu, W. 'CardioFi: Enabling heart rate monitoring on unmodified COTS WiFi devices', *Proceedings of the 15th EAI International Conference on Mobile and Ubiquitous Systems: Computing, Networking and Services (MobiQuitous'18)*. New York, NY, USA: ACM, 2018, pp. 97–106.

[123] Zhang, D., Wang, H., Wang, Y., and Ma, J., 'Anti-fall: A non-intrusive and real-time fall detector leveraging CSI from commodity WiFi devices. ICOST', *Lecture Notes in Computer Science* (Vol. 9102). Cham: Springer, 2015.

[124] Zhang, D., Hu, Y., Chen, Y., and Zeng, B., 'BreathTrack: Tracking Indoor Human Breath Status via Commodity WiFi', *IEEE Internet of Things Journal*, 2019;**6**(2):3899–3911.

[125] Wang, P., Guo, B., Xin, T., Wang, Z., and Yu, Z., 'TinySense: Multi-user respiration detection using Wi-Fi CSI signals', *2017 IEEE 19th International Conference on e-Health Networking, Applications and Services (Healthcom)*, Dalian, 2017, pp. 1–6.

[126] Liu, J., Chen, Y., Wang, Y., Chen, X., Cheng, J., and Yang, J., 'Monitoring Vital Signs and Postures During Sleep Using WiFi Signals', *IEEE Internet of Things Journal*, 2018;**5**(3):2071–2084.

[127] Gu, Y., Liu, T., Li, J., *et al.*, 'EmoSense: Data-driven emotion sensing via off-the-shelf WiFi devices', *IEEE International Conference on Communications (ICC)*, Kansas City, USA, 2018, pp. 1–6.

[128] Lee, S., Park, Y., Suh, Y., and Jeon, S., 'Design and implementation of monitoring system for breathing and heart rate pattern using WiFi signals', *15th IEEE Annual Consumer Communications & Networking Conference (CCNC)*, Las Vegas, NV, 2018, pp. 1–7.

[129] Chen, Q., Chetty, K., Woodbridge, K., and Tan, B., 'Signs of life detection using wireless passive radar', *IEEE Radar Conference (RadarConf)*, Philadelphia, PA, 2016, pp. 1–5.

[130] Yang, Y., Cao, J., Liu, X., and Xing, K., 'Multi-person sleeping respiration monitoring with COTS WiFi devices', *IEEE 15th International Conference on Mobile Ad Hoc and Sensor Systems (MASS)*, Chengdu, 2018, pp. 37–45.

[131] Gu, Y., Zhang, Y., Li, J., Ji, Y., An, X., and Ren, F., 'Sleepy: Wireless Channel Data Driven Sleep Monitoring via Commodity WiFi Devices', *IEEE Transactions on Big Data*, doi: 10.1109/TBDATA.2018.2851201.

[132] Li, S., Li, X., Lv, Q., Tian, G., and Zhang, D., 'WiFit: Ubiquitous bodyweight exercise monitoring with commodity Wi-Fi devices', *2018 IEEE SmartWorld, Ubiquitous Intelligence & Computing, Advanced & Trusted Computing, Scalable Computing & Communications, Cloud & Big Data Computing, Internet of People and Smart City Innovation (SmartWorld/SCALCOM/ UIC/ATC/CBDCom/IOP/SCI)*, Guangzhou, 2018, pp. 530–537.

[133] Guo, X., Liu, J., Shi, C., Liu, H., Chen, Y., and Chuah, M.C., 'Device-Free Personalized Fitness Assistant Using WiFi', *Proceedings of the ACM on Interactive, Mobile, Wearable and Ubiquitous Technologies*, 2018;**2**(4), Article 165.

[134] Xiao, F., Chen, J., Xie, X., Gui, L., Sun, L., and Wang, R., 'SEARE: A System for Exercise Activity Recognition and Quality Evaluation Based on Green Sensing', *IEEE Transactions on Emerging Topics in Computing*, doi: 10.1109/TETC.2018.2790080.

[135] Zhang, Y., Li, X., Li, S., Xiong, J., and Zhang, D., 'A training-free contactless human vitality monitoring platform using commodity Wi-Fi devices', *International Symposium on Pervasive and Ubiquitous Computing and Wearable Computers (UbiComp '18)*. New York, NY, USA: ACM, 2018, pp. 488–491.

[136] Ramezani, R., Xiao, Y., and Naeim, A., 'Sensing-Fi: Wi-Fi CSI and accelerometer fusion system for fall detection', *2018 IEEE EMBS International Conference on Biomedical & Health Informatics (BHI)*, Las Vegas, NV, 2018, pp. 402–405.

[137] Shu, Y., Chen, C., Shu, K., and Zhang, H., 'Research on human motion recognition based on Wi-Fi and inertial sensor signal fusion', *2018 IEEE SmartWorld, Ubiquitous Intelligence & Computing, Advanced & Trusted Computing, Scalable Computing & Communications, Cloud & Big Data Computing, Internet of People and Smart City Innovation (SmartWorld/SCALCOM/ UIC/ATC/CBDCom/IOP/SCI)*, Guangzhou, 2018, pp. 496–504.

[138] Fang, B., Lane, N.D., Zhang, M., and Kawsar, F., 'HeadScan: A wearable system for radio-based sensing of head and mouth-related activities', *2016 15th ACM/IEEE International Conference on Information Processing in Sensor Networks (IPSN)*, Vienna, 2016, pp. 1–12.

[139] Fang, B., Lane, N.D., Zhang, M., Boran, A., and Kawsar, F., 'BodyScan: Enabling radio-based sensing on wearable devices for contactless activity and vital sign monitoring', *Proceedings of the 14th Annual International Conference on Mobile Systems, Applications, and Services (MobiSys'16)*. New York, NY, USA: ACM, 2016, pp. 97–110.

[140] Venkatnarayan, R.H., Page, G., and Shahzad, M., 'Multi-user gesture recognition using WiFi', *Proceedings of the 16th Annual International Conference on Mobile Systems, Applications, and Services (MobiSys'18)*. New York, NY, USA: ACM, 2018, pp. 401–413.

[141] Lofgren, S., 'IoT: Future-Proofing Device Communications', *IEEE Internet of Things Newsletter*, 2015.

[142] https://www.cnet.com/news/ces-2019-hive-wants-to-help-you-keep-trackof-aging-loved-ones/.

[143] Burrows, A., Coyle, D., and Gooberman-Hill, R., 'Privacy, Boundaries and Smart Homes for Health: An Ethnographic Study', *Health & Place*, 2018;**50**:112–118.

[144] Birchley, G., Huxtable, M., Murtagh, M., Ter Meulen, R., Flach, P., and Gooberman-Hill, R., 'Smart Homes, Private Homes? An Empirical Study of Technology Researchers' Perceptions of Ethical Issues in Developing Smart-Home Health Technologies', *BMC Medical Ethics*.

[145] Gupta, N., Fischer, A.R., and Frewer, L.J., 'Socio-Psychological Determinants of Public Acceptance of Technologies: A Review', *Public Understanding of Science (Bristol, England)*, 2012;**21**(7):782–795.

[146] United Nations Population DivisionWorld Population Ageing: 1950–2050.

Chapter 3

Sparsity-driven methods for micro-Doppler detection and classification

Gang Li[1] and Shobha Sundar Ram[2]

3.1 Introduction

A signal is sparse if only a few entries are dominant and others are zero. Compressed sensing (CS) refers to the idea that a sparse signal can be accurately recovered from a small number of measurements [1–3], which provides a perspective for data reduction without compromising performance. In many radar applications, the micro-Doppler signals are sparse or approximately sparse in appropriate transform domains. This allows us to extract micro-Doppler features from a limited number of measurements by using CS-based or sparsity-driven recovery approaches. In [4], the features of micro-Doppler signals reflected from rotating scatterers are extracted by the orthogonal matching pursuit (OMP) algorithm, which is a typical sparse recovery algorithm. In [5], the joint estimation of the spatial distribution of scatterers and the rotational speed of a rotating target is achieved by a pruned OMP algorithm. In [6,7], the sparse signal-processing technique is combined with the time–frequency analysis to improve the accuracy of helicopter classification. In [8], the time–frequency track features of dynamic hand gestures, which are difficult to formulate analytically, are extracted by sparse decomposition of micro-Doppler signals. In this chapter, some sparsity-driven methods for micro-Doppler detection and classification and their application to human activity sensing will be introduced.

In some scenarios, data-independent basis functions or transforms (such as Fourier and wavelets) may not fit the received signals. For instance, if there are multiple dynamic movers in the same channel, the Fourier coefficients of the micro-Doppler signals will overlap, and distinguishing between them becomes challenging. Another scenario is when there is some variation in the sensor parameters during test data collection from the training scenario. In those scenarios, it may be beneficial to consider to "learn basis functions for dictionaries" to accurately represent the radar signals of each class as opposed to relying on traditional data-independent transforms. The dictionaries are learned, usually by imposing sparsity constraints on the trained

[1]Department of Electronic Engineering, Tsinghua University, Beijing, China
[2]Department of Electronics and Communication Engineering, Indraprastha Institute of Information Technology, New Delhi, India

data representations. The advantage of the data-dependent dictionaries is that they can be tuned to the underlying physics of the data as opposed to the sensor's parameters. They can further be tuned so as to increase the discriminativeness between signals belonging to different classes. As a result, the data-dependent dictionaries can be used in scenarios where there are significant differences between training and test scenarios due to either variation in the sensor parameters or be used to extract and separate signals from multiple movers.

The remainder of this chapter is organised as follows. In Section 3.2, we briefly review the fundamentals of sparse signal recovery. In Section 3.3, a sparsity-driven micro-Doppler feature extraction method for dynamic hand gesture recognition is presented. In Section 3.4, we discuss the principles of learning dictionaries to represent micro-Doppler signals for both detection and classification of periodic human motions such as walking and boxing. Concluding remarks are given in Section 3.5.

3.2   Fundamentals of sparse signal recovery

In this section, we briefly review the fundamentals of sparse signal recovery, including signal model, typical algorithms, and dictionary learning.

3.2.1   Signal model

The signal model of sparse signal recovery can be expressed as

$$\mathbf{y} = \mathbf{\Phi}\mathbf{x} + \mathbf{w} \tag{3.1}$$

where $\mathbf{y} \in \mathbb{C}^{M \times 1}$ is the measurement vector, $\mathbf{\Phi} \in \mathbb{C}^{M \times N}$ with $M < N$ is an over-complete dictionary, $\mathbf{x} \in \mathbb{C}^{N \times 1}$ is the sparse vector containing a few nonzero coefficients, and $\mathbf{w} \in \mathbb{C}^{M \times 1}$ is the additive noise. The goal of sparse signal recovery is to reconstruct $\mathbf{x}$ from (3.1). The signal $\mathbf{x}$ is said to be K-sparse if the number of nonzero elements in $\mathbf{x}$ is smaller than or equal to K, where $K \ll N$. The value of K is defined as the sparsity degree of signal $\mathbf{x}$. The set of locations of nonzero elements in $\mathbf{x}$ is referred to as the support set of $\mathbf{x}$ and is denoted by

$$U(\mathbf{x}) = \{i | x_i \neq 0, \quad i = 1, 2, \dots, N\} \tag{3.2}$$

where x_i is the ith entry of $\mathbf{x}$. The sparsity can also be defined by the l-norm, i.e., $\|\mathbf{x}\|_0 \overset{\Delta}{=} |U(\mathbf{x})| \leq K$

Given the received signal $\mathbf{y}$, it is obvious that the sparse solution $\mathbf{x}$ depends on the dictionary $\mathbf{\Phi}$. The formulation of $\mathbf{\Phi}$ varies with different applications. Some widely used dictionaries in radar applications include

* the Fourier dictionary [9], where the frequency domain coefficients of the received signal are assumed to be sparse;
* the array steering vector dictionary [9], where the spatial distribution of the signal sources is assumed to be sparse;
* the chirp dictionary [10] and Gabor dictionary [8], where the time–frequency distribution of the received signal is assumed to be sparse.

In all of these examples, the columns of the dictionary Φ, which are called atoms or basis-signals, are expected to be capable of representing the received signal $\mathbf{y}$ in a sparse fashion. If a predesigned dictionary does not perform well for sparse representation due to the mismatch between the basis-signals and the received signal, one may consider dictionary learning, i.e., to learn an appropriate form of the dictionary from the received signal. In what follows we will first introduce some typical algorithms of sparse signal recovery with a known dictionary and then discuss the dictionary learning problem.

3.2.2 Typical algorithms for sparse signal recovery

3.2.2.1 Convex optimisation

A sufficient condition of sparse signal recovery with convex optimisation is the restricted isometry property (RIP).

Definition 3.1 [11]: The dictionary Φ is said to satisfy the restricted RIP of order K if there exists a constant $\delta_K \in (0, 1)$ such that

$$(1 - \delta_K) \|\mathbf{x}\|_2^2 \le \|\Phi\mathbf{x}\|_2^2 \le (1 + \delta_K) \|\mathbf{x}\|_2^2 \tag{3.3}$$

holds for all the K-sparse signals.

Assume that the dictionary Φ satisfies RIP of order $2K$ with $\delta_{2K} < 0.4651$, the stable solution of (3.1) can be obtained by [12]:

$$\hat{\mathbf{x}} = \arg\min_{\mathbf{x}} \|\mathbf{x}\|_1, \quad \text{s.t.} \ \|\mathbf{y} - \Phi\mathbf{x}\|_2^2 \le \eta \tag{3.4}$$

where $\|\mathbf{x}\|_1 = \sum_{i=1}^{N} |x_i|$ is the l_1-norm of a vector, and η is the expected noise level. The problem in (3.4) is also known as the basis pursuit denoising [13].

Solving the problem in (3.4) is equivalent to finding the solution of the following optimisation problem:

$$\hat{\mathbf{x}} = \arg\min_{\mathbf{x}} \{ \|\mathbf{x}\|_1 + \gamma \|\mathbf{y} - \Phi\mathbf{x}\|_2^2 \} \tag{3.5}$$

where γ is a regularisation parameter controlling the balance between the sparsity of the solution and the recovery error. There are a number of algorithms/toolboxes of convex optimisation available for solving (3.5), such as l_1-Magic [14], CVX [15], SPGL$_1$ [16], and YALL1 [17]. The computational burden of convex optimisation-based methods is $O(MN^2)$ or $O(N^3)$ [18].

3.2.2.2 Bayesian sparse signal recovery

Without loss of generality, here we consider a real-value case, which can be easily extended to the complex-value case. Suppose that the entries of $\mathbf{w}$ are independent and identically distributed (i.i.d.) variables with the Gaussian probability distribution function (PDF) $\mathcal{N}(\cdot \ ; \mu, \sigma^2)$, where μ is the mean value and σ^2 is the variance. One can calculate that $f(\mathbf{y}|\mathbf{x}) = \mathcal{N}(\mathbf{y}; \ \Phi\mathbf{x}, \sigma_w^2 \mathbf{I}_{M \times M})$. The sparse signal $\mathbf{x}$ is regarded as a realisation of a stochastic process with the joint PDF $f(\mathbf{x})$, which is known *a priori*. According to Bayes' theorem, we have

$$\log f(\mathbf{x}|\mathbf{y}) \propto -\|\mathbf{y} - \Phi\mathbf{x}\|_2^2 + 2\sigma_w^2 \log f(\mathbf{x}) \tag{3.6}$$

Then one can maximise $\log f(\mathbf{x}|\mathbf{y})$ in (3.6) to obtain the maximum a posteriori (MAP) estimation of $\mathbf{x}$. The MAP estimation problem can be formulated as

$$\hat{\mathbf{x}} = \arg\min_{\mathbf{x}}\left\{\|\mathbf{y} - \mathbf{\Phi}\mathbf{x}\|_2^2 - 2\sigma_w^2 \log f(\mathbf{x})\right\} \tag{3.7}$$

Different forms of $f(\mathbf{x})$ have been considered in the existing literature. In [19], the entries of $\mathbf{x}$ are assumed to follow the i.i.d. zero-mean generalised Gaussian distribution (GGD):

$$f(\mathbf{x}) = \prod_{i=1}^{N} \frac{\alpha\beta}{\Gamma(1/\beta)} \exp\left[(-\alpha|x_i|)^{\beta}\right] \tag{3.8}$$

where α and β are the scale and shape parameters, respectively, and $\Gamma(\cdot)$ denotes the gamma function. In [20], the entries of $\mathbf{x}$ are assumed to be independent but nonidentically distributed Gaussian variables:

$$f(\mathbf{x};\boldsymbol{\theta}) = \prod_{i=1}^{N} \mathcal{N}\left(x_i; 0, \theta_i\right) \tag{3.9}$$

where $x_i = 0$ if $\theta_i = 0$ and x_i is a nonzero variable if θ_i is a large value. With different forms of $f(\mathbf{x})$, the problem in (3.7) can be solved by majorisation–minimisation (MM) [21] or expectation–maximisation (EM) [20] algorithms. The computational complexity of Bayesian sparse signal recovery is $O\left(M^2N\right)$ [22], which is lower than that of convex optimisation-based algorithms.

3.2.2.3   Greedy algorithms

The greedy algorithms aim to estimate the support set $U(\mathbf{x})$ of the K-sparse signal $\mathbf{x}$ iteratively. Once the support set $U(\mathbf{x})$ is determined, one can easily recover the nonzero coefficients of $\mathbf{x}$ by using least squares estimation. A typical greedy algorithm is OMP [23]. At each iteration, OMP finds one column of $\mathbf{\Phi}$ most correlated with the recovery residual, and then the residual is updated by orthogonal projection. The detailed steps

Table 3.1   Orthogonal matching pursuit (OMP) algorithm

Input: Compressed data $\mathbf{y}$, sensing matrix $\mathbf{\Phi}$ (with normalised columns),
  sparsity degree K
Initialisation: $\mathbf{r}^{(0)} = \mathbf{y}$, $\hat{\mathbf{x}}^{(0)} = \mathbf{0}^{N\times1}$, $U^{(0)} = \emptyset$
For $k = 1 : K$

  (1) $\mathbf{g}^{(k)} = \mathbf{\Phi}^H \mathbf{r}^{(k-1)}$

  (2) $U^{(k)} = U^{(k-1)} \cup \left\{\arg\max_{i \in \{1,2,\cdots,N\}}\left|g_i^{(k)}\right|\right\}$

  (3) $\hat{\mathbf{x}}_{U^{(k)}}^{(k)} = \mathbf{\Phi}_{U^{(k)}}^{\dagger}\mathbf{y}$

  (4) $\mathbf{r}^{(k)} = \mathbf{y} - \mathbf{\Phi}\hat{\mathbf{x}}^{(k)}$

End For
Output: $\hat{\mathbf{x}}^{(k)}$

of OMP are summarised in Table 3.1, where $(\cdot)^{\dagger} = [(\cdot)^{H}(\cdot)]^{-1}(\cdot)^{H}$, $(\cdot)^{H}$ denotes the conjugate transpose, and $(\cdot)^{-1}$ denotes the inverse of a matrix. Other greedy algorithms include subspace pursuit (SP) [23], compressive sampling matching pursuit (CoSaMP) [24], and iterative hard thresholding (IHT) [25], etc. Compared to convex optimisation-based and Bayesian methods, greedy algorithms have much less computational complexity [26].

3.2.3 Dictionary learning

In the previous discussions, it was assumed that the atoms of $\boldsymbol{\Phi}$ are based on standard transforms, such as Fourier, wavelets, and Gabor. The advantages of these traditional off-the-shelf bases are that they are readily available for use. More recent researches suggest that discriminative codes obtained from data driven dictionaries can significantly outperform the patterns from traditional bases in the areas of classification and recognition, since such dictionaries can be tuned to uniquely represent the signal characteristics of each class [27]. The disadvantage, however, is that learning dictionaries are computationally expensive and require large volumes of training data.

There are two types of dictionary learning frameworks—the synthesis and analysis frameworks. In the *synthesis framework* (3.1) is modified as

$$\mathbf{y}_c = \boldsymbol{\Phi}_c \mathbf{x}_c + \mathbf{w} \tag{3.10}$$

Note that the dictionaries or bases $\boldsymbol{\Phi}_c$ must be learnt for each target class c based on training data $\mathbf{y}_c$ using the optimisation function:

$$\langle \boldsymbol{\Phi}_c, \mathbf{x}_c \rangle = \underset{\boldsymbol{\Phi}_c, \mathbf{x}_c}{\arg\min} \left\| \mathbf{y}_c - \boldsymbol{\Phi}_c \mathbf{x}_c \right\|_2^2, \text{ s.t.} \|\mathbf{x}_c\|_0 < \tau \tag{3.11}$$

where the l-norm is the cardinality function that indicates the number of nonzero elements in $\mathbf{x}_c$ The optimisation function constrains the representation accuracy of the formulation by enforcing sparsity (τ) on each of the columns of the coefficient matrix. The function can be solved using an iterative approach called K-SVD [27]. In order to reduce its computational complexity, the l_0-minimisation operation can be reduced to the convex l_1-minimisation operation as shown in

$$\langle \boldsymbol{\Phi}_c, \mathbf{x}_c \rangle = \underset{\boldsymbol{\Phi}_c, \mathbf{x}_c}{\arg\min} \left\{ \left\| \mathbf{y}_c - \boldsymbol{\Phi}_c \mathbf{x}_c \right\|_2^2 + \gamma \|\mathbf{x}_c\|_1 \right\} \tag{3.12}$$

where γ is a regularisation parameter that trades off between the reconstruction accuracy and the sparsity. There are numerous methods for solving l_1-minimisation problems such as iterative soft thresholding [28]. The algorithm solving (3.12) is hitherto referred to as *synthesis dictionary learning (SDL)*.

These dictionaries can be further refined by adding depth to their representations. This formulation is called the *deep dictionary learning (DDL)* [29]. It is well known to the machine-learning community that the addition of depth facilitates representation of higher order features of signals belonging to a class. The deep representation of the signal is shown in the following equation:

$$\begin{aligned}
\mathbf{y}_c &= \boldsymbol{\Phi}_c^1 \mathbf{x}_c^1 \\
\mathbf{x}_c^1 &= \boldsymbol{\Phi}_c^2 \mathbf{x}_c^2 \\
&\vdots \\
\mathbf{x}_c^m &= \boldsymbol{\Phi}_c^{m+1} \mathbf{x}_c^{m+1}
\end{aligned} \tag{3.13}$$

The first layer representation is identical to (3.10). Subsequently, the coefficient matrix to the first layer is represented with the second layer dictionary bases and the corresponding coefficients. This is continued up to M layers while the dimensionality of each successive layer is reduced. The dictionary learning algorithm involves solving for the dictionary atoms and coefficients at each mth layer as shown in the following equation:

$$\begin{aligned}
\langle \boldsymbol{\Phi}_c^1, \mathbf{x}_c^1 \rangle &= \arg \min_{\boldsymbol{\Phi}_c^1, \mathbf{x}_c^1} \left\{ \left\| \mathbf{y}_c - \boldsymbol{\Phi}_c^1 \mathbf{x}_c^1 \right\|_2^2 + \gamma^1 \left\| \mathbf{x}_c^1 \right\|_1 \right\} \\
&\vdots \\
\langle \boldsymbol{\Phi}_c^{m+1}, \mathbf{x}_c^{m+1} \rangle &= \arg \min_{\boldsymbol{\Phi}_c^{m+1}, \mathbf{x}_c^{m+1}} \left\{ \left\| \mathbf{x}_c^m - \boldsymbol{\Phi}_c^{m+1} \mathbf{x}_c^{m+1} \right\|_2^2 + \gamma^{m+1} \left\| \mathbf{x}_c^{m+1} \right\|_1 \right\}
\end{aligned} \tag{3.14}$$

Note that regularisation parameters ($\gamma^m, m = 1 \cdots M$) have been introduced for each optimisation step. The dictionary for a class is obtained by the matrix multiplication of all the layer dictionaries corresponding to that class:

$$\boldsymbol{\Phi}_c = \boldsymbol{\Phi}_c^1 \times \boldsymbol{\Phi}_c^2 \times \cdots \times \boldsymbol{\Phi}_c^M \tag{3.15}$$

Note that the final representation of the signal is given by

$$\mathbf{y}_c = \boldsymbol{\Phi}_c \mathbf{x}_c^M \tag{3.16}$$

which involves the representation of the signal with a dimensionally small code $\mathbf{x}_c^M$.

During the training stage, the dictionaries for each class can be learnt individually on the basis of the previous formulation. Alternatively, the dictionaries of all the classes can be learnt jointly by enforcing sparsity on each of the coefficient classes as well as a penalty to increase the discrimination across multiple classes as [30]:

$$\langle \boldsymbol{\Phi}, w, x \rangle = \arg \min_{\langle \boldsymbol{\Phi}, \mathbf{w}, \mathbf{x} \rangle} \left\{ \left\| \mathbf{y} - \boldsymbol{\Phi} \mathbf{x} \right\|_2^2 + \alpha \left\| \mathbf{q} - \mathbf{w} \mathbf{x} \right\|_2^2 \right\} \quad \text{s.t.} \quad \left\| \mathbf{x} \right\|_0 < \tau \tag{3.17}$$

where $\mathbf{y}$ is a matrix whose columns consist of signals from different classes while the columns of $\mathbf{q}$ are the corresponding discriminative sparse codes. The regularisation parameter, α, trades off between the discrimination error and the representation accuracy. The above optimisation problem can be similarly handled by the K-SVD algorithm. Hence, this method is called the labelled consistent K-SVD algorithm [30]. The main advantage of the LC-KSVD is that the training of multiple classes is carried out simultaneously. However, the computational complexity of the dictionary learning is significant.

The alternate mechanism to the synthesis framework is the *analysis framework* [31] where $\mathbf{y}_c$ is analysed with the basis function to generate its representation:

$$\mathbf{x}_c = \mathbf{\Phi}_c \mathbf{y}_c \tag{3.18}$$

where both the analysis dictionary, $\mathbf{\Phi}_c$, and its representation vector, $\mathbf{x}_c$, are learnt for each cth class of signals. They are learnt using the optimisation function:

$$\langle \mathbf{\Phi}_c, \mathbf{x}_c \rangle = \min_{\mathbf{\Phi}_c, \mathbf{x}_c} \| \mathbf{x}_c - \mathbf{\Phi}_c \mathbf{y}_c \|_2^2 \quad \text{s.t.} \quad \| \mathbf{x}_c \|_0 < \tau \tag{3.19}$$

Again the l_0-norm can be reduced to the convex l_1-minimisation operation as shown in

$$\langle \mathbf{\Phi}_c, \mathbf{x}_c \rangle = \arg\min_{\mathbf{\Phi}_c, \mathbf{x}_c} \left\{ \| \mathbf{x}_c - \mathbf{\Phi}_c \mathbf{y}_c \|_2^2 + \gamma_1 \| \mathbf{x}_c \|_1 + \gamma_2 \| \mathbf{\Phi}_c \|_F^2 \right\} \tag{3.20}$$

An additional constraint is imposed on the Frobenius norm of the dictionary to ensure that it does not become zero or an ill-conditioned matrix. The advantage of the analysis framework is the reduction in the computational complexity during test phase since the test features can be generated through a matrix multiplication operation as shown in (3.18) as opposed to an inversion operation in (3.11). The formulation in (3.20) is hitherto referred to as *analysis dictionary learning (ADL)*.

The sparse discriminative codes $\mathbf{x}_c$ obtained from synthesis or ADL can be directly used for classification. Alternately, the dictionaries can be used for signal reconstruction in the case of the presence of signals from multiple classes. Here, the dictionaries from all the C classes are combined together to form a single dictionary matrix as

$$\mathbf{\Phi} = [\mathbf{\Phi}_1, \ldots, \mathbf{\Phi}_C] \tag{3.21}$$

Now, if a test signal, $\mathbf{y}$, consists of signals from multiple classes, the reconstructed signal from a particular class is obtained by

$$\tilde{\mathbf{x}}_{1:C} = \min_{\tilde{\mathbf{x}}_{1:C}} \left\{ \| \mathbf{y} - \mathbf{\Phi} \tilde{\mathbf{x}}_{1:C} \|_2^2 + \gamma \| \tilde{\mathbf{x}}_{1:C} \|_1 \right\} \tag{3.22}$$

where $\tilde{\mathbf{x}}_{1:C}$ consists of the representation coefficients matrices corresponding to all the classes arranged in a column-wise manner. If the energy corresponding to a class c, $\| \tilde{\mathbf{x}}_c \|_2^2$ is above a predefined threshold, we can conclude that the signal belonging to class c is present and can be reconstructed as

$$\tilde{\mathbf{y}}_c = \mathbf{\Phi}_c \mathbf{x}_c \tag{3.23}$$

In this manner, signals from multiple classes can be separated from an aggregate signal and reconstructed.

3.3   Sparsity-driven micro-Doppler feature extraction for dynamic hand gesture recognition

In this section, we present a sparsity-driven method of micro-Doppler feature extraction for dynamic hand gesture recognition [8]. First, the radar echoes reflected from dynamic hand gestures are mapped into the time–frequency domain through the

Gabor dictionary. Then, the micro-Doppler features of the dynamic hand gestures are extracted via the OMP algorithm and fed into the modified-Hausdorff-distance-based nearest neighbour (NN) classifier for recognition. Experiments with real data collected by a K-band radar show that (1) the recognition accuracy produced by the sparsity-driven method exceeds 96% under moderate noise, and (2) the sparsity-driven method outperforms the methods based on principal component analysis (PCA) and deep convolutional neural network (DCNN) in conditions of small training dataset.

3.3.1 Measurement data collection

The measurement data are collected by a K-band CW radar system. The carrier frequency and the baseband sampling frequency are 25 GHz and 1 kHz, respectively. The radar antenna is oriented directly to the human hand at a distance of 0.3 m. The following four dynamic hand gestures are considered: (a) hand rotation, (b) beckoning, (c) snapping fingers, and (d) flipping fingers. The illustrations and descriptions of the four dynamic hand gestures are shown in Figure 3.1 and Table 3.2, respectively. The data are collected from three personnel targets: two males and one female. Each person repeats a particular dynamic hand gesture for 20 times. Each 0.6 s time interval containing a complete dynamic hand gesture is recorded as a signal segment. The total number of the signal segments is (4 gestures) × (3 personnel targets) × (20 repeats) = 240.

To visualise the time-varying characteristics of the dynamic hand gestures, the short time Fourier transform with a Kaiser window is applied to the received signals to obtain the corresponding spectrograms. The resulting spectrograms of the four dynamic hand gestures from one personnel target are shown in Figure 3.2. It is clear that the time–frequency trajectories of these dynamic hand gestures are different from each other. The Doppler shifts corresponding to the gesture "hand rotation" continuously change along the time axis because the velocity of the hand continuously changes during the rotation process. The echo of the gesture "beckoning" contains a negative Doppler shift and a positive Doppler shift, which are corresponding to the back and forth movements of the fingers, respectively. The negative Doppler shift of the gesture "snapping fingers" is larger than its positive Doppler shift, since the velocity corresponding to the retreating movement of the fingers is much larger than

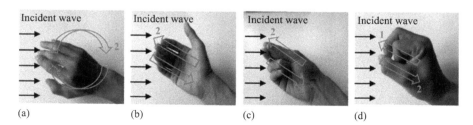

(a)   (b)   (c)   (d)

Figure 3.1   Illustrations of four dynamic hand gestures: (a) hand rotation, (b) beckoning, (c) snapping fingers, and (d) flipping fingers

Table 3.2   Four dynamic hand gestures under study

Gesture	Description
(a) Hand rotation	The gesture of rotating the right hand for a cycle. The hand moves away from the radar in the first half cycle and toward the radar in the second half
(b) Beckoning	The gesture of beckoning someone with the fingers swinging back and forth for one time
(c) Snapping fingers	The gesture of pressing the middle finger and the thumb together and then flinging the middle finger onto the palm while the thumb sliding forward quickly. After snapping fingers, pressing the middle finger and the thumb together again
(d) Flipping fingers	The gesture of bucking the middle finger under the thumb and then flipping the middle finger forward quickly. After flipping fingers, bucking the middle finger under the thumb again

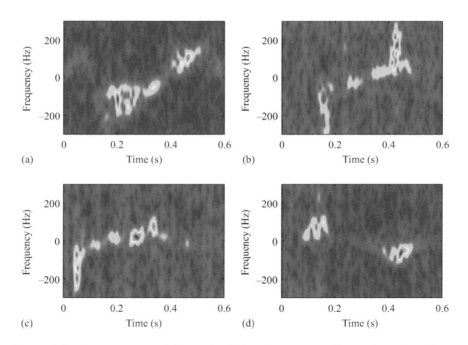

Figure 3.2   Spectrograms of the received signals corresponding to four dynamic hand gestures from one personnel target: (a) hand rotation, (b) beckoning, (c) snapping fingers, and (d) flipping fingers

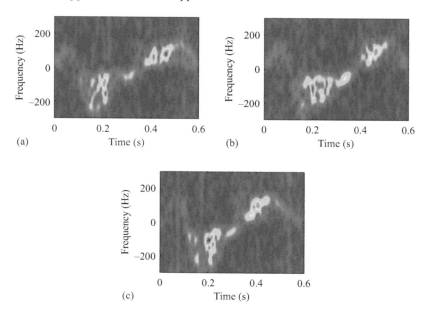

Figure 3.3   Spectrograms corresponding to dynamic hand gesture "hand rotation" from three personnel targets: (a) Target 1, (b) Target 2, and (c) Target 3

the velocity corresponding to the returning movement. The time–frequency trajectory of the gesture "flipping fingers" starts with a positive Doppler shift that corresponds to the middle finger flipping toward the radar. The differences among the time–frequency trajectories imply the potential to distinguish different dynamic hand gestures. From Figure 3.2, we can also see that most of the power of the dynamic hand gesture signals is distributed in limited areas in the time–frequency domain. This allows us to use sparse signal recovery techniques to extract micro-Doppler features of dynamic hand gestures. Figure 3.3 shows the spectrograms of received signals corresponding to dynamic hand gesture "hand rotation" from three personnel targets. It can be seen that the time–frequency spectrograms of the same gesture from different personnel targets have similar patterns.

3.3.2   Sparsity-driven dynamic hand gesture recognition

The scheme of the presented method is illustrated in Figure 3.4. This method contains two subprocesses, the training process and the testing process. The training process is composed of two steps. First, the time–frequency trajectory of each training signal is extracted using the Gabor dictionary and the OMP algorithm. Next the K-means algorithm is employed to cluster the time–frequency trajectories of all training signals and generate the central trajectory corresponding to each dynamic hand gesture. In the testing process, the modified Hausdorff distances [32,33] between the time–frequency trajectory of the testing signal and the central trajectories of dynamic hand

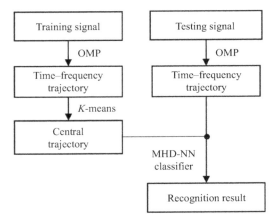

Figure 3.4 Processing scheme of sparsity-driven dynamic hand recognition method

gestures are computed and inputted into the NN classifier to determine the type of the dynamic hand gesture under test.

3.3.2.1 Extracting time–frequency trajectory

As shown in Figure 3.2, the time–frequency distributions of the dynamic hand gesture signals are generally sparse. The signal model is the same as in (3.1), where $\mathbf{y}$ is the received signal, $\mathbf{x}$ is the coefficient vector in the time–frequency domain, and $\boldsymbol{\Phi}$ is a Gabor dictionary [34] and its nth column (i.e., the basis-signal) can be expressed as

$$\Phi[:,n] = [\phi_n(1), \phi_n(2), \ldots, \phi_n(M)]^T \tag{3.24}$$

where

$$\phi_n(m) = \left(\pi\sigma^2\right)^{-(1/4)} \exp\left(-\frac{(m-t_n)^2}{\sigma^2}\right) \exp(j2\pi f_n m) \tag{3.25}$$

where t_n and f_n represent the discrete time shift and the frequency shift of the basis-signal, respectively, σ is the variance of the Gaussian window.

With the Gabor dictionary in (3.24), the sparse solution of $\mathbf{x}$ can be obtained by the OMP algorithm as

$$\hat{\mathbf{x}} = \text{OMP}(\mathbf{y}, \Phi, K) = \left(0, \ldots, \hat{x}_{i_1}, 0, \ldots, \hat{x}_{i_2}, 0, \ldots, \hat{x}_{i_K}, \ldots\right)^T \tag{3.26}$$

where K is the sparsity and $\{\hat{x}_{i_1}, \hat{x}_{i_2}, \ldots, \hat{x}_{i_K}\}$ are the nonzero entries. This implies that the time–frequency characteristics of $\mathbf{y}$ can be described by a group of basis signals at the time–frequency position (t_{i_k}, f_{i_k}) and with the corresponding intensity $A_{i_k} \triangleq |\hat{x}_{i_k}|$, for $k = 1, 2, \ldots, K$. Based on this observation, we define the time–frequency trajectory of the received signal $\mathbf{y}$ as

$$T(\mathbf{y}) = \left\{(t_{i_k}, f_{i_k}, A_{i_k}), \quad k = 1, 2, \ldots, K\right\} \tag{3.27}$$

which are utilised as the features for dynamic hand gesture recognition.

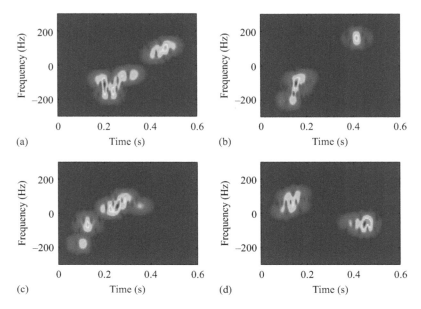

Figure 3.5   Spectrograms of reconstructed signals yielded by the OMP algorithm with K = 10: (a) hand rotation, (b) beckoning, (c) snapping fingers, and (d) flipping fingers

To explain the sparse signal representation clearer, the OMP algorithm is applied to analyse the measured signals presented in Figure 3.2. The length of each dynamic hand gesture signal is 0.6 s and the sampling frequency is 1 kHz, which means that the value of N is 600. The sparsity K is set to be 10. The variance σ of the Gaussian window is set to be 32. The dictionary $\boldsymbol{\Phi}$ is designed as discussed above and its size is $600 \times 2,400$. The OMP algorithm is used to solve the sparse vector , and then the reconstructed signal is obtained by $\mathbf{y}_{\text{rec}} = \boldsymbol{\Phi}\hat{\mathbf{x}}$. The time–frequency spectrograms of the reconstructed signals $\mathbf{y}_{\text{rec}}$ are plotted in Figure 3.5. By comparing Figures 3.2 and 3.5, we can find that the reconstructed signals contain exactly the majority part of the original time–frequency features. In addition, it is clear that the background noise has been significantly suppressed in the reconstructed signals, which is beneficial to dynamic hand gesture recognition. The locations of time–frequency trajectory, i.e. ($k = 1, 2, \ldots, K$) extracted by the OMP algorithms are plotted in Figure 3.6. By comparing Figures 3.2 and 3.6, we can see that the extracted time–frequency trajectories are capable of representing the time–frequency patterns of the corresponding dynamic hand gestures.

3.3.2.2   Clustering for central time–frequency trajectory

In the training process, a central time–frequency trajectory is clustered for each dynamic hand gesture using the K-means algorithm based on the time–frequency

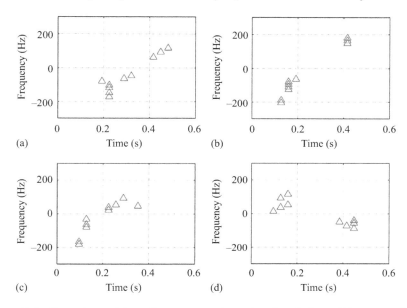

Figure 3.6 *Locations of time–frequency trajectories (k = 1, 2, ..., K) extracted by the OMP algorithms with K = 10: (a) hand rotation, (b) beckoning, (c) snapping fingers, and (d) flipping fingers*

trajectories of training signals. The details of the clustering process are presented as next.

We assume there are S segments of training signals for each dynamic hand gesture, denoted as $\mathbf{y}_g^{(s)}$ ($s = 1, 2, \ldots, S$), where s and g are the indexes of training segments and dynamic hand gestures, respectively. The time–frequency trajectory of $\mathbf{y}_g^{(s)}$ is denoted as $T(\mathbf{y}_g^{(s)})$, which is composed of K time–frequency positions as presented in (3.27). In the ideal case, different realisations of a certain dynamic hand gesture are expected to have the same time–frequency trajectory. However, in realistic scenarios, a human can hardly repeat one dynamic hand gesture in a completely same way. Therefore, there are minor differences among the time–frequency trajectories extracted from different realisations of one dynamic hand gesture. In order to explain this phenomenon more clearly, we plot the locations of the time–frequency trajectories extracted from the eight segments of signals corresponding to the gesture "snapping fingers" in Figure 3.7(a) for an example. It is clear that the time–frequency trajectories of different signal segments are distributed closely to each other with slight differences. In order to extract the main pattern from the time–frequency trajectories of training data, the K-means algorithm, which is a clustering technique widely used in pattern recognition [32,33], is employed to generate the central time–frequency trajectory of each dynamic hand gesture. The inputs of the K-means algorithm are the time–frequency trajectories of S training signals and the total number of input time–frequency positions is $K \times S$. With the K-means algorithm, K central time–frequency positions are produced to minimise the mean squared distance from each input time–frequency position to its nearest central position. The K-means algorithm is capable of compressing data and suppressing disturbances while retaining the major pattern of input data. More details about the K-means algorithm can be found in [33]. We denote the central time–frequency trajectory of dynamic hand gesture g generated by the K-means algorithm as

$$
\begin{aligned}
T_{c,g} &= K\text{-means}\left(T(\mathbf{y}_g^{(1)}), T(\mathbf{y}_g^{(1)}), \ldots, T(\mathbf{y}_g^{(S)})\right) \\
&= \left\{\left(t_{c,g}^{(k)}, f_{c,g}^{(k)}, A_{c,g}^{(k)}\right), k = 1, 2, \ldots, K\right\}
\end{aligned}
\tag{3.28}
$$

where $t_{c,g}^{(k)}$, $f_{c,g}^{(k)}$, and $A_{c,g}^{(k)}$ denote the time shift, the frequency shift, and the magnitude of the kth time–frequency position on the central time–frequency trajectory, respectively, and the superscript g is the dynamic hand gesture index. Figure 3.7(b) shows the location of the central time–frequency trajectory generated by the K-means algorithm using the time–frequency positions in Figure 3.7(a). It is clear that the majority of time–frequency positions in Figure 3.7(a) are located around the central time–frequency trajectory in Figure 3.7(b), which implies that the central time–frequency trajectory is capable of representing the major time–frequency pattern of a dynamic hand gesture.

3.3.2.3   Nearest neighbour classifier based on modified Hausdorff distance

In the testing process, the type of dynamic hand gesture corresponding to a given testing signal is determined by the NN classifier. The modified Hausdorff distance, which is widely used in the fields of pattern recognition [35,36], is used to measure the similarity between the time–frequency trajectory of the testing signal and the central time–frequency trajectory of each dynamic hand gesture.

For a testing signal $\mathbf{y}^{(*)}$, the classification process can be divided into three steps.

1. The time–frequency trajectory $T\left(\mathbf{y}^{(*)}\right)$ is extracted using the OMP algorithm as described in Section 3.3.2.1.
2. The modified Hausdorff distances between $T\left(\mathbf{y}^{(*)}\right)$ and the central time–frequency trajectories $T_{c,g}$ $(g = 1, 2, \ldots, G)$ are computed according to the following formula [35]:

$$
\text{MHD}\left(T(\mathbf{y}^{(*)}), T_{c,g}\right) = \sum_{\tau^{(*)} \in T(\mathbf{y}^{(*)})} d_H\left(\tau^{(*)}, T_{c,g}\right)
\tag{3.29}
$$

where τ is an element in set $T\left(\mathbf{y}^{(*)}\right)$, i.e., is a parameter set composed of the time shift, the frequency shift, and the amplitude as described in (3.7), and $d_H(\cdot, \cdot)$ represents the Hausdorff distance:

$$
d_H\left(\tau^{(*)}, T_{c,g}\right) = \min_{\tau \in T_{c,g}} \left\| \tau^{(*)} - \tau \right\|_2
\tag{3.30}
$$

where is an element in set $T_{c,g}$.

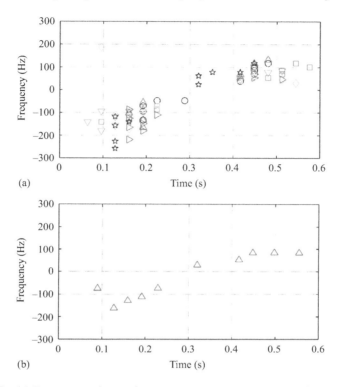

(a)

(b)

Figure 3.7   *(a) Locations of time–frequency trajectories extracted from eight segments of signals corresponding to the gesture "snapping fingers" with K = 10, where each type of marker indicates the time–frequency trajectory of a certain signal segment. (b) Locations of the clustered time–frequency trajectory generated by the K-means algorithm*

3.   The type of the dynamic hand gesture corresponding to $\mathbf{y}^{(*)}$ is determined by the following NN classifier:

$$g^{(*)} = \arg\min_{g \in \{1,2,\dots,G\}} \text{MHD}\left(T(\mathbf{y}^{(*)}), T_{c,g}\right) \tag{3.31}$$

where G represents the total number of dynamic hand gestures and $g^{(*)}$ indexes the recognition result.

3.3.3   Experimental results

In this section, the real data measured with the K-band radar are used to validate the presented method in terms of recognition accuracy, which is defined as the proportion of correctly recognised dynamic hand gesture signals among all the testing signals.

3.3.3.1   The recognition accuracy versus the sparsity

In this experiment, the recognition accuracies of the presented method with different values of sparsity K is evaluated. The performance of the presented method is compared with that of the sparse-SVM (support vector machine) method [37]. As for the sparse-SVM method in [37], the time–frequency trajectories of the dynamic hand gestures are extracted by the OMP algorithm as described in Section 3.3.2.1 and inputted into SVM for recognition. The sparsity K is varied from 7 to 21 with a step size of 2, and the recognition accuracies are computed by cross-validation. For each value of sparsity K, a certain proportion of measured signals are randomly selected for training, and the remaining data are used for testing. The recognition accuracies are averaged over 50 trials with randomly selected training data. The variance σ of the Gaussian window is set to be 32. The recognition accuracies yielded by the presented method and the sparse-SVM method using 30% and 70% of data for training are illustrated in Figure 3.8, and the confusion matrix yielded by the sparsity-driven method with $K = 17$ with 30% training data is presented in Table 3.3.

It is clear from Figure 3.8 that the recognition accuracies of the presented method increases as the sparsity K increases when $K \leq 15$. The reason is that more features of the dynamic hand gestures are extracted as the sparsity K increases. The recognition accuracies change slightly as the sparsity K increases when $K \geq 17$. This means that 17 features are sufficient to distinguish the gestures in this dataset. The training scheme proposed in [38] can be used to determine the proper value of sparsity K. That is, the sparsity-driven method is evaluated under different values of sparsity K by conducting multifold validation within the training dataset off-line, and the value of sparsity K corresponding to the highest recognition accuracy is selected in the final recognition system.

In addition, it can be seen from Figure 3.8 that the sparsity-driven method outperforms the sparse-SVM method under each value of sparsity. Furthermore, the recognition accuracies corresponding to 70% of training data are higher than that corresponding to 30% training data.

Table 3.3   Confusion matrix yielded by the sparsity-driven method with $K = 17$ and 30% of data for training

	Hand rotation (%)	Beckoning (%)	Snapping fingers (%)	Flipping fingers (%)
Hand rotation	96.67	2.32	0.72	0
Beckoning	3.21	95.42	3.45	0
Snapping fingers	0.12	2.26	95.71	0
Flipping fingers	0	0	0.12	100

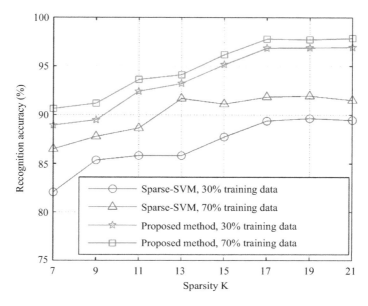

Figure 3.8   Recognition accuracies of the sparsity-driven method and the sparse-SVM method versus different the values of sparsity K

3.3.3.2   The recognition accuracy versus the size of training dataset

In this experiment, the performance of the sparsity-driven method is analysed with different sizes of training dataset. We compare the recognition accuracies yielded by the presented method with that yielded by the sparse-SVM method, the PCA-based methods, and the DCNN-based method. In the PCA-based methods, the micro-Doppler features of dynamic hand gestures are obtained by extracting the principal components of the received signals and inputted into SVM for recognition. Two kinds of PCA-based methods are considered here: (1) the PCA in the time domain, which extracts the features in the time-domain data [39] and (2) the PCA in the time–frequency domain, which extracts the features in the time–frequency domain as presented in [40,41]. As for the DCNN-based method, the time–frequency spectrograms are fed into a DCNN, where the micro-Doppler features are extracted using convolutional filters and the recognition is performed through fully connected perceptron functions. The structure of the DCNN used here is similar to that used in [42]. The proportions of training data are set to be varied from 10% to 90% with a step size of 10%, the sparsity K is set to be 17, and the variance σ of the Gaussian window is set to be 32. The resulting recognition accuracies are shown in Figure 3.9, where the sparsity-driven method obtains the highest recognition accuracies under different sizes of training set. In addition, the advantages of the presented method over the PCA-based and the DCNN-based methods are remarkable especially when the proportion of the training data is less

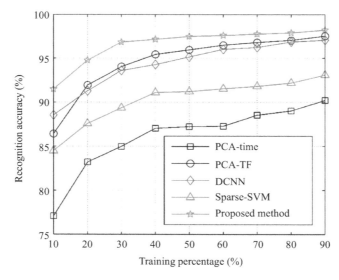

Figure 3.9   *Recognition accuracies yielded by the sparsity-driven method, the sparse-SVM method, the time-domain PCA-based method (denoted as PCA-time), the time–frequency domain PCA-based method (denoted as PCA-TF), and the DCNN-based method (denoted as DCNN) under different sizes of training set*

than 50%. This implies that the sparsity-driven method is more applicable than the PCA-based and the DCNN-based methods when the training set is small.

3.3.3.3   Recognition accuracy for unknown personnel targets

As described in Section 3.3.1, the dynamic hand gesture signals are measured from three personnel targets, denoted as Target 1, 2, and 3, respectively. In previous two experiments, the data measured from Target 1, 2, and 3 are mixed together, and a part of the data are used for training and the remaining data are used for testing. In this experiment, the data measured from one of Targets 1, 2, and 3 are used for training, and the data measured from the other two personnel targets are used for testing. This experiment aims to validate the sparsity-driven method in the condition of recognising the dynamic hand gestures of unknown personnel targets. Cross-validation is employed in this experiment. We randomly select 70% of the data measured from one of Targets 1, 2, and 3 for training and 70% of the data measured from the other two personnel targets for testing. The recognition accuracies are averaged over 50 trials with randomly selected training data and testing data. The sparsity K is set to be 17, and the variance σ of the Gaussian window is set to be 32. The resulting recognition accuracies are listed in Table 3.4. It can be seen that the sparsity-driven method obtains the highest recognition accuracies under all conditions. This implies that the presented method is superior to the sparse-SVM method, the PCA-based methods and

Table 3.4   Recognition accuracies for unknown personnel targets

	Training data from		
	Target 1 (%)	Target 2 (%)	Target 3 (%)
Sparsity-driven method	96.96	96.88	95.48
Sparse-SVM	90.54	90.54	88.21
DCNN	94.38	95.25	91.87
PCA-TF	92.14	91.80	91.14
PCA-time	84.68	84.59	81.07

Table 3.5   Time consumption of the dynamic hand gesture recognition methods

	Training time for one sample (ms)	Testing time for one sample (ms)
Sparsity-driven method	650	220
Sparse-SVM	130	154
DCNN	850	18
PCA-TF	4	11
PCA-time	0.1	0.5

the DCNN-based method in terms of recognising dynamic hand gestures of unknown personnel targets.

3.3.3.4   Computational costs considerations

In the sparsity-driven method for dynamic hand gesture recognition, the OMP algorithm performs K inner iterations for each received signal to extract the micro-Doppler feature, where K is the sparsity of the received signal. As demonstrated by the experimental results in Section 3.3.3.1, a moderate value of sparsity K can yield satisfying recognition accuracy.

The computational time consumed by the dynamic hand gesture recognition methods are measured in this subsection. The hardware platform is a laptop with an Intel® Core™ i5-4200M CPU inside, and the CPU clock frequency and the memory size are 2.5 GHz and 3.7 GB, respectively. The software platform is MATLAB® 2014a and the operating system is Windows 10. For each method, the running time for training and classifying are measured by averaging over 100 trials. The results of running time are presented in Table 3.5. It can be seen that the presented method and the DCNN-based method need more time for classifying and training, respectively, among all the tested approaches. In realistic applications, the dynamic hand gesture recognition needs to be real-time processing. Considering that the training process can be accomplished off-line, the bottleneck of real-time processing is the time consumption for classifying. Since the running time for classifying one hand gesture by the sparsity-driven sparsity-driven method with the non-optimised MATLAB® code is only 0.22 s, it is promising to achieve real-time processing with optimised code on DSP or FPGA platforms in practical applications.

3.4   Sparsity-based dictionary learning of human micro-Dopplers

Algorithms for micro-Doppler-based classification of both periodic human motions such as walking, running, crawling and boxing and aperiodic motions such as sitting, standing and falling have been extensively researched. However, in these works, the common assumption is that the propagation channel consists of only a single target or class of targets. This assumption is often violated in both indoor and outdoor environments. In indoor scenarios, there may be more than one moving human along with other dynamic targets such as fans. In outdoor scenarios, a pedestrian may share the propagation channel with animals and ground moving vehicles. When multiple dynamic targets coexist in a propagation channel, their radar returns interfere. Therefore, before we consider the problem of classification of micro-Dopplers, we must first separate or resolve the micro-Dopplers from distinct targets [43]. This problem falls under the category of single channel source separation that has been extensively researched in multiple domains such as speech recognition. Traditionally, returns from multiple targets are separated by incorporating hardware complexity into the radar. For example, broadband coherent radars resolve targets along the range and Doppler dimensions. Alternatively, narrowband sensors are fitted with multiple antenna elements and corresponding receiver channel in order to resolve targets in the Doppler and direction-of-arrival space. However, when resources are constrained, it is useful to rely on signal-processing methods to separate multiple target returns solely along the Doppler domain. Traditional data independent bases such as Fourier and wavelets have been extensively applied on human radar data. But micro-Doppler returns significantly overlap in the frequency domain and are difficult to separate. In this section, we demonstrate that data driven dictionaries learned directly from the time domain human radar data result in discriminative representations of the motion characteristics of the target behaviour. Due to the sparsity constraint imposed on the dictionary learning methods, the resultant bases are distinct and well resolved and useful for separating and reconstructing the micro-Dopplers from each of the multiple movers.

Next, we consider the classical classification problem. We consider situations where the presence of significant clutter or interference may result in weak the detection of targets resulting in incorrect classification. For example, many indoor radars use the ISM band that is shared with several other applications such as Wi-Fi and Bluetooth. Therefore, a degree of reconfigurability is desirable in indoor radars which may be facilitated by the implementation of the radar on software defined radio platforms. The hardware reconfigurability must, likewise, be supported by algorithms that can handle diversity of sensor parameter data. Since the sparsity-based data driven dictionaries represent motion characteristics, they can be used in these situations where there are considerable variations in the sensor parameters. We demonstrate the efficacy of dictionary learning by considering one such sensor parameter—the carrier frequency. We classify radar signals gathered at different radar carrier frequencies from that used during training [44].

3.4.1 Measurement data collection

The experimental set up in both applications consists of a monostatic radar configuration consisting of a N9926A Fieldfox vector network analyser (VNA) and two double ridged horn antennas. The VNA is configured to make narrowband time-domain two-port scattering parameter measurements (transmission parameter S_{21}) of different types of human motions along different trajectories in indoor line-of-sight conditions. Additional data are gathered for a table fan (TF) with moving blades. This target is chosen since the rotating blades' frequency modulates the carrier signal giving rise to distinct micro-Doppler patterns that clutter the Doppler signatures of humans in indoor environments.

3.4.2 Single channel source separation of micro-Dopplers from multiple movers

Here, we consider a scenario where the channel may consist of one or more targets moving simultaneously but along distinct trajectories. They are human walking toward the radar (FH), human walking away from the radar (BH), human walking tangentially across the path of the radar (SH) and a TF. Individual dictionaries for each of the cases are learnt using the formulations presented in (3.11). Training narrowband radar data at 7.5 GHz consists of only a single mover in the channel. However, the test scenario may have one or more movers. Therefore, the test radar signal consists of the superposition of scattered returns from multiple targets. The individual radar signals of the four classes of targets are reconstructed from the aggregate signal based on (3.22). Based on the energy of the reconstructed signals, we determine if that target class is present in the channel.

We show an example where the channel consists of three targets—two human movers walking in opposite directions (FH and BH) and a TF. After disaggregation, the signals from each of the targets are reconstructed.

Figure 3.10(a) shows the spectrogram generated from the superposed signals from all three targets. The remaining figures show the spectrograms reconstructed from the disaggregated signals from each of the targets. Figure 3.10(b) shows the positive Dopplers from the torso and limbs of the human walking toward the radar, while Figure 3.10(c) shows the negative Dopplers from the human walking away from the radar. The human walking away from the radar is nearer than the human walking toward the radar, and hence the signals are stronger for BH. Figure 3.10(d) shows the very distinctive patterns arising from the returns from the fan. These figures illustrate that the method is able to disaggregate signals that have considerable overlap along both time and frequency axes.

In Table 3.6, we show the false alarm and true detections for single and multiple target scenarios. The results show that the method is able to detect multiple movers

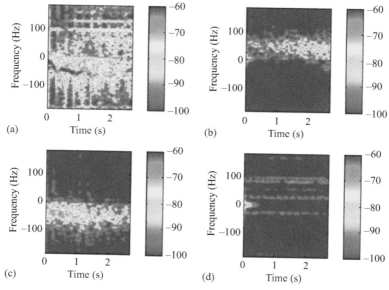

Figure 3.10   *Doppler spectrograms of (a) combined signal from three targets*
(b) reconstructed/disaggregated signal from a human walking toward
a radar, (c) reconstructed/disaggregated signal of a human walking
away from radar, and (d) reconstructed/disaggregated signal of
table fan

Table 3.6   *True detection and false alarm rates for single or more dynamic movers*
using dictionary learning methods

Target scenarios	Average true detections	Average false alarms
Single target		
FH	100	9
BH	100	10
SH	94	2
TF	100	0
Two targets		
FH+TF	88 + 94	12
BH+TF	94 + 96	28
SH+TF	100 + 98	22
FH+BH	88 + 80	56
Three targets		
FH+BH+TF	95 + 90 + 90	0
FH+SH+TF	95 + 100 + 100	0
BH+SH+TF	95 + 100 + 100	25
Four targets		
FH+BH+SH+TF	88 + 68 + 100 + 100	Not applicable

with significantly overlapping micro-Doppler patterns (in the time–frequency representations). The TF features are spread over the entire joint time–frequency space and hence have considerable overlap with the other micro-Doppler signals. However, their unique patterns are captured in the dictionary learning framework and hence, they are nicely segregated from the radar signals scattered from the humans. When there are multiple humans moving simultaneously, there are instances when one of them shadows the other from the radar. Hence, the detection performance deteriorates in these circumstances. When the human approaches the radar, the Dopplers are mostly positive while they are negative when the subject moves away. Therefore the SH category consists of positive Dopplers for a certain duration following which the Dopplers become negative. Also, these Doppler tend to be low due to low radial velocity components. Hence, the radar signatures again overlap significantly with the human walking toward and away from the radar. However, the algorithm succeeds in separating the radar signals from these classes.

We observe that the problem, in this case, becomes more challenging as the number of movers in the channel increases. The algorithm confuses between the different categories of human motions.

3.4.3 Target classification based on dictionaries

Next, we consider the scenario when there are training radar data collected at different sensor parameters when compared to the test data. We consider training data collected at four of five different carrier frequencies (2.4, 3, 4, 4.5, and 5.8 GHz) while the test data are collected at the fifth carrier frequency. We consider four categories of micro-Doppler signals: two humans walking before the radar along any trajectory (TH), a human standing and boxing (HB), a single human walking before the radar (HW), and a TF. In Table 3.7, we present the classification accuracy for the four dictionary

Table 3.7   Accuracy for classification of radar micro-Doppler from two humans (TH), boxing human (HB), walking human (HW), and table fan (TF) for different dictionary learning algorithms

Algorithm	TH (%)	HB (%)	HW (%)	TF (%)
Synthesis dictionary learning (SDL)	67	88	93	95
Deep dictionary learning (DDL)	68	98	97	99
Analysis dictionary learning (ADL)	54	42	58	68
Label consistent KSVD (LC-KSVD)	15	75	19	92

Table 3.8   Training and test time for different dictionary learning algorithms

Algorithm	Training time (s)	Testing time (s)
SDL	415	3.4
DDL	631	0.2
ADL	291	0.07
LC-KSVD	1,815	0.2

learning algorithms—SDL, DDL, ADL, and LC-KSVD. The formulations for each of these were described in detail in Section 3.2.3 in (3.12), (3.14), (3.17), and (3.20), respectively. The results show that both the synthesis dictionaries perform fairly well at classifying these radar signals despite the variation in the sensor parameters. In other words, the dictionaries capture the physics of the motions instead of being limited by the sensor parameters. As anticipated, the deeper representations result in the better classification performance than the shallow representations.

Table 3.8 shows the test and training time for the four algorithms. It is observed that the LC-KSVD algorithm has the highest training time. However, in real world deployments, the test times are far more important than training times. Here, we observe that the ADL has the lowest test time due to the straight forward matrix multiplication operation during test. The DDL has a much lower test time than SDL due to the dimensionality reduction that is introduced with multiple layer representations.

3.5   Conclusions

In this chapter, we have demonstrated the feasibility of sparsity-driven methods for micro-Doppler detection and classification. Two scenarios have been considered: (1) the dictionary is pre-defined and fixed during the processing. Taking advantage of the sparse property of radar echo reflected from dynamic hand gestures, the Gabor dictionary and the OMP algorithm are used to extract the micro-Doppler features. The extracted features are inputted into modified-Hausdorff-distance-based nearest neighbour classifier to determine the type of dynamic hand gestures. Experimental results based on measured data show that the sparsity-driven method obtains recognition accuracy higher than 96% and outperforms the PCA-based and the DCNN-based methods in the conditions of small training dataset. (2) The dictionary is learnt from the training data using dictionary learning methods. We used these sparse codes to detect multiple movers (humans moving along different trajectories) in the channel as well as to classify different types of human motions.

References

[1] E. J. Candes and T. Tao, "Decoding by linear programming," *IEEE Transactions on Information Theory*, vol. 51, no. 12, pp. 4203–4215, 2005.

[2] E. Candes and M. B. Wakin, "An introduction to compressive sampling," *IEEE Signal Processing Magazine*, vol. 25, no. 2, pp. 21–30, 2008.

[3] R. G. Baraniuk, "Compressive sensing," *IEEE Signal Processing Magazine*, vol. 24, no. 4, pp. 118–121, 2007.

[4] Y. Luo, Q. Zhang, C. Qiu, S. Li, and T. S. Yeo, "Micro-Doppler feature extraction for wideband imaging radar based on complex image orthogonal matching pursuit decomposition," *IET Radar, Sonar & Navigation*, vol. 7, no. 8, pp. 914–924, 2013.

[5] G. Li and P. K. Varshney, "Micro-Doppler parameter estimation via parametric sparse representation and pruned orthogonal matching pursuit," *IEEE Journal of Selected Topics in Applied Earth Observations and Remote Sensing*, vol. 7, no. 12, pp. 4937–4948, 2014.

[6] D. Gaglione, C. Clemente, F. Coutts, G. Li, and J. J. Soraghan, "Model-based sparse recovery method for automatic classification of helicopters," In: Proc. 2015 IEEE Radar Conf., May 2015, pp. 1161–1165.

[7] F. K. Coutts, D. Gaglione, C. Clemente, G. Li, I. K. Proudler, and J. J. Soraghan, "Label consistent K-SVD for sparse micro-Doppler classification," In: Proc. 2015 IEEE Int. Conf. Digit. Signal Process., Jul. 2015, pp. 90–94.

[8] G. Li, R. Zhang, M. Ritchie, and H. Griffiths, "Sparsity-driven micro-Doppler feature extraction for dynamic hand gesture recognition," *IEEE Transactions on Aerospace and Electronic Systems*, vol. 54, no. 2, pp. 655–665, 2018.

[9] J. H. G. Ender, "On compressive sensing applied to radar," *Signal Processing*, vol. 90, pp. 1402–1414, 2010.

[10] M. G. Amin, B. Jokanovic, Y. D. Zhang, and F. Ahmad, "A sparsity-perspective to quadratic time–frequency distributions," *Digital Signal Processing*, vol. 46, pp. 175–190, 2015.

[11] E. J. Candes and T. Tao, "Decoding by linear programming," *IEEE Transactions on Information Theory*, vol. 51, no. 12, pp. 4203–4215, 2005.

[12] S. Foucart, "A note on guaranteed sparse recovery via L_1-minimization," *Applied and Computational Harmonic Analysis*, vol. 29, no. 1, pp. 97–103, 2010.

[13] S. S. Chen, D. L. Donoho, and M. A. Saunders, "Atomic decomposition by basis pursuit," *SIAM Review*, vol. 43, no. 1, pp. 129–159, 2001.

[14] E. Candes and J. Romberg, "L_1-Magic: Recovery of Sparse Signals Via Convex Programming," 4, 2005, p. 14, URL: www.acm.caltech.edu/l1magic/downloads/l1magic.pdf

[15] M. Grant, S. Boyd, and Y. Ye, "CVX: Matlab Software for Disciplined Convex Programming," 2008.

[16] SPGL1, "A Solver for Large-Scale Sparse Reconstruction," https://friedlander.io/spgl1/

[17] Y. Zhang, J. Yang, and W. Yin, http://yall1.blogs.rice.edu/

[18]   S. Boyd and L. Vandenberghe, *Convex Optimization*. Cambridge, UK: Cambridge University Press, 2004.

[19]   R. G. Baraniuk, V. Cevher, and M. B. Wakin, "Low-dimensional models for dimensionality reduction and signal recovery: A geometric perspective," *Proceedings of the IEEE*, vol. 98, no. 6, pp. 959–971, 2010.

[20]   D. P. Wipf and B. D. Rao, "Sparse Bayesian learning for basis selection," *IEEE Transactions on Signal Processing*, vol. 52, no. 8, pp. 2153–2164, 2004.

[21]   M. A. Figueiredo, J. M. Bioucas-Dias, and R. D. Nowak, "Majorization-minimization algorithms for wavelet-based image restoration," *IEEE Transactions on Image Processing*, vol. 16, no. 12, pp. 2980–2991, 2007.

[22]   S. Ji, Y. Xue, and L. Carin, "Bayesian compressive sensing," *IEEE Transactions on Signal Processing*, vol. 56, no. 6, pp. 2346–2356, 2008.

[23]   J. Tropp and A. C. Gilbert, "Signal recovery from partial information via orthogonal matching pursuit," *IEEE Transactions on Information Theory*, vol. 53, no. 12, pp. 4655–4666, 2007.

[24]   D. Needell and J. A. Tropp, "CoSaMP: Iterative signal recovery from incomplete and inaccurate samples," *Applied and Computational Harmonic Analysis*, vol. 26, no. 3, pp. 301–321, 2009.

[25]   T. Blumensath and M. Davies, "Iterative hard thresholding for compressed sensing," *Applied and Computational Harmonic Analysis*, vol. 27, no. 3, pp. 265–274, 2009.

[26]   Y. C. Eldar and G. Kutyniok, *Compressed Sensing: Theory and Applications*. Cambridge, UK: Cambridge University Press, 2012.

[27]   M. Aharon, M. Elad, and A. Bruckstein, "K-SVD: An algorithm for designing overcomplete dictionaries for sparse representation," *IEEE Transactions on Signal Processing*, vol. 54. no. 11, pp. 4311–4322, 2006.

[28]   I. Daubechies, M. Defrise, and C. De Mol, "An iterative thresholding algorithm for linear inverse problems with a sparsity constraint," *Communications on Pure and Applied Mathematics*, vol. 57, no. 11, pp. 1413–1457, 2004.

[29]   S. Tariyal, A. Majumdar, R. Singh, and M. Vatsa, "Deep dictionary learning," *IEEE Access*, vol. 4, pp. 10096–10109, 2016.

[30]   J. Zhuolin, Z. Lin, and L. S. Davis, "Label consistent K-SVD: Learning a discriminative dictionary for recognition," *IEEE Transactions on Pattern Analysis and Machine Intelligence*, vol. 35, no. 11, pp. 2651–2664, 2013.

[31]   R. Rubinstein, T. Peleg, and M. Elad, "Analysis K-SVD: A dictionary-learning algorithm for the analysis sparse model," *IEEE Transactions on Signal Processing*, vol. 61, no. 3, pp. 661–677, 2013.

[32]   T. Hastie, R. Tibshirani, and J. Friedman, *The Elements of Statistical Learning: Data Mining, Inference, and Prediction*. Berlin, Germany: Springer Series in Statistics, 2009.

[33]   T. Kanungo, D. M. Mount, N. S. Netanyahu, C. D. Piatko, R. Silverman, and A. Y. Wu, "An efficient k-means clustering algorithm: Analysis and implementation," *IEEE Transactions on Pattern Analysis and Machine Intelligence*, vol. 24, no. 7, pp. 881–892, 2002.

[34] S. G. Mallat and Z. Zhang, "Matching pursuits with time–frequency dictionaries," *IEEE Transactions on Signal Processing*, vol. 41, no. 12, pp. 3397–3415, 1993.

[35] M. P. Dubuisson and A. K. Jain, "A modified Hausdorff distance for object matching," In: Proceedings of the International Conference on Pattern Recognition (ICPR'94), Oct. 1994, pp. 566–568.

[36] D. P. Huttenlocher, G. A. Klanderman, and W. J. Rucklidge, "Comparing images using the Hausdorff distance," *IEEE Transactions on Pattern Analysis and Machine Intelligence*, vol. 15, no. 9, pp. 850–863, 1993.

[37] G. Li, R. Zhang, M. Ritchie, and H., Griffiths, "Sparsity-based dynamic hand gesture recognition using micro-Doppler signatures," In: Proceeding of 2017 IEEE Radar Conference, May 2017, pp. 0928–0931.

[38] P. Molchanov, *Radar Target Classification by Micro-Doppler Contributions*. Tampere, Finland: Tampere University of Technology, 2014.

[39] A. Balleri, K. Chetty, and K. Woodbridge, "Classification of personnel targets by acoustic micro-Doppler signatures," *IET Radar, Sonar & Navigation*, vol. 5, no. 9, pp. 943–951, 2011.

[40] Q. Wu, Y. D. Zhang, W. Tao, and M. G. Amin, "Radar-based fall detection based on Doppler time–frequency signatures for assisted living," *IET Radar, Sonar & Navigation*, vol. 9, no. 2, pp. 164–172, 2015.

[41] B. Jokanovic, M. Amin, F. Ahmad, and B. Boashash, "Radar fall detection using principal component analysis," In: Proc. SPIE 9829, Radar Sensor Technology XX, 982919, May 2016.

[42] Y. Kim and B. Toomajian, "Hand gesture recognition using micro-Doppler signatures with convolutional neural network," *IEEE Access*, vol. 4, pp. 7125–7130, 2016.

[43] S. Vishwakarma and S. S. Ram, "Detection of multiple movers based on single channel source separation of their micro-Dopplers," *IEEE Transactions on Aerospace and Electronic Systems*, vol. 54, no. 1 pp. 159–169, 2018.

[44] S. Vishwakarma and S. S. Ram, "Dictionary learning with low computational complexity for classification of human micro-Dopplers across multiple carrier frequencies," *IEEE Access*, vol. 6, pp. 29793–29805, 2018.

Chapter 4

Deep neural networks for radar micro-Doppler signature classification

Sevgi Zubeyde Gurbuz[1] and Moeness G. Amin[2]

In recent years, advances in machine learning, parallelisation, and the speed of graphics processing units, combined with the availability of open, easily accessible implementations, have brought deep neural networks (DNNs) to the forefront of research in many fields. DNNs are a key enabler of driverless vehicles, and have made possible amazing feats in image and natural language processing, including object recognition and classification, colourisation of black and white images, automatic text and handwriting generation, and automatic game playing, to name a few. DNNs are now able to harvest extremely large amounts of training data to achieve record breaking performance. For example, in 2015, the winner of the ImageNet Large Scale Visual Recognition Challenge (ILSVRC) was a 152-layer residual neural network (ResNet [1]), which categorised 100,000 images into 1,000 categories.

Likewise, deep learning has offered significant performance gains is the classification of radar micro-Doppler signatures, paving the way for new civilian applications of radio frequency (RF) technologies that require a greater ability to recognise a larger number of classes that are similar in nature. Current applications of radar micro-Doppler analysis include not just military applications of border control and security, perimeter defence, and surveillance, but also applications impacting the daily life, such as the monitoring daily activity as part of a smart environment, home security (intruder detection), assisted living, remote health monitoring, telemedicine, and human–computer interfaces.

The primary focus of assisted living and remote health applications has been non-contact measurement [2] of heart rate [3,4] and respiration [5,6], including detection of related conditions, such as sleep apnea [7] or sudden infant death syndrome [8], as well as fall detection [9–13]. Falling remains a leading cause of mortality among the elderly, even in nursing homes, where it is estimated that as much as 20% of falls result in serious injury and 1,800 elderly die each year as a result of falls [14]. The prognosis after a fall is dependent upon the amount of time that elapses until medical care for fall-related injuries is provided. Thus, rapid response is critical to minimising long-term debilitation, and ensuring senior citizens maintain their independence and

[1]Department of Electrical and Computer Engineering, The University of Alabama, Tuscaloosa, AL, USA
[2]Center for Advanced Communications, Villanova University, Villanova, PA, USA

a high quality of life. As such, fall detection specifically has become a dominant focus in noncontact, remote gait analysis and sensing technologies for telemedicine and smart homes. However, radar has shown great potential to assess not just falls, but to detect gait abnormalities [15] for fall risk prediction, monitor neuromuscular disorders, evaluate balance, identify concussions [16], and monitor patient response to treatments, operations, physical therapy, and rehabilitation [17].

In the context of home security and energy-efficient smart homes, occupancy (or conversely, vacancy) sensing [18] can be used to intelligently operate home systems, such as lighting or heating, ventilation, and air conditioning units. Beyond simple motion detectors that could be triggered by fine motion, such as the turning of the pages of a book, radar Doppler sensors can detect and classify larger-scale activity. The unique patterns in the micro-Doppler signature of individuals can also be exploited to identify specific people [19,20], count the number of people in a room [21], recognise individuals within groups of 4 to 20 people [22], and detect and discriminate intruders from among known residents of a house [23]. In fact, it has even been shown that the placement of objects, e.g. hand grenades, about the waist, or carrying a rifle can affect changes in the micro-Doppler signature, which can be used for active shooter recognition [24], suicide bomber detection [25], and armed/unarmed personnel classification [26].

In all of the aforementioned applications, DNNs have been an enabling technology in the advancement of radar micro-Doppler classification algorithms [27]. Conversely, the unique properties of RF data have been motivating research into novel approaches to DNN design specific to micro-Doppler classification. RF data have a fundamentally different phenomenology from optical imagery or speech signals. Rather, the signal received by a radar is typically a complex time-series, whose amplitude and phase are related to the physics of electromagnetic scattering and kinematics of target motion. It is only after a number of preprocessing stages, such as clutter filtering, pulse compression, and time–frequency (TF) analysis, that one-, two-, or three-dimensional (1D, 2D, or 3D) data can be generated and presented to a DNN. As a result, DNN approaches, popular in the image processing domain, do not directly translate to the RF domain, and their benefits and performance must therefore be separately reassessed. For example, although GoogleNet [28] surpassed VGGnet [29] in terms of classification accuracy on the ImageNet database, when utilised with transfer learning on radar micro-Doppler signatures, in fact the reverse was found—VGGnet outperformed GoogleNet [30]. In a related fashion, sometimes conventional wisdom in radar signal processing can prove detrimental when applied to DNN-based classification of radar data. For example, while pre-processing techniques, such as high-pass filtering or moving target indication (MTI) has generally been advantageous when extracting handcrafted features, such filtering of ground clutter can be detrimental to DNN performance [21].

A related challenge, which further necessitates the exploitation of a physics-based knowledge-aided signal processing approach with DNN design and training, is the small amount of measured data that is available in RF applications. Whereas in image processing millions of samples can be easily available, it is both costly and time-consuming to acquire radar data. Thus, the depth and performance of DNNs used in radar applications are challenged by only having several hundred or a thousand samples available for training. Moreover, these samples may also not be completely representative of the broad range of operational scenarios and target types the radar is expected to classify.

In this chapter, we focus on the aspects of DNN design that are specific to radar micro-Doppler data. First, we will describe preprocessing steps applied to the raw complex time stream of RF measurements, the derivation of different data representations, and how this affects the subsequent DNN architecture used for classification and its performance. Next, methods for addressing the challenges involved in training DNNs for radar micro-Doppler classification will be presented. A performance comparison of several DNN architectures will be presented using benchmarking on a common micro-Doppler data set. Finally, current challenges and future directions for DNN research will be discussed.

4.1   Radar signal model and preprocessing

The signal received by a radar is, in general, a time-delayed, frequency-shifted version of the transmitted signal. While the structure of a vehicle may be represented by the scattering of a discrete point target, for a high resolution, linear frequency-modulated continuous wave (FMCW) radar, the human body behaves as a distributed target. Finite element models of the electromagnetic scattering from the body [31] have been used to determine Swerling models that approximate the radar cross section (RCS) of the human body during different activities. However, for many practical radar scenarios, multipath reflections can be ignored, and it has been shown that the scattering from the entire human body, $x[n]$, may be approximated using superposition of the returns from K body parts [32,33]. Thus, for an FMCW radar:

$$x[n] = \sum_{i=1}^{K} a_i \exp\left\{-j\frac{4\pi f_c}{c}R_{n,i}\right\} \tag{4.1}$$

where $R_{n,i}$ is the range to the ith body part at time n, f_c is the transmit frequency, c is the speed of light, and the amplitude a_i is the square root of the power of the received signal as given by

$$a_i = \frac{\sqrt{G_{tx}G_{rx}}\lambda\sqrt{P_t\sigma_i}}{(4\pi)^{3/2}R_i^2\sqrt{L_s}\sqrt{L_a}} \tag{4.2}$$

Here, G_{tx} and G_{rx} are the gains of the transmit and receive antennas, respectively; λ and P_t are the wavelength and power of the transmitted signal, respectively; σ_i is the RCS of the ith body part; and L_s and L_a are system and atmospheric losses, respectively.

The effect of target kinematics is predominantly reflected in the frequency modulations of the received signal. While the distance of the target to the radar does affect the received signal power, these variations are not significant for human subjects given typical body sizes. Rather, the amplitude of the received signal is primarily dependent upon the transmit frequency being used and RCS of target observed. Thus, at higher frequencies, such as 77 GHz, the received signal power from a person will be much lower than that of a radar at lower frequencies, such as 24 or 5.8 GHz. Not just transmit frequency, but also bandwidth, pulse repetition interval, observation duration, and aspect angle have been shown to affect the radar received signal and the performance of machine learning algorithms applied to the data [34].

A fundamental question is then how to best make accessible those signal properties that are most critical to classification. One approach could be to directly classify the 1D complex time stream acquired by the radar [35]. This relies on the DNN to automatically learn suitable features from the raw in-phase and quadrature (I/Q) data, from which it is not possible to directly compute handcrafted features that have physical meaning [36], such as bandwidth or stride rate, often used in supervised classification with machine learning. As the number of classes increases, direct classification of raw radar data requires the DNN to learn an increasingly high-dimensional mapping between the raw data and motion classes. This drives up the computation cost and complexity of the resulting network, while also precluding utilisation of radar signal processing approaches that can mitigate clutter and simplify the DNN architecture—these are techniques which typically also result in networks with more robust performance, simpler training, and optimisation.

Thus, a much more common approach is to apply some TF analysis as well as range processing to generate 2D and 3D representations of the radar data, depicted in Figure 4.1 and thereby bring out the wealth of information contained in the amplitude and frequency modulations of the received signal.

4.1.1   2D Input representations

Four different types of 2D inputs can be generated from raw radar data: micro-Doppler signatures, range maps, range-Doppler (RD) images, and cadence velocity diagrams (CVDs) [37]. The range map shows the variation of target position with slow time, while RD images show the relationship between target position and velocity. Micro-Doppler signatures reflect the variation of frequency with time, and may be computed using a variety of TF transforms. The CVD is computed by taking the fast Fourier transform (FFT) of a given frequency bin over time.

TF analysis methods can be classified into linear transforms and quadratic TF distributions (QTFDs) [38,39]. The former includes the short-time Fourier transform (STFT) [38], Gabor transform [40], fractional Fourier transform [41], and wavelet transform [42]. The most commonly applied TF transform is the spectrogram, denoted by $S(t, \omega)$, which is the square modulus of the STFT, and a special case of (4.5). It can be expressed in terms of the employed window function, $w(t)$, as

$$S(t, \omega) = \left| \int_{-\infty}^{\infty} w(t - u)x(u)e^{-j\omega t} du \right|^2, \qquad (4.3)$$

where $x(t)$ is the continuous-time input signal.

While reaping the benefits of linearity, STFTs must trade-off time and frequency resolution due to fixed length windows. In contrast, alternative transforms can capture local behaviour of the signal in differing ways.

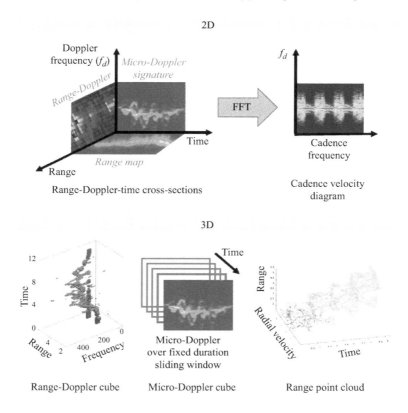

Figure 4.1 Multidimensional input representations of radar data

The instantaneous autocorrelation function (IAF) is [38]:

$$C(t, \tau) = x(t + \tau)x^*(t - \tau) \tag{4.4}$$

where τ is the time lag, and (:)* denotes the complex conjugate. The Wigner–Ville distribution (WVD) is the Fourier transform of the IAF with respect to τ, whereas the Fourier transform over the time delay t yields the ambiguity function, $A(\theta, \tau)$. The WVD has improved time and frequency resolution in comparison to the STFT; however, the presence of cross-term interference limits its practical application to micro-Doppler analysis. As a result, the Gabor–Wigner transform, defined as the product of the STFT with the Wigner distribution, has been proposed as a means for removing cross-term interference while maintaining the clarity and resolution of the result [43].

QTFDs, on the other hand, aim at concentrating the signal power along the instantaneous frequency of each signal component. In the case of radar backscattering from

a moving target, such concentration accurately reveals the target velocity, acceleration, and higher-order terms. The Cohen's class of QTFDs of signal $x(t)$ is defined
as [44,45]:

$$D(t, \omega) = FT_2[\Phi(\theta, \tau)A(\theta, \tau)] \int_{-\infty}^{\infty} \int_{-\infty}^{\infty} \Phi(\theta, \tau)A(\theta, \tau)e^{-j(\omega\tau + \theta t)}d\tau d\theta \quad (4.5)$$

where $FT_2[:]$ denotes 2D Fourier transform. The kernel $\Phi(\theta, \tau)$ typically has low-
pass filter characteristics to suppress signal cross-terms and preserve auto-terms,
leading to reduced interference distributions [44]. The backscattered radar signal
from humans in motion are viewed as deterministic signal in additive noise rather
than random processes [46].

4.1.2   3D Input representations

The availability of different joint domain representations of radar measurements has
led to not just the proposal of fusing classification results achieved with different
2D representations [47,48], but also to 3D representations in terms of time, range,
and frequency. The most common approach to forming a 3D data cube is to form a
sequence of frames as a function of time that would effectively resemble video, but
for radar data. Both RD video sequences [49–52] as well as micro-Doppler video
sequences [53] have been applied to radar micro-Doppler classification of human
activities and gestures. A single RD frame is obtained by applying a Fourier transform
at each range bin over a non-overlapping coherent processing interval of slow time on
the range map. Additionally, a radar point cloud [54] may be derived from RD videos
by applying constant false alarm rage (CFAR) detection on the data, computing an
iso-surface mesh from points with similar intensity [55] and then applying farthest
point sampling to arrive at a 3D, discrete representation [56].

4.1.3   Data preprocessing

4.1.3.1   Dimensionality reduction

Often times the signal processing used to generate suitable 2D or 3D representations
results in large structures, which when flattened, translate into vectors that have too
high a dimensionality. For example, a radar data cube of size $64 \times 64 \times 64$ results
in a vector of size $262,144 \times 1 \times 1$, upon flattening. In contrast, due to the curse of
dimensionality, DNNs often take much smaller input representations. For example, in
the CIFAR-10 database, In CIFAR-10, images are only of size $32 \times 32 \times 3$ (32 wide,
32 high, 3 colour channels), so a single, fully connected neuron in the first layer of a
DNN would have 3,072 weights.

Thus, dimensionality reduction (DR) is often applied on data prior to input to
DNNs, both in the case of 2D and 3D data. However, the first step of some DR
approaches, such as conventional principal component analysis (PCA), involves vectorisation, or flattening, of the data. This reshaping operation breaks down the natural
structure and correlation in the original data [57]. One alternative approach is to
represent the images as matrices (second-order tensors) and find the image covariance matrix. Spatial correlation of the image pixels within a localised neighbourhood

is then used to create a less restrictive PCA. Alternatively, multi-subspace learning techniques, such as multilinear PCA and multilinear discriminant analysis, have been proposed for dimension reduction of 3D radar data cubes [49]. It should be noted that image resizing techniques, which are applied after computation of the TF transform, tend to also degrade the data so that it yields lower classification accuracy than would have otherwise been achieved had a smaller image been originally generated through manipulation of the window length and overlap parameters of the spectrogram.

4.1.3.2   Mitigation of clutter, interference, and noise

The received signal model presented in (4.1) considers only reflections from the target of interest. In fact, the radar also receives reflections from other objects or surfaces in its vicinity, known as clutter, as well as signals from other sources of unwanted RF interference, such as from jammers or communications signals that may fall in the same receive frequency band. A complete received signal model that includes both interference and clutter is $x_r[n] = x[n] + x_c[n] + x_i[n]$, where $x[n]$ is the target return and $x_c[n]$ and $x_i[n]$ are random signals representing the noise/clutter and interference, respectively. While reflections from stationary objects result in a narrow clutter spectrum centered on 0 Hz, the rotational motion of objects such as a swivel chair, table, pedestal, or ceiling fan can result in non-zero micro-Doppler with its own distinct features. In both cases, the clutter returns can mask the low-frequency components of the target micro-Doppler, especially in the case of slow-moving targets, such as typical indoor human motion.

Oftentimes, pre-processing can be applied to isolate the micro-Doppler frequency components corresponding only to the motion of interest. One of the most basic ways to minimise noise is to apply an image-processing-based approach, where a threshold is determined based on a histogram of the pixel values in the image. The eCLEAN [55] algorithm implements a more advanced thresholding scheme, where the 2D histogram is first downsampled and normalised before the threshold is determined. The eCLEAN algorithm scans across the time axis, determining the number of points that should be extracted based on the threshold. Mask functions are created to extract the signal for each slice of time. This method has been shown to yield more robust results than predefined thresholding when applied to micro-Doppler signatures [49]. Radar signal processing approaches for mitigating noise and clutter include high-pass filtering or MTI filtering to remove ground clutter, or range gating to mitigate both noise and clutter by isolating those data samples corresponding to the target location. Figure 4.2 gives examples of clutter and noise mitigation techniques applied on micro-Doppler signatures.

An important open question in the application of DNNs on radar data, however, is whether it is always necessary to apply such conventional pre-processing techniques to the data. It has been reported in the literature [21] that filtering to remove clutter has in fact had a detrimental effect on DNN performance, because such filters do not just remove noise, but signal components as well, which can supply useful information for learning effective features for classification. Moreover, different DNNs respond to the presence of spurious signals in different ways. Consider the activation layers for different DNNs in response to an unwanted spike caused by a buffering error in

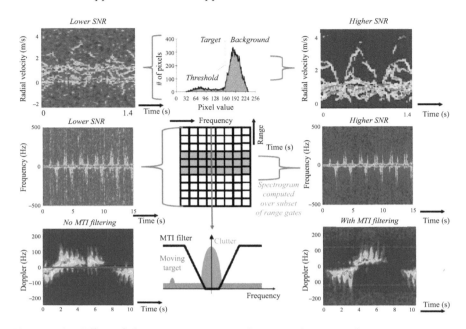

Figure 4.2   Effect of clutter suppression and noise reduction techniques on spectrogram of radar data

the software-defined radio used to acquire an observation of a person falling, shown in Figure 4.3. Transfer learning using VGGNet and a three-layer convolutional neural network (CNN) appear to preserve this spike, regarding it as signal, whereas this same spike is rejected by both the convolutional autoencoder (CAE) and transfer learning from GoogleNet. Why one network is more susceptible to such unwanted artefacts while others appear immune is an open question that requires more investigation if robust DNNs are to be designed with high performance in real-world environments with high clutter and interference.

4.2   Classification with data-driven learning

The typical approach for supervised classification involves first learning a set of predefined or handcrafted features from a training data set, and then applying the same feature computation process on the test data to decide the class of each test sample. In contrast, data-driven feature learning methods can modify or adapt the process of extracting features by exploiting knowledge gained from analysis of the training data set. A block diagram illustrating the main steps in classification with data-driven learning is shown in Figure 4.4. Deep learning is currently one of the most popular approaches for data-driven learning; however, other data-driven approaches include PCA and optimisation of feature extraction using genetic algorithms.

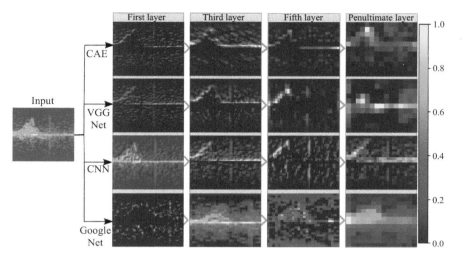

Figure 4.3   Visualisation of activation layers for different DNNs when data include unwanted artefacts [30]

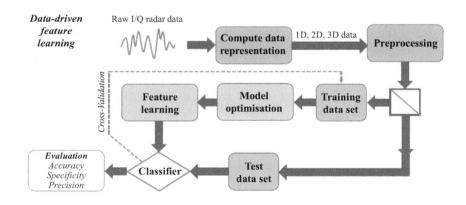

Figure 4.4   Flowchart of data-driven classification approach

4.2.1   Principal component analysis

An example of an unsupervised approach to data-driven feature learning is PCA. PCA is a dimension reduction technique which seeks to apply a linear transformation on the data such that the variance of the components in the resulting representation is maximised. PCA begins by first removing the mean and normalising the variance of correlated data. The direction for which the variance is maximised corresponds to the principal eigenvector of the covariance matrix Σ of the data. To project the data to a

k-dimensional subspace, the new components should be selected as the top k eigenvectors of Σ, which therefore also have the property of being mutually orthogonal. In this way, application of PCA results in a lower-dimensional representation where the transformed data are now uncorrelated. To apply PCA to micro-Doppler analysis, the spectrogram of the data is first vectorised and M corresponding vectors x_i of the same class are used to form an $N \times M$ input data matrix $X = (x_1|x_2|\ldots x_M)$. The eigendecomposition of the unbiased covariance matrix can be computed to find the matrix of principal components (W): $X^\top X = W \Lambda W^\top$, resulting in a new representation of the data by $z = W^\top x$. PCA can be extended to operate on 2D structures with 2D-PCA and generalised 2D-PCA [58] which compute the correlation in both rows and columns and remove any redundancies across both variables. Both 1D-PCA and 2D-PCA have been proven to be a computationally efficient approach to feature extraction, and are thus especially well suited for embedded implementations [59]. Interestingly, one interpretation of DNNs [60] is that they are a nonlinear extension of PCA, implemented using a one-layer neural network with linear hidden and activation units.

4.2.2   Genetic algorithm-optimised frequency-warped cepstral coefficients

Another approach to data-driven classification that has been recently proposed is the application of genetic algorithms to optimise the design of the filter bank used to compute frequency-warped cepstral coefficients. The method is essentially a modification of mel-frequency cepstral analysis, originally developed for speech processing, but refined so as to match the micro-Doppler frequencies represented in the training data. The mel-scale is a frequency scale based upon the perception of equal distance between pitches, which mimics the perception of the human auditory system. Although mel-frequency cepstral coefficients (MFCCs) have been applied to micro-Doppler classification [61], the frequencies spanned by radar signatures have no relation to auditory perception, and the MFCC features are sub-optimal for radar classification:

1.   Radar micro-Doppler signatures often contain negative frequency components due to target motion away from the radar; however, mel-frequency filter banks only pass positive frequencies.
2.   The frequencies of interest in human activity recognition are oftentimes higher frequencies, because that is where the richness in human motion tends to reside. Lower frequencies map to average, translational motions. However, the mel-frequency filters are spaced much more closely at low frequencies than high. This is exactly opposite what is needed to distinguish micro-Doppler.
3.   The frequency range spanned by micro-Doppler depends on the transmit waveform parameters, such as transmit frequency and pulse repetition frequency (PRF) in addition to target motion. It is likely therefore that very little of the RF signal would fall within the frequencies spanned by the mel-scale.

As a result of these sources of mismatch, MFCC features, when extracted from RF signals, will not be able to capture qualities important for distinguishing radar micro-Doppler.

This problem may be remedied, however, by redesigning the filter bank used during the cepstral coefficient computation process to match the spectral range of the signal received by the radar. Training data may be utilised to determine the optimum placement of the centre frequencies and bandwidth of each filter so as to maximise classification accuracy. Figure 4.5 shows the resulting placement of filter banks, as optimised by a genetic algorithm for 5 filter banks and classification of 11 different human motions. Notice how the placements are no longer at regular intervals or bandwidths. Instead, the optimisation results in more filters being concentrated in frequencies corresponding to the envelopes, while only two filters are broad and map global features. As will be shown in the case study presented in Section 1.4, this approach, while being purely rooted in frequency analysis of the received signal, yields classification accuracies comparable to that of DNNs.

4.2.3 Autoencoders

Autoencoders (AEs) represent a special case of neural networks in which unsupervised learning is used to reconstruct the input at the output. In other words, for a given input vector x, the AE aims to approximate the identity operation $h_w(x) \approx x$. This is accomplished by using a symmetric encoder–decoder structure in the AE, as illustrated in Figure 4.6. First, the encoder computes a nonlinear mapping of the inputs as $e_i = \sigma(Wx_i + b)$, where σ denotes a nonlinear activation function, W denotes weights and b denotes the biases of the encoder. The encoded features are then decoded to reconstruct the input vector x using $z_i = \sigma(\tilde{W}x_i + \tilde{b})$, where $\tilde{W}$ and $\tilde{b}$ are the weights and biases of the decoder, respectively. In unsupervised pre-training, the AE tries to minimise the reconstruction error between output and input:

$$argmin_\theta \, J(\theta) = \frac{1}{N} \sum_{i=1}^{N} (x_i - z_i)^2 + \beta \sum_{j=1}^{h} KL(p\|p_j) \tag{4.6}$$

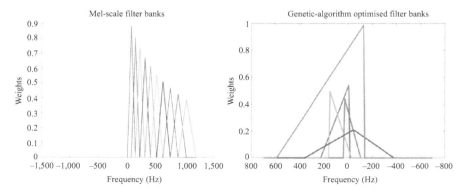

Figure 4.5 *Comparison of mel-scale filter bank with filter bank specified as a result of optimisation with a genetic algorithm*

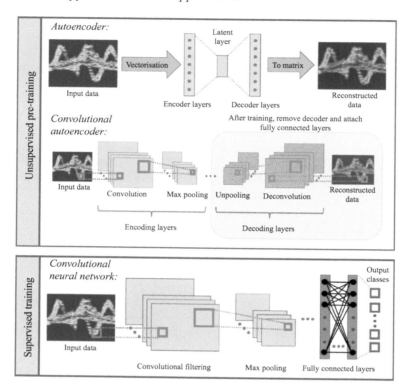

Figure 4.6   Commonly used DNN architectures

Here, θ is a parameter vector that includes the weights and biases of both the encoder and decoder: $\theta = [W, b, \tilde{W}, \tilde{b}]$; here h denotes the number of neurons in the hidden layer; β denotes sparsity proportion, and $\sum_{j=1}^{h} KL(p\|p_j)$ denotes the Kullback–Leibler (KL) divergence between the Bernoulli random variables with mean p and p_j, respectively. The sparsity parameter β functions as a regulariser and prevents the network from collapsing into the identity function.

If the decoder is removed from the network, the remaining encoder components can be fine-tuned in a supervised manner by adding a softmax classifier after the encoder. An important advantage of AEs is that the unsupervised pre-training step minimises requirements for labelled training data sets and is thus effective when only a small amount of data is available. This is particularly relevant for radar applications, where it is time-consuming and costly to collect a large number of measurements, and often not feasible to conduct experiments that span all expected scenarios and target profiles. For example, typical studies of radar micro-Doppler classification involve just one or two thousand measured data samples—this pales in comparison to the 1.28 million optical images used to train the 152-layer ResNet [1].

4.2.4 Convolutional neural networks

CNNs are perhaps the most popular DNNs architecture currently in use today. CNNs use spatially localised convolutional filters to capture the local features of images. Distinct features, such as lines, edges, and corners are learned in the initial layers, while more abstract features are learned as layers go deeper. CNNs are comprised of convolutional, pooling, and fully connected layers (see Figure 4.6). For a given matrix P, the mth neuron in the CNN calculates:

$$M[i,j] = \sigma \left(\sum_{x=-2k-1}^{2k+1} \sum_{y=-2k-1}^{2k+1} f_m[x,y]P[i-x,j-y] + b \right) \tag{4.7}$$

where M is the activation map of the input P, $2k+1$ is the dimension of the square input image, f_m is the mth convolution filter, σ is the nonlinear activation function, and b is a bias. After each convolutional layer, the output is downsampled using a pooling process. Although averaging can also be used, it is more typical to apply max pooling, where subregions are binned and only the local maxima are carried forward. Max pooling thus not only reduces computational complexity, but also adds translational invariancy to the network. The final stage of a CNN is comprised of fully connected layers which apply nonlinear combinations of the outputs of the previous layers to extract features.

The training of CNNs is a supervised process, where large amounts of training data are required for optimisation of network weights. Because the objective function of CNNs are highly non-convex, the parameter space of the model contains many local minima. As a result, the random initialisation typically used at the start of training does not necessarily guarantee that the local minimum to which the optimisation algorithms converge will also be a global optimum. In fact, if there is not much training data available, as is often the case in RF applications, the likelihood of converging to a good, if not optimum, solution decreases. As a result, alternative approaches to training under low sample support are required, e.g. CAEs and transfer learning, discussed next.

4.2.5 Convolutional autoencoders

CAEs reap the advantages of local spatial processing using convolutional filtering, and minimise training data requirements by implementing unsupervised pre-training for network weight initialisation. Thus, CAEs have an encoder–decoder structure that includes convolutional and deconvolutional filters. Deconvolutional filters may be learned from scratch, or they may simply be defined as transposed versions of the convolutional filters used in the encoder. In symmetry with the max pooling layers used in the encoder, the deconvolutional layers of the decoder are followed by unpooling layers. During pooling, the locations of the maximum values are stored, so that during unpooling the values at these locations are preserved, while the remaining values are set to zero. After unsupervised pre-training of the network weights, the decoder is disconnected from the network and supervised fine-tuning with labelled measured data is performed to optimise the final values of the network weights. The initialisation provided by unsupervised pre-training is typically closer to the global optimum; thus, better performance is typically achieved on small training data sets.

4.3   The radar challenge: training with few data samples

Deep learning is a data greedy approach to automatic feature learning. Unfortunately, acquisition of radar data is often costly and time-consuming. Thus, most micro-Doppler studies utilise about a thousand samples or less for training [30,62–69] with the notable exception of Trommel *el al.* [67], who collected 17,580 samples over 6 classes to train a 14-layered DNN. The lack of large amounts of data limits the depth, and, hence, classification accuracy attainable by DNNs. Another negative implication of low training sample support is that the diversity or degree to which the available data represents probable target presentations is also low—this hinders the ability of a DNN to generalise within a class; for example, it would be difficult to recognise the motion of people with significantly different height than that represented by the training samples. Thus, an essential step in classification is the acquisition or generation of a sufficient amount of adequately diverse and statistically sound training data set.

4.3.1   Transfer learning from optical imagery

One way of minimising real data requirements is to transfer knowledge gained from a different domain to initialise the weights of the DNN. Examples include using audio data for acoustic scene classification or optical imagery to classify ultrasound images. Because the trained model from which the weights are transferred is related to the given task, once the network is initialised using many samples of data set A, fine-tuning of the network with a few data samples from data set B can yield improved classification results. Thus, transfer learning can be a powerful technique when the amount of training data is insufficient. For example, Park *et al.* [70] used deep CNN models, namely AlexNet [71] and VGGnet [29], pre-trained on the ImageNet database, composed of 1.5 million RGB images, to classify 625 data samples of five different aquatic activities. The ImageNet database contains a sufficient diversity of shapes that a classification performance of 80.3% was reported. This represents a 13.6% improvement over training from scratch and starting with randomly initialised weights.

However, the phenomenology of radar signals fundamentally differs from that of optical signals, and thus, how well generic images may be able to represent micro-Doppler features can be questioned. As an alternative to transfer learning, a CAE, which uses unsupervised pre-training to minimise data requirements, could be used. In a study that compared transfer learning to unsupervised pre-training [30], two interesting results were reported:

1.   Although in ILSVRC, GoogleNet outperformed VGGnet, on radar micro-Doppler data, the exact reverse was found. VGGnet surpassed GoogleNet by nearly 5%.

2. When the number of training samples is roughly below 650, the performance of transfer learning with VGGnet surpasses that of the CAE, but above that threshold, the CAE surpasses VGGnet in performance.

This analysis demonstrates the important point that results from other applications, such as computer vision or speech processing, do not necessarily translate into radar signal processing, and, moreover, that a moderate amount of radar data is sufficient to surpass the results achieved with initialisation on optical imagery. This, in turn, has motivated much research into the generation of simulated micro-Doppler signatures, and using transfer learning to map between the simulated and real domains as opposed to bridging the gap between different sensors.

4.3.2 Synthetic signature generation from MOCAP data

Synthetic radar data generation requires two models: (1) a model for target motion, including position, velocity, and acceleration, and (2) a model for the electromagnetic scattering from the target. Equation (4.1) provides a point target approximation to target scattering and relies on superposition to compute the overall target response. More advanced models may be achieved by computing the return from a distributed target, finite element model. For most applications of micro-Doppler analysis, however, the point target model has been found to be of sufficient accuracy.

Kinematic modelling is more challenging, especially for the case of modelling human micro-Doppler signatures, which can exhibit high diversity. One often used kinematic model is the global human walking model [72] proposed by R. Boulic, N.M. Thalmann, and D. Thalmann. The Boulic model is based on an empirical mathematical parameterisation using a large amount of experimental biomechanical data, and it provides 3D spatial positions and orientations of normal human walking as a function of time. This model is only applicable to walking, and is limited in the degree of variability it can represent because its only two input parameters are height and speed. Thus, it cannot take into account individual differences in gait and generates signatures with separate and distinct trajectories for each limb.

As a result, in recent years, the use of motion capture (MOCAP) data to synthesise micro-Doppler data has become more popular. For instance, the Graphics Library of the Carnegie Mellon University has used multiple infrared cameras to capture motion by placing many retroreflective markers on a body suit worn by the subject [73]. The markers' positions and joint orientations in 3D space are tracked and stored in a MOCAP database. This MOCAP data can then be used to animate a skeleton model of the human body and enable it to move in accordance with the MOCAP data. Thus, this technique can be applied to simulate almost any desired human motion. While complete MOCAP sensor systems can be quite costly, markerless systems that exploit RGB-D cameras, such as Microsoft's Kinect sensor, can also be used for skeleton tracking.

However, because these MOCAP data sets are dependent upon obtaining real infrared and optical sensor data from test subjects, as with radar data, the amount of synthetic data generated is still limited by human effort, time, and cost of data collections. One way to increase the amount of data is to apply transformations to the measurements, known as data augmentation. Examples of data augmentation techniques from image processing include translation, rotation, cropping, and flipping. However, application of these techniques to radar micro-Doppler has a corrupting effect on the training data, because such transformations result in signatures that map to kinematically impossible target behaviour. Instead, it has been recently proposed that just a small set of MOCAP data be used to generate a large number of synthetic micro-Doppler signatures by applying transformations on the underlying skeletal structure tracked by a Kinect sensor. Thus, scaling the skeletal dimensions can model different body sizes, while scaling the time axis can model different motion speeds. Individualised gaits can even be synthesises by adding perturbations to a parametric model of the time-varying joint trajectories. An overview of the diversification transformations applied to the skeletal model is given in Figure 4.7. Although there is no way to guarantee that the result of such skeletal transformations is fully compatible with the kinematic constraints of human motion, the extent of any discrepancies are greatly limited by the skeletal model. Thus, diversified MOCAP data provide an alternative source domain for transfer learning that is much more related to the target (RF) domain than optical imagery.

Recent results have shown that 55 MOCAP measurements of 5 test subjects performing 7 different human activities could be used to generate diversified MOCAP data set of 32,000 samples [74]. These data were then used to initialise the weights of a CNN which was later fine-tuned with 474 real data measurements. It was found that the classification accuracy of two- and seven-layer CNNs increased by 9% and 19%, respectively. More significantly, the increased amount of data available for training also facilitated the construction of a deeper, 15-layer residual neural network (DivNet-15), which raised classification accuracies after fine-tuning from 86% using a CNN to 97%.

4.3.3 *Synthetic signature generation using adversarial learning*

An alternative approach for generating synthetic data, which is gaining increased attention in the radar community, is adversarial learning. Generative adversarial networks (GANs) have been proposed for synthesising realistic images in a variety of applications [75], including synthetic aperture radar (SAR) [76]. In [77], a 4% increase in the classification accuracy of 50 running, walking, and jumping signatures was achieved using a GAN that augmented a training data set of 150 signatures with 1,000 GAN-generated signatures.

The efficacy of different generative models to accurately synthesise radar micro-Doppler signatures varies. A recent study [78] shows, for example, that auxiliary condition GANs (ACGANs) generate crisper signatures than conditional variational autoencoders (CVAEs). For radar micro-Doppler applications, however, fidelity cannot be evaluated by considering image quality alone. A critical challenge is the possibility for generative models to provide misleading synthetic signatures. The characteristics of the micro-Doppler signature resulting from human activity are constrained not only by the physics of electromagnetic scattering, but also by human kinematics. The skeleton physically constrains the possible variations of the

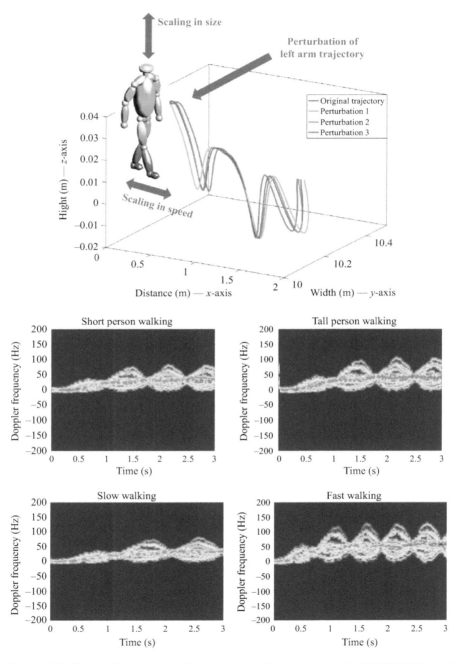

Figure 4.7 Diversification transformations used to augment initial MOCAP-based training data set

spectrogram corresponding to a given class. However, generative models have no knowledge of these constraints. It is possible, therefore, for GANs to generate synthetic samples that may appear visually similar, but are in fact incompatible with possible human motion.

4.3.3.1   Case study: auxiliary conditional GANs

To evaluate the quality of micro-Doppler signatures synthesised using generative models, let us consider the efficacy of ACGANs as a case study. As with all adversarial networks, the ACGAN employs a game-theoretic approach in which two players (the generator and the discriminator) compete in a zero-sum game. The generator seeks to generate samples that are of the same distribution as the input data. Meanwhile, the discriminator aims at classifying whether the resulting image is real or fake. GANs are trained by maximising the probability that the discriminator makes a mistake. The loss function of adversarial networks can thus be expressed as

$$\min_G \max_D E_{x \sim p_{data}} \log(D(x)) + E_{z \sim p_z}[\log(1 - D(G(z)))]) \tag{4.8}$$

where $G(z)$ is the synthetic image generated and $D(x)$ is the probability of the image being real. The discriminator is trained to maximise $D(x)$ for images with $x \sim p_{data}$, while the generator tries to produce images such that $D(G(z)) \sim p_{data}$. ACGANs attempt to improve the quality of the generated images by tasking the discriminator directly with reconstructing labels, as opposed to feed it class information. This is done by modifying the discriminator to include an auxiliary decoder network, which outputs the class labels for the training data [79]. This modification results in the objective function having two parts: the log-likelihood of the correct source, L_s, and the log-likelihood of the correct class, L_y. Thus, the discriminator is trained to maximise $L_s + L_Y$, while the generator is trained to maximise $L_Y - L_s$, where

$$L_s = E[\log p(s = real | x_{real})] + E[\log p(s = fake | x_{fake})] \tag{4.9}$$

$$L_y = E[\log p(Y = y | x_{real})] + E[\log p(Y = y | x_{fake})] \tag{4.10}$$

and s are the generated images.

In this example, an ACGAN is implemented with the following parameters. The input to the generator is drawn from uniformly distributed $N(0, 2)$ random noise supplied as a 100×1 vector, while the output is a $64 \times 64 \times 1$ spectrogram. The generator itself is comprised of a fully connected dense layer reshaped to size $4 \times 4 \times 128$ and four convolutional layers with the first three layers having 256, 128, and 64 filters, respectively, while the last layer contains only one filter due to the grey-scale channel size. The kernel size for each filter is 3×3. Batch normalisation with a momentum of 0.8 and 2D upsampling (kernel size 2×2 with strides of 2) are applied to each layer (including the dense layer) of the generator network, except for the output layer. In addition to the batch normalisation, dropout of 0.15 is also applied in every even layer due to the small amount of real training data available. This serves to regularise the generator as well as prevent overfitting and mode collapsing. Rectified linear unit (ReLU) activation functions are used throughout the network, except in the output layer, which employs a *tanh* activation function instead.

The discriminator network consists of seven convolutional layers, which have 64, 128, 128, 256, 256, 512, and 512 filters in each layer, respectively, and apply a kernel size of 3 × 3. A LeakyReLU activation function is used everywhere, with the exception of the last layer, for which the slope of the leak is 0.2. Max-pooling is only included in the first layer with a filter size of 2 × 2 and strides of 2. Downsampling is done in every odd convolutional layer with a stride rate of 2. Batch normalisation with momentum 0.8 is utilised in every layer except the first. The validity of the generated images is tested using a sigmoid function in the last layer of the discriminator, followed by a softmax classifier to reconstruct the class labels.

The overall architecture of the ACGAN is shown in Figure 4.8. As preprocessing, the eCLEAN algorithm is applied, followed by scaling between $(-1, 1)$ for the *tanh* activation function. A 19-layer CNN is trained using the synthetic data generated by the ACGAN, and optimised using adaptive moment estimation (ADAM) with a learning rate of 0.0002, $\beta_1 = 0.5$, $\beta_2 = 0.999$ for 3,000 epochs, minibatch size of 16. Some examples generated by the proposed ACGAN are depicted in Figure 4.9. A total of 40,000 synthetic spectrograms are generated using the ACGAN (5,000 for each class).

4.3.3.2 Kinematic fidelity of ACGAN-synthesised signatures

In this section, we examine the kinematic properties of synthetic walking and falling signatures, in comparison to their real counterparts. Walking and falling represent challenging cases for the ACGAN due to the great diversity within the real training samples as well as the greater richness of frequencies comprising the signature. The kinematic fidelity of the synthetic data can be evaluated by considering its consistency with certain properties that are known to be common to all samples within a class. For example, walking micro-Doppler exhibits the following properties:

1. Periodicity: The cyclic motion of the left and right legs results in a periodic micro-Doppler signature.
2. Leg velocity > torso velocity: Walking requires the legs to move much faster than the average speed of the torso; thus, the maximum micro-Doppler frequency of the signature should be greater than the maximum frequency of the strongest return (given by the torso).
3. Frequency composition of signature should match direction of motion: If moving towards the radar, most of the micro-Doppler frequencies should be positive, with any negative frequency components having a very small Doppler frequency. The reverse will be true if moving away from the radar.

The properties can be used to establish a set of rules with which each synthetic signature can be checked. To test whether a walking signature satisfies these rules, first the upper and lower envelopes are extracted and their maximum and minimum values compared. To illustrate the results of assessing these kinematic rules, 12 synthetic walking signatures generated by the ACGAN are randomly selected and shown in Figure 4.10. The green labels indicate that the synthetic images passed all three kinematic rules, while orange indicates minor issues (i.e. only one or two rules failed), and red indicates that the image fails all rules. Inspecting the images that violate one

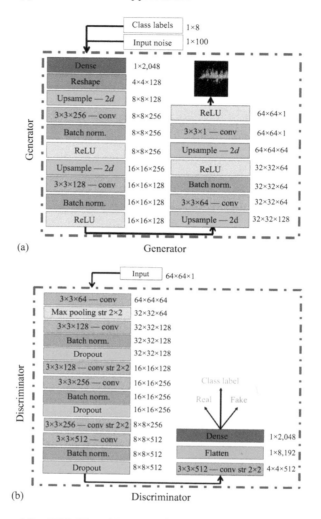

Figure 4.8   ACGAN architecture: (a) generator; (b) discriminator

or more of the rules, it may be observed that Sub-images (2,3) and (4,3) fail because they degenerate with virtually no target signal, and only a faint clutter component. Sub-images (1,3) and (3,3) also fail all rules because they are not periodic and have an inconsistent distribution of positive and negative frequencies, and the strongest return is not consistent with typical torso motion. Rule 1 fails twice because of lack of periodicity in the synthetic signatures. Rule 2 is violated by Subimage (1,2) because the negative frequency component is almost as large as the positive frequency component, which is inconsistent with directed human motion towards or away from the radar. Overall, when these rules are applied to the 5,000 synthetic walking spectrograms generated by ACGAN, 15% of the signatures failed all three of the kinematic rules.

Synthetic samples generated with ACGAN

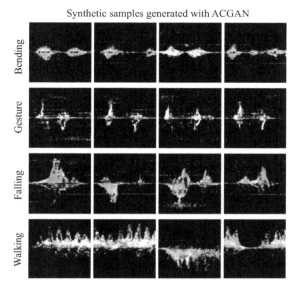

Figure 4.9   Samples of synthetic signatures generated by ACGAN for the classes of bending, gesturing, falling, and walking

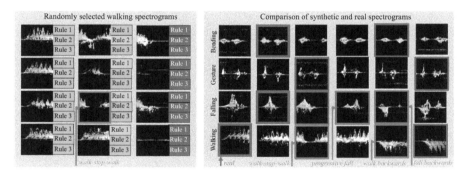

Figure 4.10   Interpretation of kinematics reflected in ACGAN-generated signatures: (left) results of kinematic rule test; (right) annotation of synthetic signatures in comparison with real signatures

In the case of falling, only the second and third rules are applicable—falling is an impulsive motion, not a period one. When these rules are enforced on the synthetic falling samples, 10% of the generated images fail.

Figure 4.10 also shows a comparison of synthetic and real signatures. For each class (row), two out of the five samples are real (shown in red), while the remaining are synthetic. Notice that the synthetic images in many cases can capture not just signal components, but also noise or interference components in the data. However, some of the synthetic images can also be interpreted as belonging to different classes of motion, as highlighted in blue. Thus, in trying to emulate a fall, the ACGAN may generate a signature that in fact more closely resembles a progressive fall—first falling to the knees, then to the ground. Or the ACGAN may generate variances in which the direction of motion is reversed, or the person stops in between. These are striking results which showcase the challenge in using adversarial learning to simulate radar micro-Doppler: due to the physical meaning belying the patterns in the signatures, generating data that simply resembles the training data is not sufficient. The ACGAN must (somehow) also be constrained from generating misleading samples, which while similar, in fact correspond to different human motions. These inaccurate signatures will actually mislead the classifier and result in degraded performance, as shown in Section 1.3.3.4.

4.3.3.3   Diversity of ACGAN-synthesised signatures

One risk in using GANs is a phenomenon known as mode collapse: the generative model collapses by outputting just a single prototype that maximally fools the discriminator [80]. This is an undesirable outcome as many samples of the same image cannot be used for training any DNN. To evaluate potential mode collapse, we checked the similarity of synthetic signatures using a quantitative similarity measured called MS-SSIM [81]. MS-SSIM attempts to weight different parts of the image according to its importance for human perception of similarity. Defined over the interval of 0 to 1, the MS-SSIM metric takes on higher values when two images are very similar. Mathematically, it is defined as

$$\text{SSIM}(x, y) = [l_M(x, y)]^{\alpha_M} \prod_{j=1}^{M} [c_j(x, y)]^{\beta_j} [s_j(x, y)]^{\gamma_j} \tag{4.11}$$

where x and y are the compared images; α_M, β_j, and γ_j are used to adjust the relative importance of different components; $l((x, y))$, $c((x, y))$, and $s((x, y))$ are defined as the luminance, contrast, and structural comparison metrics, respectively; and M depicts the scale that will be used in the iterative filtering and downsampling.

In generating a database of synthetic samples for training, it is critical to consider not just the fidelity of the data to the underlying class model, but also to ensure sufficient diversity so that entire range of possible samples within a class is represented. MS-SSIM is used to evaluate the diversity of the ACGAN-generated synthetic signatures, and the resulting box plot for 100 randomly selected image pairs in each class is shown in Figure 4.11. The walking class exhibits the most diversity, which is indicative of the number of potential variations and kinematic complexity of the motion. As falling is an uncontrolled motion and gesturing is a very broad class, open to participant interpretation during experiments, these results match expectations.

It is important to also observe from Figure 4.11(b) that the degree of similarity between synthetic samples depends strongly on the amount of real data used to train the ACGAN, but that beyond 100 samples the MS-SSIM levels off. Thus, from a diversity perspective, 100 real data samples are sufficient to train the ACGAN. From

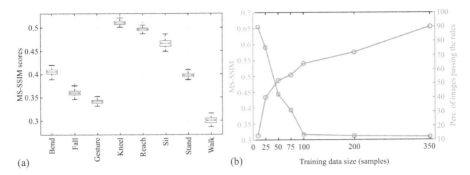

(a)   (b)   Training data size (samples)

Figure 4.11   *Diversity measures: (a) box plots of the intra-class diversities measured by MS-SSIM; (b) MS-SSIM diversity values and percentages of kinematically correct images as a function of the real training samples used in the training*

the perspective of kinematic fidelity, however, the more the training samples, the higher the percentage of synthetic samples that pass the kinematic rules defined earlier. It is only when 350 training samples are utilised that 90% of synthetic samples pass all kinematic rules.

4.3.3.4   PCA-based kinematic sifting algorithm

To improve the efficacy of the ACGAN-synthesised training data set, a PCA-based kinematic sifting algorithm has been developed to discard samples that deviate from human motion typical of the class. To generalise beyond just the walking and falling classes, generalised PCA [82] is used to establish the feature space spanned by each class, based on the training data. The boundaries of this feature space is determined using the convex hull algorithm. To allow for some intra-class variations, this region can be expanded according to a predefined allowable tolerance. Then each synthetic sample can be mapped to this lower-dimensional, expanded feature space and tested to see whether its features represent that of a sample that is reasonable considering the allowed intra-class variations. If the synthetic sample falls outside the region in the feature space, that sample is deemed to have poor kinematic consistency with its class and discarded.

The performance improvement achieved by throwing out kinematically inconsistent samples can be demonstrating by considering an eight-class experiment, in which the training samples for the ACGAN are collected from a line-of-sight configuration, while the test samples are collected from a through-the-wall configuration. Note that the signatures shown in Figure 4.9 include horizontal lines outside the region where the target micro-Doppler resides—these are artefacts generated in the electronics of the RF transceiver used to acquire the data and are sensor-specific. Their presence in the synthesised data demonstrates the ability of the ACGAN to capture not just target components, but also clutter, interference, and noise components as well. A total of 385 test samples were collected across multiple aspect angles ($0°$, $30°$, $45°$, and $60°$).

A total of 40,000 synthetic signatures are generated using the ACGAN. When all synthetic signatures are used to train a 19-layer CNN to classify the test data, a classification accuracy of 82.5% was achieved. Alternatively, if the PCA-based kinematic sifting method is first applied on the training data set generated by ACGAN, 11% of bending, 18% of falling, 8% of gesture, 33% of kneeling, 7% of reaching, 15% of sitting, 36% of standing, and 22% of walking class samples are discarded. This results in the training data set size for the CNN being reduced from 40,000 samples to 31,133 samples. However, the classification accuracy *increases* to 93.2%—an improvement of over 10%.

4.4   Performance comparison of DNN architectures

We compare the classification accuracy achieved by several DNN architectures using a common data set for the classification of 12 daily activities: (1) walking, (2) using a wheelchair, (3) limping, (4) walking with a cane, (5) walking with a walker, (6) falling after tripping, (7) walking with crutches, (8) creeping while dragging abdomen on the ground, (9) crawling on hands and knees, (10) jogging, (11) sitting, and (12) falling off a chair. A total of 1,007 measurements were collected using a 4 GHz continuous wave software-defined radar positioned 1 metre above the ground from 11 participants, who performed each activity while moving along the radar line of sight. The classification approaches listed below were each tested using tenfold cross validation, and specified so as to try to achieve the best possible performance for each approach:

1. *Supervised classification with handcrafted features (mSVM-50):* A total of 127 features were extracted using a wide variety of methods, after which sequential backward elimination was used to optimally select just 50 features. The types and number of features computed were based on prior studies [83]; ultimately, the features utilised were (1) the bandwidth of the torso response, (2) mean torso frequency, (3) mean of the upper envelope, (4) mean of the lower envelope, (5) 2 features from the CVD [37], (6) 2 cepstral coefficients [61], (7) 37 linear predictive coding coefficients [84], and (8) 5 discrete cosine transform coefficients [85]. A multi-class support vector machine (SVM) classifier with linear kernel outperformed polynomial and radial basis function kernels, as well as random forest and xgboost classifiers.

2. *2D-PCA*: Generalised 2D-PCA [58] is implemented on a spectrogram of size 128-by-128. The right (rows) and left (columns) projection matrices are of dimension 128-by-15 and 15-by-128, respectively, describing 15 principal components. A 3-kNN classifier is implemented with the size of training, validation, and testing sets, respectively, chosen to be 80%, 10%, and 10% of the data.

3. *AE:* A three-layer AE with layers of 200-100-50 neurons, respectively, was found to yield the best results considering depth and width of the network. ADAM algorithm [86] was used for optimisation with a learning rate of 0.0001. The KL

divergence term for regularisation and sparsity parameter β were chosen as 2 and 0.1, respectively.

4. *CNN:* Two different convolutional filter sizes of 3×3 and 9×9 were applied and concatenated in each convolutional layer. A total of three convolutional layers using 2×2 max-pooling followed by two fully connected layers of 150 neurons each yielded the highest accuracy. To mitigate the potential for overfitting, 50% dropout was implemented.

5. *CAE:* The convolutional and deconvolutional layers of a three-convolutional layer CAE were populated with thirty 3×3 and 9×9 convolutional filters. After unsupervised pre-training, two fully connected layers with 150 neurons each were utilised with a softmax classifier.

6. *Transfer learning from VGGnet pre-trained on ImageNet database:* Although there are many possible pre-trained models that could be exploited to illustrate transfer learning, VGGnet has been shown to outperform GoogleNet and give similar performance to ResNet-50 on radar micro-Doppler data [74].

7. *Residual neural network:* Deep residual networks are comprised of building blocks that compute the residual mapping of

$$y_l = h(x_l) + F(x_l, W_l) \tag{4.12}$$

$$x_{l+1} = f(f_l) \tag{4.13}$$

where x_l is the input to the *l*th residual unit (RU); $W_{l,k|1 <= k <= K}$ is a set of weights and biases for the *l*th RU; K is the number of layers in an RU; F is a residual function (e.g. a stack of two convolutional layers); $h(x_l) = x_l$ is an identity mapping that is performed by using a shortcut path; and f is an activation function (e.g. ReLU). A variant of the RU, known as a modified RU [87], which places the activation function before the final addition, is often used as it is more easily trained and possess improved generalisation properties. Residual networks enable the design of very DNNs by permitting the creation of short connections within the network via identity mappings. This drives new layers to learn something different from what has already been encoded. Moreover, it provides a mechanism for mitigating the vanishing gradient problem seen in very DNNs by simply learning an identity mapping if adding layers gives no benefit. Using diversified MOCAP signatures (see Section 1.3.2) as a source of a large amount of training data, a 15-convolutional layer residual neural network (DivNet-15) was constructed, as shown in Figure 4.12.

8. *Genetic algorithm-optimised frequency-warped cepstral coefficient (GA-FWCC):* A genetic algorithm with a population size of 128, crossover rate of 0.8, and mutation rate of 0.01 was implemented over 30 generations to optimise the centre frequency and bandwidth of each filter used to calculate the cepstral features. GA-FWCC features were then supplied a two-layer shallow neural network with ten neurons for classification.

A performance comparison of these techniques is given in Figure 4.13. The benefits of deep learning can be clearly seen as SVM offers just 76.9% accuracy

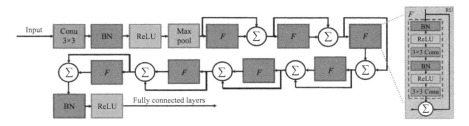

*Figure 4.12   DivNet-15: architecture for residual neural network with 15
convolutional layers, trained on diversified MOCAP signatures [74]*

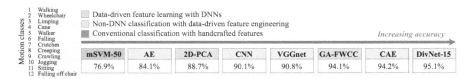

*Figure 4.13   Comparison of conventional, data-driven, and deep learning
classification approaches*

in sharp contrast to the CAE at 94.2%. Some important observations include the
following:

1. Transfer learning from optical imagery yields only a slight improvement in per-
    formance beyond that of a CNN trained on a small amount of real radar data,
    despite VGGnet being a deeper network. VGGnet is a 16-layer CNN, while the
    CNN used in this work has only 3 convolutional layers—a significantly less
    complex network.
2. The GA-FWCC algorithm yields performance comparable to that of DNNs,
    in particular CAEs. This underscores the importance of developing machine
    learning algorithms that are consistent with radar phenomenology.
3. The highest performance is achieved by DivNet-15, which exploits diversified
    kinematic models to enable the training of much deeper DNNs than otherwise
    possible.

4.5   RNNs and sequential classification

The DNNs surveyed so far have all processed the radar micro-Doppler measurements as 2D snapshots computed over a finite window of time. However, typically,
the motion of targets is time-varying, cycling through a sequence of many different
motion classes. Dynamic sequential classification remains an important open problem

in micro-Doppler analysis; however, recent efforts have focused on recurrent neural networks (RNNs) as an important mechanism towards a solution. RNNs model temporal behaviour using connections between nodes that form a directed graph along a sequence. At each time step, an output is produced. Long short-term memory (LSTM) RNNs are able to model longer-term behaviour through the inclusion of a memory block that consists of a cell, input, output, and forget gates. Studies on gesture recognition [52,53] have proposed using LSTM RNNs; however, sequential data were not processed, with only one motion class being present in each data sample. Gated recurrent units (GRUs) are a variant of RNNs, which have a simpler structure and better performance on smaller data sets. In one recent study, snapshots of the micro-Doppler signatures for different activities were concatenated to form a sequence, and classified using stacked GRUs [88]. Concatenating spectrograms, however, does not accurately represent the overall measurement acquired from continuous observation of motion sequences, because of the transitional motions that occur as part of the natural kinematic flow of the human body. In [89], continuous observational data are processed using an SVM classifier on a sliding window of data, but the classification accuracy achieved is lower than those typical of deep learning when applied to the micro-Doppler signature of a snapshot.

Thus, although great progress has been made with deep learning over the past few years, micro-Doppler classifiers that are dynamic, robust to noise and clutter, pervasive, and highly accurate remain in need of the new innovations of the future.

References

[1]   He K, Zhang X, Ren S, and Sun J. Deep Residual Learning for Image Recognition. In: 2016 IEEE Conference on Computer Vision and Pattern Recognition, CVPR 2016, Las Vegas, NV, USA, June 27–30, 2016; 2016. pp. 770–778.

[2]   Li C, Lubecke VM, Boric-Lubecke O, *et al.* A Review on Recent Advances in Doppler Radar Sensors for Noncontact Healthcare Monitoring. *IEEE Transactions on Microwave Theory and Techniques.* 2013;61(5):2046–2060.

[3]   Massagram W, Lubecke VM, Hst-Madsen A, *et al.* Assessment of Heart Rate Variability and Respiratory Sinus Arrhythmia via Doppler Radar. *IEEE Transactions on Microwave Theory and Techniques.* 2009;57(10):2542–2549.

[4]   Hu W, Zhao Z, Wang Y, *et al.* Noncontact Accurate Measurement of Cardiopulmonary Activity Using a Compact Quadrature Doppler Radar Sensor. *IEEE Transactions on Biomedical Engineering.* 2014;61(3):725–735.

[5]   Rahman A, Lubecke VM, BoricLubecke O, *et al.* Doppler Radar Techniques for Accurate Respiration Characterization and Subject Identification. *IEEE Journal on Emerging and Selected Topics in Circuits and Systems.* 2018;8(2):350–359.

[6]   Dell'Aversano A, Natale A, Buonanno A, *et al.* Through the Wall Breathing Detection by Means of a Doppler Radar and MUSIC Algorithm. *IEEE Sensors Letters.* 2017;1(3):1–4.

[7]   Lee YS, Pathirana PN, Steinfort CL, *et al.* Monitoring and Analysis of Respiratory Patterns Using Microwave Doppler Radar. *IEEE Journal of Translational Engineering in Health and Medicine.* 2014;2:1–12.

[8]   Ziganshin EG, Numerov MA, and Vygolov SA. UWB Baby Monitor. In: 2010 5th Int. Conf. on UWB and Ultrashort Impulse Signals; 2010. pp. 159–161.

[9]   Amin MG, editor. *Radar for Indoor Monitoring: Detection, Classification, and Assessment.* Boca Raton, FL: CRC Press; 2017.

[10] Mercuri M, Soh PJ, Pandey G, *et al.* Analysis of an Indoor Biomedical Radar-Based System for Health Monitoring. *IEEE Transactions on Microwave Theory and Techniques.* 2013;61(5):2061–2068.

[11] Garripoli C, Mercuri M, Karsmakers P, *et al.* Embedded DSP-based Telehealth Radar System for Remote In-door Fall Detection. *IEEE Journal of Biomedical and Health Informatics.* 2015;19(1):92–101.

[12] Su BY, Ho KC, Rantz MJ, *et al.* Doppler Radar Fall Activity Detection Using the Wavelet Transform. *IEEE Transactions on Biomedical Engineering.* 2015;62(3):865–875.

[13] Amin MG, Zhang YD, Ahmad F, *et al.* Radar Signal Processing for Elderly Fall Detection: The Future for In-home Monitoring. *IEEE Signal Processing Magazine.* 2016;33(2):71–80.

[14] Falls and Fractures in the Home. Accessed September 14, 2019. https://www.nursinghomeabusecenter.com/nursing-home-injuries/falls-fractures/.

[15] Seifert A, Zoubir AM, and Amin MG. Radar Classification of Human Gait Abnormality Based on Sum-of-harmonics Analysis. In: Proc. IEEE Radar Conference; 2018. pp. 0940–0945.

[16] Palmer JW, Bing KF, Sharma AC, *et al.* Detecting Concussion Impairment with Radar Using Gait Analysis Techniques. In: Proc. IEEE Radar Conf.; 2011. pp. 222–225.

[17] Postolache O, Pereira JMD, Viegas V, *et al.* Gait Rehabilitation Assessment Based on Microwave Doppler Radars Embedded in Walkers. In: Proc. IEEE Int. Symp. on Medical Measurements and Appl.; 2015. pp. 208–213.

[18] Yavari E, Song C, Lubecke V, *et al.* Is There Anybody in There? Intelligent Radar Occupancy Sensors. *IEEE Microwave Magazine.* 2014;15(2): 57–64.

[19] Bjorklund S, Petersson H, and Hendeby G. On Distinguishing Between Human Individuals in Micro-Doppler Signatures. In: 2013 14th International Radar Symposium (IRS). vol. 2; 2013. pp. 865–870.

[20] Chen Z, Li G, Fioranelli F, *et al.* Personnel Recognition and Gait Classification Based on Multistatic Micro-Doppler Signatures Using Deep Convolutional Neural Networks. *IEEE Geoscience and Remote Sensing Letters.* 2018;15(5):669–673.

[21] Yang X, Yin W, and Zhang L. People Counting Based on CNN Using IR-UWB radar. In: 2017 IEEE/CIC Int. Conf. on Comm. in China; 2017. pp. 1–5.

[22] Cao P, Xia W, Ye M, *et al.* Radar-ID: Human Identification Based on Radar micro-Doppler Signatures Using Deep Convolutional Neural Networks. *IET Radar, Sonar and Navigation.* 2018;12(7):729–734.

[23] Vandersmissen B, Knudde N, Jalalvand A, *et al.* Indoor Person Identification Using a Low-power FMCW Radar. *IEEE Transactions on Geoscience and Remote Sensing.* 2018;56(7):3941–3952.

[24] Li Y, Peng Z, and Li C. Potential Active Shooter Detection Using a Portable Radar Sensor with Micro-Doppler and Range-Doppler Analysis. In: International Applied Computational Electromagnetics Society Symposium; 2017. pp. 1–2.

[25] Greneker G. Very Low-cost Stand-off Suicide Bomber Detection System Using Human Gait Analysis to Screen Potential Bomb Carrying Individuals. In: Proceedings of the SPIE, Radar Sensor Technology IX. vol. 5788;2005.

[26] Fioranelli F, Ritchie M, and Griffiths H. Classification of Unarmed/Armed Personnel Using the NetRAD Multistatic Radar for Micro-Doppler and Singular Value Decomposition Features. *IEEE Geoscience and Remote Sensing Letters.* 2015;12(9):1933–1937.

[27] Gurbuz SZ, and Amin MG. Radar-based Human-Motion Recognition with Deep Learning: Promising Applications for Indoor Monitoring. *IEEE Signal Processing Magazine.* 2019;36(4):16–28.

[28] Szegedy C, Liu W, Jia Y, *et al.* Going Deeper with Convolutions. CoRR. 2014;abs/1409.4842. Available from: http://arxiv.org/abs/1409.4842.

[29] Simonyan K, and Zisserman A. Very Deep Convolutional Networks for Large-scale Image Recognition; 2014. Available from: https://arxiv.org/abs/1409.1556.

[30] Seyfioglu MS, and Gurbuz SZ. Deep Neural Network Initialization Methods for Micro-Doppler Classification with Low Training Sample Support. *IEEE Geoscience and Remote Sensing Letters.* 2017;14(12):2462–2466.

[31] Dogaru M, and Le C. Validation of Xpatch Computer Models for Human Body Radar Signature. US Army Research Laboratory Technical Report, ARL-TR-4403.

[32] Geisheimer JL, Greneker EF, and Marshall WS. High-resolution Doppler Model of the Human Gait. In: Proceedings of the AeroSense Conference; 2002.

[33] van Dorp P, and Groen FCA. Human Walking Estimation with Radar. *IET Radar, Sonar and Navigation.* 2003;150(5):356–365.

[34] Gurbuz SZ, Erol B, Cagliyan B, *et al.* Operational Assessment and Adaptive Selection of micro-Doppler Features. *IET Radar, Sonar and Navigation.* 2015;9(9):1196–1204.

[35] Loukas C, Fioranelli F, Le Kernec J, *et al.* Activity Classification Using Raw Range and I & Q Radar Data with Long Short Term Memory Layers. In: IEEE 16th Intl Conf on Dependable, Autonomic and Secure Computing, 16th Intl Conf on Pervasive Intelligence and Computing, 4th Intl Conf on Big Data Intelligence and Computing and Cyber Science and Technology Congress(DASC/PiCom/DataCom/CyberSciTech); 2018. pp. 441–445.

[36] Kim Y, Ha S, and Kwon J. Human Detection Using Doppler Radar Based on Physical Characteristics of Targets. *IEEE Geoscience and Remote Sensing Letters.* 2015;12(2):289–293.

[37]   Otero M. Application of a Continuous Wave Radar for Human Gait Recognition. In: Proceedings of SPIE; 2005.

[38]   Cohen L. Time-frequency Distributions—A Review. *Proceedings of the IEEE.* 1989;77(7):941–981.

[39]   Chen V, and Ling H. *Time-frequency Transforms for Radar Imaging and Signal Analysis.* Boston, MA: Artech House; 2002.

[40]   Tivive FHC, Phung SL, and Bouzerdoum A. Classification of Micro-Doppler Signatures of Human Motions Using Log-Gabor Filters. *IET Radar, Sonar and Navigation.* 2015;9(9):1188–1195.

[41]   Almeida LB. The Fractional Fourier Transform and Time-frequency Representations. *IEEE Transactions on Signal Processing.* 1994;42(11):3084–3091.

[42]   Daubechies I. The Wavelet Transform, Time-frequency Localization and Signal Analysis. *IEEE Transactions on Information Theory.* 1990;36(5):961–1005.

[43]   Sridharan K, Thayaparan T, SivaSankaraSai S, *et al.* Gabor–Wigner Transform for Micro-Doppler Analysis. In: 9th Int. Radar Symp. India; 2013.

[44]   Choi H, and Williams WJ. Improved Time-frequency Representation of Multicomponent Signals Using Exponential Kernels. *IEEE Transactions on Acoustics, Speech, and Signal Processing.* 1989;37(6):862–871.

[45]   Amin MG. Spectral Decomposition of Time-frequency Distribution Kernels. *IEEE Transactions on Signal Processing.* 1994;42(5):1156–1165.

[46]   Amin MG. Time-frequency Spectrum Analysis and Estimation for Nonstationary Random Processes. In: Boashash B, editor. *Time-frequency Signal Analysis: Methods and Applications.* Melbourne: Longman Cheshire; 1992.

[47]   Jokanovic B, Amin M, and Erol B. Multiple Joint-variable Domains Recognition of Human Motion. In: 2017 IEEE Radar Conference; 2017. pp. 0948–0952.

[48]   Li Y, Peng Z, Pal R, *et al.* Potential Active Shooter Detection Based on Radar Micro-Doppler and Range-Doppler Analysis Using Artificial Neural Network. *IEEE Sensors Journal.* 2019;19(3):1052–1063.

[49]   Erol B, and Amin MG. Radar Data Cube Processing for Human Activity Recognition Using Multisubspace Learning. *IEEE Transactions on Aerospace and Electronic Systems.* 2019;55(6):3617–3628.

[50]   He Y, Le Chevalier F, and Yarovoy AG. Range-Doppler processing for indoor human tracking by multistatic ultra-wideband radar. In: 2012 13th International Radar Symposium; 2012. pp. 250–253.

[51]   He Y. Human Target Tracking in Multistatic Ultra-Wideband Radar. In: PhD Dissertation, Delft University of Technology; 2014.

[52]   Wang S, Song J, Lien J, *et al.* Interacting with Soli: Exploring Fine-grained Dynamic Gesture Recognition in the Radio-frequency Spectrum. In: Proc. of the 29th Annual Symposium on User Interface Software and Technology. New York, NY, USA: ACM; 2016. pp. 851–860.

[53]   Zhang Z, Tian Z, and Zhou M. Latern: Dynamic Continuous Hand Gesture Recognition Using FMCW Radar Sensor. *IEEE Sensors Journal.* 2018;18(8): 3278–3289.

[54] Schumann O, Hahn M, Dickmann J, *et al.* Semantic Segmentation on Radar Point Clouds. In: 21st Int. Conf. on Information Fusion; 2018. pp. 2179–2186.

[55] He Y, Molchanov P, Sakamoto T, *et al.* Range-Doppler Surface: A Tool to Analyse Human Target in Ultra-wideband Radar. *IET Radar, Sonar and Navigation.* 2015;9(9):1240–1250.

[56] Du H, Jin T, Song Y, Dai Y, and Li M. A Three-dimensional Deep Learning Framework for Human Behavior Analysis Using Range-Doppler Time Points. *IEEE Geoscience and Remote Sensing Letters.* 2020;17(4):611–615.

[57] Lu H, Plataniotis KN, and Venetsanopoulus AN. *Multilinear Subspace Learning: Dimensionality Reduction of Multidimensional Data.* Boca Raton, FL: Chapman and Hall/CRC; 2012.

[58] Kong H, Wang L, Teoh EK, *et al.* Generalized 2D Principal Component Analysis for Face Image Representation and Recognition. *Neural Networks.* 2005;18:585–594.

[59] Zabalza J, Clemente C, Di Caterina G, *et al.* Robust PCA Micro-Doppler Classification Using SVM on Embedded Systems. *IEEE Transactions on Aerospace and Electronic Systems.* 2014;50(3):2304–2310.

[60] Goodfellow I, Bengio Y, Courville A, *et al. Deep Learning.* Cambridge, MA: The MIT Press; 2016.

[61] Yessad D, Amrouche A, Debyeche M, *et al.* Micro-Doppler Classification for Ground Surveillance Radar Using Speech Recognition Tools. In: Proc. of the 16th Iberoamerican Congress on Progress in Pattern Recognition, Image Analysis, Computer Vision, and App. Berlin, Heidelberg: Springer-Verlag; 2011. pp. 280–287.

[62] Jokanovic B, Amin M, and Ahmad F. Radar Fall Motion Detection Using Deep Learning. In: IEEE Radar Conference; 2016. pp. 1–6.

[63] Mendis GJ, Randeny T, Wei J, *et al.* Deep Learning-based Doppler Radar for Micro UAS Detection and Classification. In: IEEE Military Communications Conference; 2016. pp. 924–929.

[64] Parashar KN, Oveneke MC, Rykunov M, *et al.* Micro-Doppler Feature Extraction Using Convolutional Auto-encoders for Low Latency Target Classification. In: IEEE Radar Conference; 2017. pp. 1739–1744.

[65] Kim Y, and Toomajian B. Hand Gesture Recognition Using Micro-Doppler Signatures with Convolutional Neural Network. *IEEE Access.* 2016;4:7125–7130.

[66] Kim Y, and Moon T. Human Detection and Activity Classification Based on Micro-Doppler Signatures Using Deep Convolutional Neural Networks. *IEEE Geoscience and Remote Sensing Letters.* 2016;13(1):8–12.

[67] Trommel R, Harmanny R, Cifola L, *et al.* Multi-target Human Gait Classification Using Deep Convolutional Neural Networks on Micro-Doppler Spectrograms. In: IEEE European Radar Conference; 2016. pp. 81–84.

[68] Kim BK, Kang HS, and Park SO. Drone Classification Using Convolutional Neural Networks with Merged Doppler Images. *IEEE Geoscience and Remote Sensing Letters.* 2017;14(1):38–42.

[69] Lombacher J, Hahn M, Dickmann J, *et al.* Potential of Radar for Static Object Classification Using Deep Learning Methods. In: IEEE MTT-S Int. Conf. on Microwaves for Intel. Mob.; 2016.

[70] Park J, Javier RJ, Moon T, *et al.* Micro-Doppler Based Classification of Human Aquatic Activities via Transfer Learning of Convolutional Neural Networks. *Sensors.* 2016;16(12).

[71] Krizhevsky A, Sutskever I, and Hinton GE. ImageNet Classification with Deep Convolutional Neural Networks. In: Pereira F, Burges CJC, Bottou L, *et al.*, editors. *Advances in Neural Information Processing Systems 25.* Curran Associates, Inc.; 2012. pp. 1097–1105.

[72] Boulic R, Thalmann NM, and Thalmann D. A Global Human Walking Model with Real-time Kinematic Personification. *The Visual Computer.* 1990;6(6): 344–358.

[73] CMU Graphics Library. Available from: http://mocap.cs.cmu.edu/ (Accessed: 9/15/19).

[74] Seyfioglu MS, Erol B, Gurbuz SZ, and Amin MG. DNN Transfer Learning from Diversified Micro-Doppler for Motion Classification. *IEEE Transactions on Aerospace and Electronic Systems.* 2019;55(5):2164–2180.

[75] Creswell A, White T, Dumoulin V, *et al.* Generative Adversarial Networks: An Overview. *IEEE Signal Processing Magazine.* 2018;35(1):53–65.

[76] Lewis B, Liu J, and Wong A. Generative Adversarial Networks for SAR Image Realism. In: Proc.SPIE. vol. 10647; 2018.

[77] Mi Y, Jing X, Mu J, *et al.* DCGAN-based Scheme for Radar Spectrogram Augmentation in Human Activity classification. In: 2018 IEEE International Symposium on Antennas and Propagation USNC/URSI National Radio Science Meeting; 2018. pp. 1973–1974.

[78] Erol B, Gurbuz SZ, and Amin MG. GAN-based Synthetic Radar Micro-Doppler Augmentations for Improved Human Activity Recognition. In: IEEE Radar Conference; 2019.

[79] Odena A, Olah C, and Shlens J. Conditional Image Synthesis with Auxiliary Classifier GANs. 2016 Oct; Available from: http://arxiv.org/abs/1610.09585.

[80] Arjovsky M, Chintala S, and Bottou L. Wasserstein GAN. 2017 Jan; Available from: https://arxiv.org/abs/1701.07875.

[81] Wang Z, Bovik AC, Sheikh HR, *et al.* Image Quality Assessment: From Error Visibility to Structural Similarity. *IEEE Transactions on Image Processing.* 2004;13(4):600–612.

[82] Erol B, Francisco M, Ravisankar A, *et al.* Realization of Radar-based Fall Detection Using Spectrograms. In: Compressive Sensing VII: From Diverse Modalities to Big Data Analytics. vol. 10658. International Society for Optics and Photonics; 2018. p. 106580B.

[83] Seyfioglu MS, Ozbayoglu AM, and Gurbuz SZ. Deep Convolutional Autoencoder for Radar-based Classification of Similar Aided and Unaided Human Activities. *IEEE Transactions on Aerospace and Electronic Systems.* 2018;54(4):1709–1723.

[84]   Javier RJ, and Kim Y. Application of Linear Predictive Coding for Human Activity Classification Based on Micro-Doppler Signatures. *IEEE Geoscience and Remote Sensing Letters*. 2014;11(10):1831–1834.

[85]   Molchanov P, Astola J, Egiazarian K, *et al.* Ground Moving Target Classification by Using DCT Coefficients Extracted from Micro-Doppler Radar Signatures and Artificial Neuron Network. In: Proc. of Microwaves, Radar and Remote Sensing Symposium; 2011. pp. 173–176.

[86]   Kingma D, and Ba J. Adam: A Method for Stochastic Optimization. arXiv preprint arXiv:14126980. 2014.

[87]   He K, Zhang X, Ren S, *et al.* Identity Mappings in Deep Residual Networks. In: Proc. of the 14th European Conference on Computer Vision, Amsterdam, The Netherlands; 2016. pp. 630–645.

[88]   Wang M, Cui G, Yang X, *et al.* Human Body and Limb Motion Recognition via Stacked Gated Recurrent Units Network. *IET Radar, Sonar and Navigation*. 2018;12(9):1046–1051.

[89]   Li H, Shrestha A, Heidari H, *et al.* Activities Recognition and Fall Detection in Continuous Data Streams Using Radar Sensor. In: 2019 IEEE MTT-S International Microwave Biomedical Conference (IMBioC); 2019. pp. 1–4.

Chapter 5

Classification of personnel for ground-based surveillance

*Ronny Harmanny[1], Lorenzo Cifola[1],
Roeland Trommel[1] and Philip van Dorp[2]*

A well-designed ground-based surveillance radar, with its all-weather and day-and-night capability, has the ability to detect and keep track of a large number of moving objects in a potentially large volume. This includes surface objects like vehicles, vessels and indeed animals and people. For all sorts of military and civilian safety and security applications, it is interesting to focus on individuals, in order to confirm that they are indeed human and to reveal what their intent might be. Radar can be of help when other means like cameras are not an option. A quick assessment on radar contacts like this would help the radar operator significantly in the context of intrusion detection, impending attacks by adversarial personnel, or during search-and-rescue situations.

In this chapter, we discuss how human activity such as walking and running can be recognised automatically by radar processing, and whether it is possible to extract additional knowledge from the individual's signatures, for instance whether someone carries a prohibited item such as a weapon.

5.1 Introduction

Radar, short for radio detection and ranging, is a measuring principle that uses active sensing through electromagnetic (EM) waves to analyse the environment. An EM signal is transmitted in a known direction, and reflected power is received and analysed by the radar system. These echoes carry information on the whereabouts of reflective objects and whether or not they move towards or away from the radar and at what relative radial velocity. This allows filtering out moving objects from irrelevant stationary reflective objects and surfaces.

A ground-based surveillance radar is a system that guards an area and monitors target that enter and exit the region of interest, thereby providing situational awareness

[1]Thales Nederland B.V., Delft, The Netherlands
[2]TNO, The Hague, The Netherlands

to the operators. Most surveillance radars scan their environment, meaning that the radar beam repeatedly scans through the defined volume, thereby illuminating the objects of interest only temporarily. Smart tracking algorithms are then capable of linking the identities of detected objects from scan to scan, hence providing the radar operators with historical information of the radar contacts, their origin, heading, behaviour, etc.

Radar systems can operate not only standalone but also in combination with other sensor types, such as optical and acoustic, using sensor fusion techniques. The sensor mix depends on the application, which in our case includes observing persons. An optical system can provide the observer with signature information thanks to its high angular resolution, allowing recognition and intent analysis which complements the surveillance function of a radar. An acoustic sensor may provide the operator with audio signatures that can be matched with audio profiles in a database in order to do additional classification. Both optical and acoustic sensors provide signature information but have important disadvantages at longer ranges where the object of interest does not occupy a sufficient amount of resolution cells or its acoustic signature is no longer detectable. In addition, targets can be camouflaged, making them hard to detect using optical sensors, especially in dark and foggy conditions. In those challenging circumstances, a radar can still perform detection and tracking, unaffected by day/night conditions and more resistant to weather. Therefore it is highly desirable to be able to extract a form of radar signature for classification, independently from optical and acoustic sensors. The radar signature consists of the measurable effects of the target's geometry and how they change over time. In particular, moving elements on the object, like arms and legs, that cause those changes over time, can be measured in two dimensions: position and radial velocity. The signature in the dimension of radial velocity is known as the micro-Doppler signature and can be exploited for classification.

Radar classification approaches can be roughly divided into model- and data-driven approaches. Model-driven classification approaches in radar can in turn be broadly divided into model-based (top-down) and feature-based (bottom-up) approaches, see [1–3]. (1) Model-based approaches estimate parameters of a model by finding the best fit between model predictions and measurements. This approach requires a radar model and a human model. It is unlikely that the human model can account for all human activities and will most likely be limited to, e.g. walking and running. (2) The feature-based approach is an inversion of the model-based approach. Recognisable characteristics, or features, are extracted from the data and associated with the model that is most similar. It will be clear that the search for reliable characteristics is difficult and the connection with the model can be vague.

Data-driven classification approaches forego the construction of explicit models and rely on feature learning, instead extracting information from the data itself. These features, which may or may not be associated with a model or physical phenomena, are then fed into a classifier of choice. Traditional learning techniques learn transformations of data that make it easier to extract useful information. Principal Component Analysis (PCA) is a prime example and widely used for dimensionality reduction as a part of a classification system. Other well-known techniques include linear discriminant analysis (LDA), independent component analysis (ICA), isometric feature mapping (Isomap) and locally linear embedding (LLE) [4]. In recent years, deep learning has emerged as one of the more successful data-driven machine-learning methods and is widely applied across many fields and disciplines. Deep learning represents a family of techniques based on artificial neural networks with many layers, where each layer can apply a non-linear transform to its inputs. The networks are able to learn arbitrarily complex functions from the data which makes manual feature engineering unnecessary. Deep learning can be supervised through the use of labelled data, often leading to state-of-the-art classification systems, or can be unsupervised in order to discover structure and meaningful patterns in the data.

Before we can address classification aspects of human motion, we first need to cover basic properties of human motion and the constraints under which radar is able to measure this. In Section 5.2, we will first introduce the most relevant types of human motion for ground surveillance radar, which is walking and running. Models that describe the most essential aspects of human gait are introduced. Besides providing a better understanding of the human gait, these models can also serve as a component within a classifier, as will be explained in Section 5.3, which deals with model-driven techniques for classification. In Section 5.4, we turn to data-driven methods that can learn models implicitly from a dataset in either a supervised or unsupervised fashion. Finally in Section 5.5, main conclusions are drawn for model-based and data-driven techniques, summarising the advantages and drawbacks. Furthermore, an outlook on future research dedicated to human gait classification for ground-based surveillance radars is provided.

5.2 Modelling and measuring human motion

In this section, we will introduce different models for the human gait, and we will address the radar parameters that are critical for measuring human gait. Then, we will present some examples of measurements on human gait through spectrograms which are of particular interest in micro-Doppler-based classification. At the end of this section, we will provide an alternative method to observe human motion through motion capturing hardware.

5.2.1 Human gait

The most relevant form of human motion is the way we move ourselves over land without external aids, i.e. through walking and running. Our goal is to have sufficiently realistic models of human gait, such that they can be used to generate a synthetic radar signature of any 'normal' walking or running individual [2]. This requires detailed information about the kinematics, size and shape of the human body. First, we will introduce the Boulic–Thalmann model used for modelling of human walking. After that, the Vignaud model used for the modelling of human running is introduced. Some alternative models for human gait also are mentioned at the end of this section.

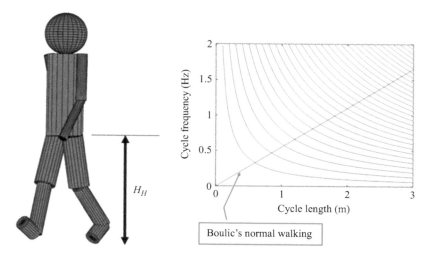

*Figure 5.1   Left: the human body size normalisation with respect to hip height.
Right: the relation between cycle length and cycle frequency.*

5.2.1.1   Human walking model

Human walking is a complex motion of swinging arms and legs. Walking is a periodic
activity, in which one single gait cycle is defined as the successive contacts of the heel
of the same foot, i.e. one step from both left and right leg. During a single gait cycle,
or 'stride', each leg undergoes two phases. In the stance phase, the corresponding foot
is on the ground. The stance phase occupies approximately 60% of the gait cycle. In
the swing phase, the foot is lifted from the ground with an acceleration or deceleration
[5]. Thalmann and Boulic developed a model for human walking based on empirical
mathematical parameterisations derived from experimental data [6]. The influence
of personalised motion is minimised by averaging the parameters from experimental
data [5]. The model consists of two aspects: the kinetics describing the time-varying
positions of the body during a gait cycle and the size and shape of the body parts,
of which there are 12. The model enables generation of normal walking humans and
abnormal walking humans with unrealistic relations between the model parameters.
Unrealistic relations give skidding or sliding feet. The Boulic–Thalmann model is
suitable for the detection of normal walking action and not for any unusual walking
motions.

Boulic normalises any observed person into a standard person with a leg-length
of 1 m, see Figure 5.1. The relative normalised velocity v (m/s) is expressed using
the ground to hip distance H_H and real body speed s (m/s), $v = s/H_H$. The relations
between the cycle length, cycle frequency and normalised velocity are depicted in
Figure 5.1. The relative normalised velocity v is the multiplication of the cycle length l^c
and the cycle frequency f^c, $v = l^c f^c$. These are the blue lines in the figure. Boulic gives

the relations of a normal walking person: $l^c = 1.346\sqrt{v}$ and after elimination $f^c = l^c/1.346^2$, which is the red line in the figure. The Boulic–Thalmann model describes realistic walking humans with only two parameters, the cycle length and the cycle frequency, both dependent on the length of the person. The positions of the body parts are calculated relative to the model's reference point which is located in the centre of the pelvis.

5.2.1.2 Human running model

There are some fundamental differences between walking human and running. First of all, the stance phase is shortened, which means that the swing phase is lengthened. Second, there is no double support, which means that during the entire gait cycle, the situation where both feet are on the ground does not occur. There is also a new phase: the non-support phase. In this phase, neither leg is weight bearing (i.e. a double float period). Compared to walking, for running a more simple model was devised from the work of Vignaud and Ghaleb [7]. They measured the kinematic behaviour of a human using 40 discrete points. From these measurements, the radial velocity as a function of average running velocity and time were modelled. In order to use the same model for the size and shape of the body parts as for a walking human, the same reference points as in the Boulic–Thalmann model were selected out of the 40 points available in the Vignaud model.

5.2.1.3 Other gait models

The advantage of the Boulic–Thalmann model is that the entire walking movement is defined with only a few parameters. Troje [8,9] studied the human gait and how we perceive others when they walk. He focused on gender and other characteristics of the human motion and extracted a five-parameter model with biological and physical attributes. The model generates different walking patterns based on gender-specific differences and emotional state such as nervous versus relaxed, and happy versus sad.

More recently, it has been shown that the body mass index has a correlation with the micro-Doppler signature of a walking human and is reasonably discriminative [10].

5.2.2 Measuring the signature of human gait

A signature contains information of an object's shape and how it evolves as function of time in one or more measurements' dimensions. Two basic radar properties are dominant for the appearance of signatures: the radar central frequency and bandwidth. Figure 5.2 presents a schematic subdivision of radar systems. A higher transmitter frequency gives higher Doppler frequency information, while a larger bandwidth gives the opportunity to add range information in the classifier. Modern surveillance radar systems scan the environment with mechanical or electronic beam control and provide azimuth information and elevation information. The additional angle information depends heavily on the radar configuration. In general, a higher carrier frequency gives better angle information to support the classification.

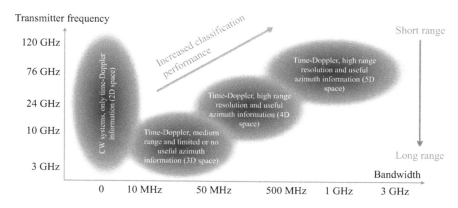

Figure 5.2   Schematic radar processing division as a function of bandwidth and transmitter frequency

If we consider an average operational-ground-based surveillance radar, we can expect a coherent, either pulsed radar or frequency-modulated continuous wave (FMCW) radar. This radar has its major design decisions determined by detection and tracking performance requirements in terms of detection probability, false track rate, scan time, etc. The carrier frequency and bandwidth of the radars we consider in this chapter are in the 3–10-GHz/10–50-MHz block in Figure 5.2.

Measuring the signature of human gait is something that the radar should do in addition to its normal detection and tracking tasks. If we look at the basic processing architecture of such radars, we can make a distinction between pre-processing, detection and post-processing. Pre-processing separates targets and clutter based on range, velocity, direction and amplitude characteristics. Detection and post-processing stages are applied to those radar cells that have crossed a certain threshold. After a target detection has been established, a track can be either initiated or updated, depending on whether the detection can be associated with any existing track.

Signature extraction is typically performed when the object's position is known. As we will see, signal extraction for human gait involves performing a dedicated measurement on the range-azimuth cells in which the object is currently moving. The objective of this dedicated measurement is to capture sufficient movements from the object's moving parts, such that the measurement contains enough information to enable successful classification.

5.2.2.1   Micro-Doppler signature

When the transmitted signal of a coherent radar system hits a moving object, the carrier frequency of the signal will be shifted by $f_d = -2v_r/\lambda$ (Hz). This is known as the Doppler effect, with $v_r = dR/dt$ (m/s) as the radial velocity of the object with respect to the radar, $\lambda = c/f_0$ (m) as the wavelength of the carrier frequency f_0, and R(m) is the range to the object.

As we have already stated, the interaction between the EM signal and the object is a complex process that involves the scattering from the different parts on the target. If some of these parts are moving with respect to the net velocity of the target, additional frequency modulations will be present on the received signal. This is commonly referred to as micro-Doppler [11] and provides a way to quantify movement, e.g. the wing beat of birds, propeller rotations of airplanes and indeed the swinging of the limbs of people and animals. Collecting a micro-Doppler signature with sufficient accuracy does however require some planning. Since we are dealing with capturing a measurement time series, things to pay attention to are

1. Radar power budget,
2. Bandwidth of the radar,
3. Radar beam time-on-target T_{tot}, and
4. Wavelength and waveform.

1. The radar power budget should be high enough to reach sufficient signal-to-noise ratio (SNR) at the range of interest for the parts that are dominant for the signature. So where the radar cross section (RCS) of an entire human body could be around 1 m^2, or 0 dBm^2, if the arms matter for our classification problem, they may produce perhaps -6 dB less return power compared to the whole body.
2. The range and angle resolution of the radar should be adapted to the size of a human body. A low resolution measurement will complicate the classification process as the micro-Doppler response from a larger area may contain multiple persons or more clutter compared to measurements from a smaller area.
3. Many micro-Doppler signatures are repetitive, like the human gait, and so a sufficient part of the signature's period has to be captured. A typical human stride takes about 1.2 s. Some of the classification techniques introduced in this chapter require at least one complete step, so half a gait cycle, which leads to a round and safe interval length of $T_{tot} \geq 1$ s. For a single beam, scanning surveillance radar, this T_{tot} could be an issue. If we assume a reasonable azimuth beamwidth of several degrees, and a revisit time T_r of several seconds to retain the running tracks, in a broad surveillance sector Υ, this $T_{tot} = \theta T_r / \Upsilon$ is much less than a second. So unless the radar can pause for a while on a detected object, or exploit multiple beams, or do something else to reach the required T_{tot}, the signature is collected in asynchronous pieces of several tens of milliseconds, which is still an unsolved challenge in the literature. In the following, we assume we have sufficient T_{tot} to do classification.
4. The human gait generates maximum velocities in the order of 10 m/s, which are generated by the feet. For classification it is preferred to collect these velocities unambiguously and with sufficient resolution.

Unambiguous velocity is obtained through a sufficiently high Doppler sampling frequency f_s (Hz), a.k.a. pulse repetition frequency (PRF) or sweep repetition frequency (SRF) for conventional pulse-Doppler or FMCW radars, respectively. The Doppler sampling frequency should be $f_s \geq 4v_{r,max}/\lambda$, where $v_{r,max}$ is the maximum unambiguously measurable velocity in either inbound or outbound direction. For a

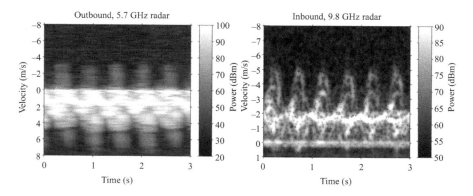

Figure 5.3   Spectrograms of a walking person

3-GHz (S-band) radar, this means $f_s \geq (4 \cdot 10)/0.1 = 400$ Hz, whilst for a 10-GHz (X-band) radar, $f_s \geq (4 \cdot 10)/0.03 \approx 1.33$ kHz. At the same time, we need our micro-Doppler signal to have sufficient velocity resolution $\delta_v = \lambda/2T_{cpi}$ (m/s), where T_{cpi} is the coherent processing interval, assuming conventional frequency analysis, i.e. by Fourier transform. Typically, $T_{cpi} \ll T_{tot}$, in order to be able to collect several traces to construct a spectrogram.

Also note that $\delta_v T_{cpi} = \lambda/2$ is constant for a certain carrier frequency, so we can exchange Doppler resolution with time resolution when we generate a micro-Doppler spectrogram.

Besides the structure and dynamics of the objects themselves, the carrier frequency of the radar and the processing parameters, the appearance of micro-Doppler depends on the aspect angle under which the measurements take place. In practice no two spectrograms are exactly the same, which is why generalisation techniques are required in order to perform human gait classification. This is true especially if we want to identify an individual amongst other individuals based on their gait induced micro-Doppler at an arbitrary aspect angle.

5.2.3   Visualisation of the micro-Doppler signature

If an object is being tracked by radar, the object's signature can be acquired and visualised. Several visualisation types are available, but most popular is the velocity–time response of the object, also referred to as a spectrogram. This is the waterfall plot of the Doppler spectra from the object's position. Depending on the coherent processing interval (CPI), it is possible to overlap integration intervals and create a smoother spectrogram in the time dimension. Typical examples for two radar bands are shown in Figure 5.3.

By assuming a radar model, it is also possible to create synthetic spectrograms from the walking and running models that were introduced in Section 5.2.1. In that case, we also need to define the reflective properties of the different human body parts, which will be addressed in the next section.

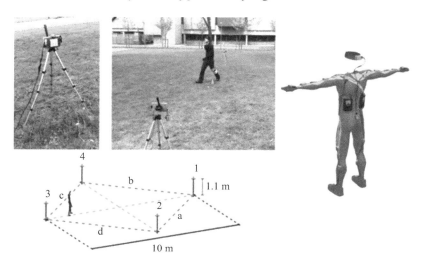

Figure 5.4 *Left three images: photos and a schematic overview of the measurement setup. Four CW radars are located on the corners of a 10 × 10 m field, recording the radar return of a test person. In addition, the motions of this person were recorded using the sensor suit as pictured on the right side*

5.2.4 Motion capturing activities on human motion

Capturing the personal traits of an individual's gait, or indeed any other motion, can be done by means of motion capturing technology. One way to do this is to have a subject wearing a motion capturing suit as depicted on the right side in Figure 5.4.

Several experiments were performed where a human subject walked back and forth in front of four industrial, scientific and medical (ISM)–band CW (continuous wave) radars, whilst his motions were recorded with a motion capturing suit from the XSens company from Enschede, The Netherlands [12]. Motion data from 17 inertial motion trackers is transferred wirelessly to the base station, which matches the data with an anatomical model.

The experimental setup from Figure 5.4 allows comparing radar data from different aspect angles to the motion data recorded by the suit. The latter was translated into a synthetic radar signal, by having a computerised mannequin, comprised mainly of ellipsoids that move synchronously according to the data recorded from the suit. Ellipsoids are used to estimate the return of body parts, specifically the aspect angle dependency, according to [13]:

$$\bar{\sigma}_{ellipsoid} = \frac{\pi a^2 b^2 c^2}{(a^2 \sin^2 \theta \cos^2 \varphi + b^2 \sin^2 \theta \sin^2 \varphi + c^2 \cos^2 \theta)^2} \tag{5.1}$$

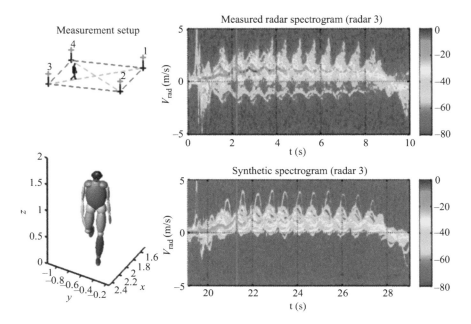

Figure 5.5   Screen capture of the animation that was used to compare the measured and the synthetic spectrograms. Note that the measured spectrogram shows a 'mirror' replica of the human gait response in the negative part of the Doppler spectrum. This is due to an IQ imbalance in the receiver.

where a, b and c fix the dimensions of the ellipsoid, and θ and φ are the two aspect angles of the ellipsoid with respect to the radar. The formula returns the ellipsoids RCS $\bar{\sigma}_{ellipsoid}$. By assuming a radar model and geometry equal to the real experimental setup, we can simulate the response and compare it with the measured spectrogram, which is shown in Figure 5.5. Note that this simulation ignores many real world EM effects, such as multipath and blocking of body parts from the radar's line of sight. Nonetheless, the synthetic spectrogram does present a satisfactorily visual similarity to the measured version.

The exercise described above allows to differentiate between the contribution from different body parts, as demonstrated in Figure 5.6. It also enables the simulation of returns from any arbitrary radar position and look angle (Figure 5.7), as well as other radar parameters, such as the carrier frequency, pulse repetition frequency, and CPI [14].

Motion capturing techniques thus enable recording any human motion, including personal traits, which can be translated into synthetic radar data. These techniques can be used for data augmentation as well as classification on an individual level.

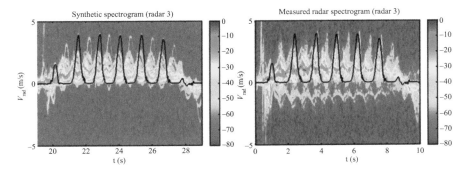

Figure 5.6 *Synthetic spectrogram (left) and measured spectrogram (right) of a person walking away from the radar. The black line is an overlay of the radial velocity of a single body part, and it serves as a highlight to accentuate the contribution of this individual part, in this case the right foot.*

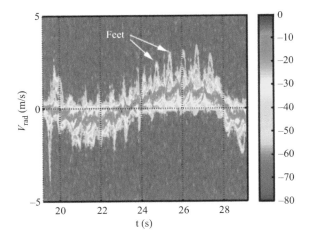

Figure 5.7 *Synthetic spectrogram for an arbitrarily chosen radar position. In this case, the 'virtual' radar is located right at the centre of the radar setup, at an altitude of 3 m, looking down. The characteristic motion of the feet is still visible.*

5.3 Model-driven classification

In this section, we will discuss model-driven techniques for classification. Section 5.3.1 introduces the basics of model-driven classification methods. Next, in Section 5.3.2, we estimate the human motion parameters by minimising the error between the data generated by the model and the measurements. Section 5.3.3 considers a

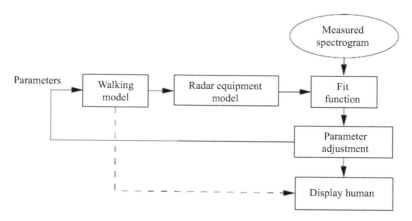

Figure 5.8   Model-based approach

particle filter to minimise the error between data generated with the model and the measurements.

5.3.1   Introduction to model-driven classification methods

The model-based approach minimises the difference between simulated data and measured data by adjusting the model parameters. Figure 5.8 presents the model-based approach. It comprises four main parts: the human model, the radar equipment model, a fit function and the adjustment of the parameters.

Human walking models have been discussed previously in Section 5.2.1. The radar equipment model contains the radar equation which relates the range of a human body part and its shape to the radar characteristics. It also includes the radar data processing model which calculates the raw radar measurements using the RCS that describes the reflection by the human body parts. This reflection depends on the shape of the body parts and the aspect angle with respect to the radar. Basic shapes such as spheres, cylinders and ellipsoids can be used to model the radar response of the body parts. More accurate modelling can be done using CAD models to extend the EM propagation modelling with shadowing and multipath. In practice, the modelling with basic shapes works well and requires less computation time.

The fit function calculates the error between the measured spectrogram and the simulated spectrogram. The result is determined by the noise properties and reflected radar signal properties. Selecting a suitable fit function depends on the error distribution: linear least-squares fit and log least-squares fit are typical choices. Estimation and adjustment of the model parameters can then be done using a variety of approaches. The parameters that belong to the best fit describe the human walking behaviour. A simulated walking person shows the connected human body parts with a specification of their relative position as a function of time. The scene of the walking person in a virtual environment is generated given the estimated walking parameters

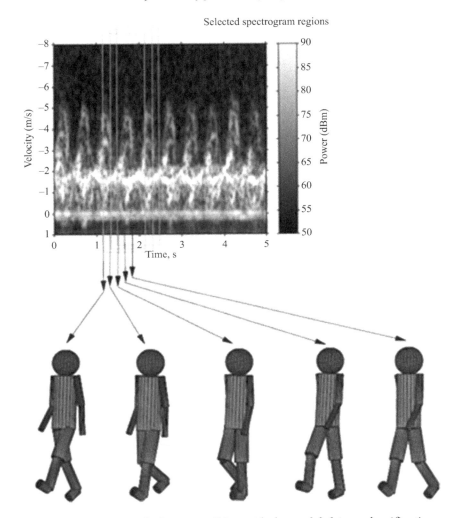

Figure 5.9   *Animation of a human walking with the model-driven classification.*
Successive spectrogram measurements give the instantaneous human
posture. Successive estimated postures give the animated human model.

and the geometric model. The simulated walking person generates an image of the
person at the measured position. Successive images of the animated person are shown
in Figure 5.9.

5.3.2   Results of model-based classification

In this section, some results of model-based classification from [2] are presented. The
Boulic–Thalmann model was used here and so the kinematic parameters that are to be

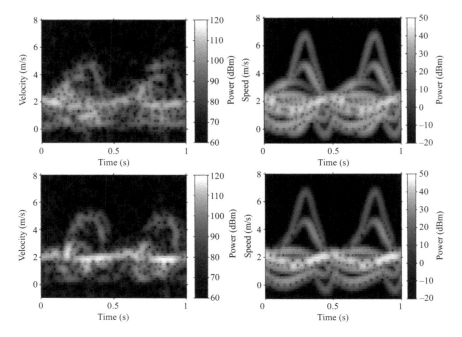

*Figure 5.10   Left: measured spectrograms of human. Right: the corresponding
            simulated spectrograms with the best fit. Top row: the walker with
            swinging arms. Bottom row: the stroller without swinging arms.*

estimated are the cycle frequency f^c, cycle length l^c and the cycle phase φ^c. The cycle
phase is an additional parameter, because the measured spectrogram has unknown
start time. Two types of human gait are considered in this section: that of a walker and
of a stroller. The walker moves his arms during motion, whereas the stroller's arms
hang by his sides. Figure 5.10 presents example spectrograms of a measurement with
1-s duration and the corresponding simulated spectrograms. One example belongs to
a normal walker, the other to the stroller. They show powerful reflections caused by
the motion of the human body parts. The mean Doppler velocity belongs to the body
of the human and is approximately 2 m/s, the cycle frequency is approximately 1 Hz.
Velocities belonging to other human body parts are also apparent in the spectrogram.
These body parts move relative to the body with velocities greater and smaller than
the average Doppler velocity. A high velocity resolution is necessary to distinguish
the individual human body parts. The simulated spectrogram of the normal walking
person shows the movement of the arms, while these are not present in the stroller
spectrogram.

The structure of the linear and log fit functions for a typical example in the cycle
length and cycle frequency plane are presented in Figure 5.11. The figure shows the
minimum fit function for a given cycle length and cycle frequency, and the best cycle
phase is estimated with a grid search. The typical example of the linear fit function

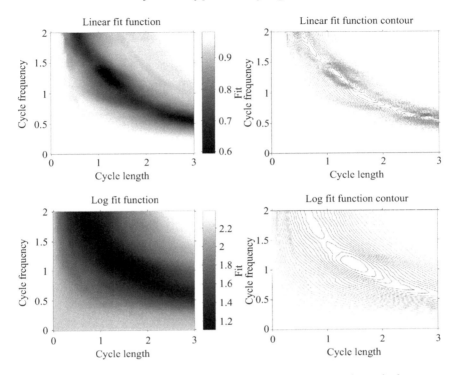

Figure 5.11   Top row: linear fit function. Bottom row: logarithmic fit function.

has many minima. These local minima lie on the line with constant relative velocity. The absolute minimum is not always the correct minimum. The typical example of the log fit function has less minima, the absolute minimum is the correct minimum. The least-squares fit counted deviant errors in the estimation of the parameters. The log transform of the RCS exponential distribution and the multiplicative noise forced the log fit function to a normal distribution without deviant errors in the estimation of parameters and gives fewer minima.

Previously, the focus has been on walking and the arms and legs movement. The classification distinction between walking and running person is easier; it is based on the average Doppler speed. Slow moving persons and little Doppler variations correspond to walking persons, and fast moving persons with high Doppler variations correspond to running persons. In general, a fast walking speed is 5–6 km/h and a slow run is 8 km/h.

In addition to the model-driven classification method discussed above, there are other model-driven techniques. One model is the periodic amplitude modulation (PAM) and frequency modulation (FM) in specific regions of the spectrogram which correspond to physical target properties, see [15–17]. This model is not specifically related to a human model but to modulation characteristics in the spectrogram. The model focuses on regions with relevant information and then extracts relatively invariant features from each region. These features are global features or related to the body response. Common features related to PAM and FM are statistical features, spectral ratio features and shape factors. The best features for classification are selected with a feature selection algorithm and then a standard classifier is trained with the selected features. The method is very dependent on the bodyline estimator with corresponding region selection. In practice, the bodyline estimate is unreliable due to the stochastic behaviour of the personal response. The theoretical background of this model is good, but in practice it is not as useful.

5.3.3   Particle-filter-based techniques for human gait classification and parameter estimation

In this section, summarised from [18,19], we discuss a motion-model-based particle filter implementation to classify human motion and to estimate key state variables, such as motion type, i.e. running or walking, as well as the subject's height. The micro-Doppler spectrum is used as the observable information. The system and measurement models of human movements are built using three parameters (relative torso velocity, height of the body and gait phase). The algorithm developed has been verified on simulated and experimental data.

5.3.3.1 Particle filter

Particle filters or sequential Monte Carlo methods are a set of Monte Carlo algorithms used to solve filtering problems arising in signal processing and Bayesian statistical inference. The principles are well explained in [20].

The idea of a particle filter is to repeatedly go through a Bayesian inference process in which a relatively large set of hypotheses on the state space, the 'particles', is extrapolated in time and then updated with new measurements. In contrast to a Kalman filter, the posterior density is then represented by the location of the particles with associated weights, whereas in a (default) Kalman filter, distributions of the current state are always assumed to be Gaussian.

By recursively making predictions for the current set of particles, and updating them according to new measurements, i.e. allowing to have more particles in likely positions and less particles in unlikely positions in the state space, in due time, the particle 'cloud' starts to describe a sampled version of the posterior probability density function.

In order to employ a particle filter, one would need the following:

- A definition of the state space x, in terms of its elements, for which we would like to find the posterior density.
- A system model, i.e. a function that describes the evolution of the state with time, to cope with the intervals in between measurements. This is a function that maps x_k to the next step x_{k+1} and accepts an additional input that quantifies the system process noise at step k.

- A measurement model, i.e. a function that maps a state x_k to a measurement value z_k. In practise though, we would measure z_k with an additional noise component that has to be taken into account.

Once we have the representation of a measurement for a particular state at time $k + 1$, we can weigh the corresponding particle, by comparing the measurement representation with the actual measurement z_{k+1}. A high resemblance means more particles are allowed in the same region of the state space. Low resemblance means the particle is likely to poorly summarise the current situation and needs to be removed.

5.3.3.2 Human gait implementation

For our human gait examples, we can fill in the constructs from the previous section as follows:

For the system state at time k, we take $x_k = [m_k \ \bar{v}_k \ h_k \ \phi_k]$, where $\bar{v}_k$ (m/s) is the relative velocity which is a scaled version of the average velocity by a dimensionless value equal to the height of the thigh of the person, h_k (m) is the height of the person and $\phi_k \in [0, 1)$ is the phase of the gait cycle. For the element m_k, we consider two different hypothesis sets. For *simulated data*, we take $m_k \in \{0, 1, 2, 3\}$ to represent to following motion model classes:

- H_0: the null-hypothesis, for motions of other origin and no motion at all
- H_1: the hypothesis for a strolling human (without arm swinging)
- H_2: the hypothesis for a walking human (with arm swinging)
- H_3: the hypothesis for a running human

For *measured data*, we will simplify $m_k \in \{0, 1, 2\}$, to represent only:

- H_0: the null-hypothesis, for motions of other origin and no motion at all
- H_1: the hypothesis for a walking human
- H_2: the hypothesis for a running human

For the system model, we only need to evolve the gait cycle in x_k towards x_{k+1}, since the other elements in the state vector can be assumed to remain constant. So, $\phi_{k+1} = \phi_k + (\Delta t / T_c(\bar{v}_k))$, where Δt is the time between k and $k + 1$, and $T_c(\bar{v}_k)$ is the time duration of one gait cycle. The process noise needs to generate sufficient variety for the whole set of particles in order to converge and adapt to the measurements. For the discrete part, the motion model, a transition probability matrix needs to be defined. More details about the process noise can be found in [18].

The measurement model is less trivial. We need to translate a point in the state space x_k to the corresponding spectrogram estimate. This translation consists of two parts:

1. From state to mannequin posture and velocities;
2. From mannequin posture and velocities to the corresponding spectrogram.

For part 1, we can use the models from Boulic–Thalmann for $m_k = 1$ and Vignaud for $m_k = 2$, as described in Sections 5.1.1.1 and 5.1.1.2, respectively. Both models give position and velocity of the relevant body parts. We can generate data from the mannequin the same way as in Section 5.2.4, with ellipsoids that give an aspect angle dependent RCS. For part 2, we need a radar model that can replicate the spectrogram. This means that the model should assume the same waveform and processing conditions. The spectrogram is generated by summing up the contributions from the different body parts.

For $m_k = 0$, i.e. the hypothesis where we assume that there is no human motion, H_0, we decide to model the spectrogram directly as a Gaussian spectrum around zero Doppler velocity, and as noise for the other velocities. This is further explained and motivated in [18]

The likelihood function is used to assign weights to the particles. For each particle, a spectrum is estimated based on the models explained above. The relation between the measured spectrum s and the estimated spectrum $\bar{s}$ is the following: $\bar{s} = s + n$, where n is an i.i.d. measurement noise process, which follows a zero-mean complex–normal distribution with variance σ_n^2. Both vectors have the same length, corresponding to the number of Doppler bins N_{bins}.

In [18], it is argued that a least squares fit can be used as a maximum likelihood estimator. This maximum likelihood estimator is subsequently used to assign weights to the particles, as follows:

$$w^j = \frac{1}{\sum_{i=1}^{N_{bins}} (s_i - \tilde{s}_i)^2} \quad \text{for } j = 1, 2, \ldots, N_s \tag{5.2}$$

where N_s represents the number of particles used. Here, $N_s = 500$ was used.

5.3.3.3 Classification and parameters estimation results

Classification results will be presented for both simulated and measured data.

Simulated data

Using the human movement models described in Section 5.2.1, micro-Doppler spectrograms have been synthesised. Three different inputs are considered. The first two inputs consider two types of walking—one where the person walks without swinging of the arms, the so-called strolling gait and the other walking hypothesis does consider the swinging of the arms. The third input considers a running person. In the top left part of Figure 5.12, a simulated spectrogram for a walking human with swinging arms is depicted. This person has a height of 1.9 m and is walking with a relative velocity of 1.5 m/s. At $t = 0$, the person starts moving, i.e. at that time, the phase in the gait cycle is 0. In the bottom left part of Figure 5.12, the classification results are indicated. The classification is performed correctly.

After analysing the distribution of the particles, the posterior distribution of the three motion parameters—the relative velocity v_{rel}, the height h and the phase gait cycle φ_{gc}—are found to be Gaussian. The mean and the standard deviation of the estimated parameters are given in Table 5.1, together with the true values of the parameters. The true value of the gait phase is evaluated by first determining the time duration of a single cycle t_c. Next, the time duration of the gait cycle is used to evaluate the phase using the measurement time at this iteration, given that the phase

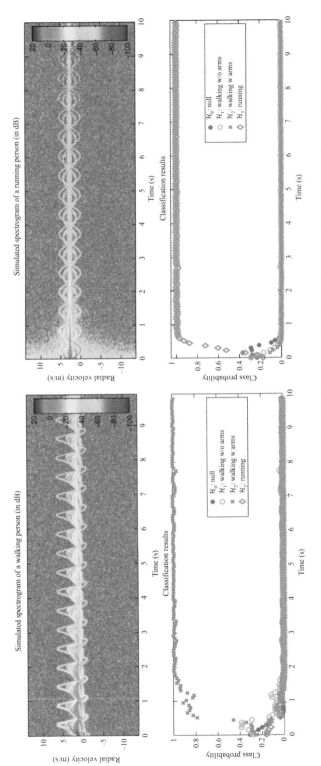

Figure 5.12 Simulated spectrogram for a walking (left) and running (right) human and its classification results

Table 5.1   Parameter estimation results in a simulation performed for both a walking and a running person

Parameters	Walking			Running		
	Estimated		True	Estimated		True
	μ	σ		μ	σ	
v_{rel} (m/s)	1.497	0.014	1.5	2.823	0.068	3.0
h (m)	1.893	0.068	1.9	1.792	0.089	1.8
φ_{gc} (−)	0.479	0.008	0.47	0.653	0.075	0.65

was 0 at $t = 0$. A similar simulation was performed for a running person. In the top right part of Figure 5.12, a simulated spectrogram of a person of 1.8 m, running with a constant relative velocity of 3.0 m/s, is depicted. The bottom right part of Figure 5.12 shows the correct classification of this running person, and the parameter estimation results are given in Table 5.1.

Measured data

Measurements of humans and animals were performed in order to test the correct performance of the algorithm on measured data. The measurement radar was positioned 0.5 m above the ground, and the targets moved in a radial position away from the radar, starting from a known position. The radar used for the measurements was a CW radar operating in the ISM band (~24 GHz). The radar video signal was sampled at 8,820 Hz. After some post processing (e.g. correcting the I–Q imbalance) the short-time Fourier transform (STFT) is performed on the radar video signal in order to obtain the spectrogram. In the STFT, a sliding window of 512 samples is used.

In the top left part of Figure 5.13, the measured spectrogram of a walking person (male) is shown. He has a height of 1.85 m. Looking at the velocity of the torso, the average walking velocity can be extracted from the spectrogram and is assumed to be constant at 1.55 m/s. The bottom left part of Figure 5.13 shows the classification result. The part of the spectrogram that contains the motion of the walking human (up to about 5.5 s) is classified correctly as human walking. When he comes to a halt, the algorithm classifies this part also correctly as the null hypothesis. The results of the estimation of the motion parameters are indicated in Table 5.2.

A similar test is performed with a measurement of a running person. The spectrogram of this person is indicated in the top right part of Figure 5.13. In the first 0.5 s of the measurement, the person is accelerating towards running, and this part of the measurement is classified as the null-hypothesis. Between around 0.5 and 2.5 s, the motion of a running person can be visually noticed in the spectrogram. The algorithm

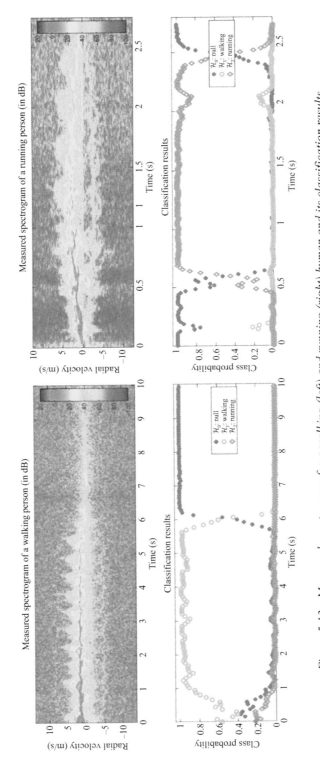

Figure 5.13 Measured spectrogram for a walking (left) and running (right) human and its classification results

Table 5.2   Parameter estimation results in a measurement performed for both a walking and a running person

Parameters	Walking			Running		
	Estimated		True	Estimated		True
	μ	σ		μ	σ	
v_{rel} (m/s)	2.91	0.04	1.71	2.91	0.03	2.97
h (m)	1.66	0.04	1.87	1.66	0.02	1.7
φ_{gc} (−)	0.10	0.03	0.14	0.10	0.05	0.11

classifies this part correctly, as is observed in the bottom right part of Figure 5.13. After 2.5 s, he is slowing down and no human motion is recognised by the algorithm. Estimated parameters at $t = 2.21$ s are indicated in Table 5.2.

To test whether the filter can recognise motions not originating from humans, measurements on moving animals were performed. In Figure 5.14, the spectrogram of a mallard duck is depicted. Since a mallard is bipedal like humans, a walking mallard shows some similarities to human walking. Especially the peaks coming from the feet and the torso component are similar to human walking. On the other hand, a mallard makes shorter and quicker steps and, therefore, explains the more spike-like nature of the spectrogram. After 4.5 s, the mallard stops moving, and hence, no motion occurs anymore. This allows the same spectrogram to be used for testing when no motion is classified correctly. In the bottom part of Figure 5.14, the classification results are indicated. From this figure, one can observe that both the movement of the duck and the part containing no motion are classified correctly.

5.3.3.4   Conclusion

Model-based classification of human movement using micro-Doppler spectrograms as inputs was demonstrated. The joint estimation of motion parameters and the classification between human motion and motion of other origin was successfully performed with two methods, a direct model fitting method and a particle filter approach. Both methods showed that the unique signature of human motion in the spectrogram can be used for jointly detecting and classifying human motion as well as for the estimation of motion parameters. The classification between human walking and human running was investigated, and the classification proved to be successful on both simulated and measured data.

Explicit gait models are very useful for radar classification algorithms. If the models are in close agreement with measurements, the presented classification algorithms based on these models are reliable and robust and can be easily used under different measurement circumstances and radar configurations. One important disadvantage is that they can only accommodate a limited set and variety of movements.

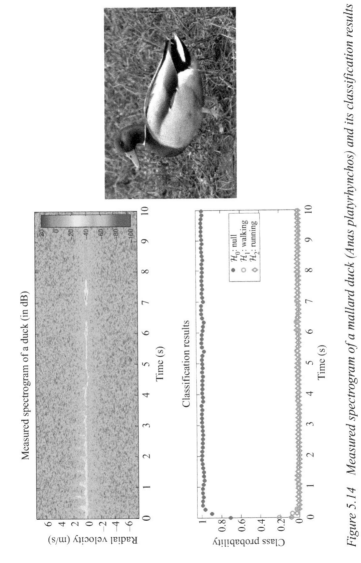

Figure 5.14 Measured spectrogram of a mallard duck (Anas platyrhynchos) and its classification results

5.4   Data-driven classification

In the previous section, examples of model-based human gait classification were discussed, where the input data was compared to a predefined model. In this section, an alternative approach is considered: the information or features, for example, needed to accomplish a classification task are extracted directly from the data. These features can be either handcrafted or learned by an algorithm during a training process. Depending on the application and available information at hand, this training can be either supervised or unsupervised. In the supervised case, input data and their corresponding correct or desired outputs (for instance, class labels for a classification problem) are provided as input to the algorithm. In unsupervised training, only raw data is used as input for the training, with the objective to find structure within the data or provide an improved initialisation for another task. For both training principles, the most effective algorithms are currently based on a hierarchical feature extraction process described by a composition of multiple non-linear transformations of the input data, known as deep learning. In the following subsections, we will discuss examples of deep neural networks applied to human gait based on previous work by the authors. In Section 5.4.1, both a convolutional neural network (CNN) and a recurrent neural network (RNN) are trained in a supervised manner to distinguish the number of human gaits in a group using micro-Doppler spectrograms as input data. Finally, in Section 5.4.2, unsupervised learning by means of Generative Adversarial Networks (GANs) is used to find structure in the micro-Doppler signature of human gaits and the semantic meaning of the automatically extracted features.

5.4.1   Deep supervised learning for human gait classification

In this section, adapted from [21], a deep convolutional neural network (DCNN) is proposed to distinguish between absence of gait, and the presence of single or multiple instances of human gait. The DCNN is applied to micro-Doppler spectrograms of both synthetic data and measured data. Next, following up on this work, a RNN is applied to this classification problem.

5.4.1.1   Convolutional neural networks

Convolutional neural networks are a special form of neural network specifically designed for image recognition, inspired by the mammalian visual system and widely recognised as the most successful network architecture for image recognition to date [22]. The DCNN in this work was implemented using the Theano-framework [23] and the add-on Lasagne library [24]. Critical network design choices were made based on hands-on experiments and experience. The designed network consists of a few main parts: a convolutional part for the feature extraction, a subsampling part to provide spatial invariance and reduce the computational complexity and a final stage formed by fully connected layers which globally integrate all the features extracted by the previous layers. The final network comprises 14 layers, including eight convolutional layers with filter size 5×5, three 2×2 maxpooling layers, two fully

Table 5.3 Number of spectrograms per class (source: [21])

Class	Model data		Measurement data	
	Train	Test	Train	Test
C0	1,800	450	1,729	1,506
C1	1,800	450	8,510	1,379
C2	1,800	450	3,504	1,448
C3	1,800	450	3,837	1,443
C4	1,800	450	N/A	N/A
C5	1,800	450	N/A	N/A

connected layers and a final softmax layer for the classification. More details can be found in [21].

Model data

The model used for generating the data is from Boulic–Thalmann as also described in Section 5.2.1.1. A dataset containing 2,000 unique time-signal examples of single walking persons varying in height between 1.65 and 1.95 m was built. The simulated subjects walked both inbound and outbound with respect to a simulated CW radar positioned around 1 m in height. Datasets have been produced at radar frequencies of 2.4, 3, 5, 7 and 10 GHz. The resulting radar signals were sampled with a Doppler sampling rate of 2 kHz. A six-class dataset was formed. Class C0 represents the absence of human gait and is simulated by white Gaussian complex noise. Classes C1 up to C5 represent 1 to 5 people walking, respectively. The samples for each class have been obtained by the summation of the appropriate number of single gait samples, randomly selected from the original datasets for each particular radar frequency. Next, varying amounts of white Gaussian complex noise were added to this composite signal to obtain datasets with SNR levels of 0, 5, 15 and 25 dB. Subsequently, by using the STFT with a 128-point FFT, Hamming windowing and an overlap of 90% between the consecutive integration intervals, spectrograms of size 128×192 Doppler and time bins were obtained, corresponding to a time duration of 1.25 s. See Table 5.3 for training and test set sizes.

Measured data

An X-band CW radar was used to measure 29 subjects (25 males, 4 females, aged 18–47) walking individually or side by side in groups of two or three people in a radial trajectory from and towards the radar. Various other targets (cars, cyclists, rustling trees, etc.) were measured to provide a non-gait class. Thus, four classes corresponding to classes C0–C3 of the model data were formed. The baseband radar signal was sampled at 8 kHz and decimated to 2 kHz, high-pass filtered to suppress static clutter, after which the spectrograms of size 128×192 were created as described for the model data above. All data belonging to a particular test subject was assigned

*Table 5.4   Classification accuracy (%) of DCNN on model
data (source: [21])*

RF (GHz)	SNR (dB)			
	25	**15**	**5**	**0**
10	96.1	96.0	91.0	86.1
7	92.9	91.8	88.8	84.5
5	91.8	91.7	86.4	82.7
3	86.9	87.1	86.0	84.5
2.4	84.9	85.7	81.3	78.2

to either test or training set. For the multiple gait classes, the training and test sets did not overlap in the unique combination of subjects. See Table 5.3 for training and test set sizes.

Results

The results of the DCNN on the model data are shown in Table 5.4 and on the measured dataset in Figure 5.15. Generally, the DCNN achieves high accuracy that increases with higher carrier frequency due to the larger Doppler shifts and hence more easily distinguishable micro-Doppler signatures. Overall, the DCNN achieved an accuracy of 86.9% on the measurement data. The DCNN could easily distinguish the non-gait from the gait classes. The single gait class was classified very well, whereas the two-gait class appeared to be somewhat difficult for the network, and it overlapped significantly with both the single gait and the three-gaits class. The performance of the DCNN on the measurement data is less than on the model data. We attribute this to the much more complex and varied appearance of the measurement data. The model assumes people swing both their arms. It does not include the effects of clothing or the blocking of body parts from view. Figure 5.17 shows typical spectrograms of the human gait classes, and they show a much more complex and varied appearance of the measurement data when compared with the model data in Figure 5.16. The contributions of the various body parts vary somewhat in strength over time and are hard to distinguish, especially for the multiple gaits as the lower Doppler frequency region of the spectrogram becomes cluttered.

By using guided back propagation [25], insight into the behaviour of the trained network is obtained. For a particular input spectrogram, this method produces a saliency map that shows which 'pixels' affect the classification the most. In Figure 5.16, an example spectrogram of multiple human gaits and its associated saliency map are shown. At first glance, the spectrogram shows three human gaits, but close inspection reveals that it contains four gaits. Note the slightly thicker curve of the leg contribution as indicated by the arrow in Figure 5.16. The DCNN classified this particular example correct, demonstrating its ability to discern subtle differences. The saliency maps for synthetic data in Figure 5.16 and for measured data in

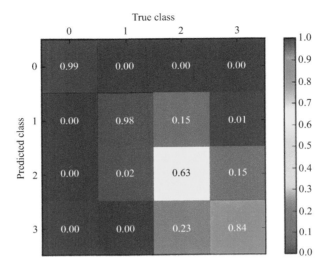

Figure 5.15 *Confusion matrix DCNN on measurement data (adapted from: [21]).*
Overall accuracy: 86.9%

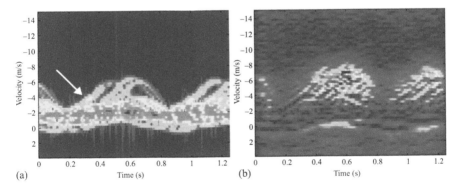

Figure 5.16 *(a) Spectrogram and (b) saliency map of model-based multiple human*
gaits (RF = 10 GHz) (source: [21])

Figure 5.17 show that for the DCNN classification, the results are determined by the
high-frequency regions of the signature, while the region containing the torso contribution is largely irrelevant. This makes sense, as the high frequency components
due to the legs and feet occur only during a small part of the gait cycle and thus high
frequency contributions occupying larger parts of the cycle indicate a higher number
of gaits. The torso contribution is nearly constant during the gait cycle and hence not
very distinctive for the number of gaits. This result shows the potential difficulty in

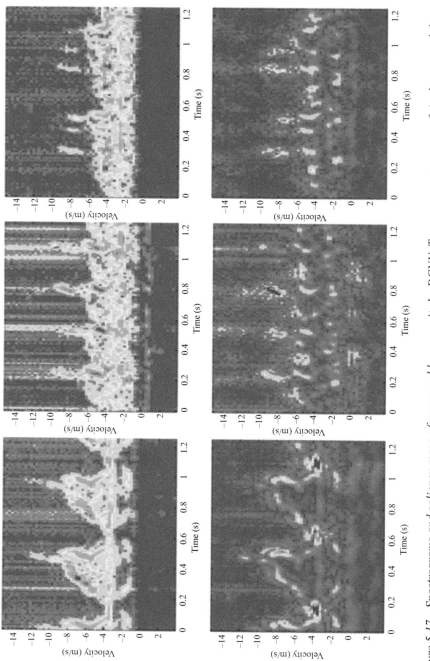

Figure 5.17 *Spectrograms and saliency maps of measured human gaits by DCNN. Top row: spectrograms of single, two and three gaits. Bottom row: saliency maps associated to their respective spectrogram (source: [21])*

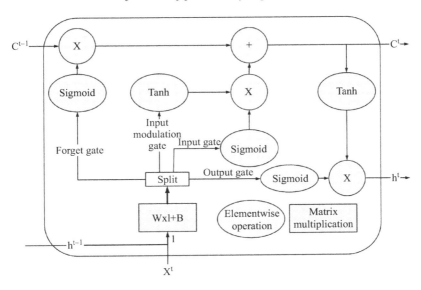

Figure 5.18 Architecture of LSTM cell (source: [26])

determining the number of gaits when the gaits are synchronised in phase and thus completely overlap one another in the spectrogram.

5.4.1.2 Recurrent neural networks

In this section, summarised from [26], another type of neural network is used to approach the same multi-target human gait classification problem as discussed in the previous section. Though successful, the CNN approach has a few drawbacks, and by using a so-called RNN, some of these drawbacks might be overcome or circumvented.

CNNs are oriented towards image-like data, and most CNN models assume a fixed size and format of the input data. During inference, it is not able to take advantage of knowledge of previous data (each prediction is made independently) because it has no memory. The CW radar data in the human gait application however is sequential in nature and directly preceding measurements of a target provide information that can be utilised. RNNs are designed with sequential data in mind and hence are more suitable. A RNN is a network with a recurrent connection that is realised by giving the RNN a state. There are many possible ways to build a RNN; the most well-known and widely used building block however is the long short-term memory (LSTM) [27], see Figure 5.18 for a schematic overview of the LSTM cell.

The LSTM cell contains a state, where four different weight matrices are used to modulate its state and output. The outputs of the four matrix multiplications are then split into four different layers. These four layers all work independently using the input of the LSTM and the output of the previous step. The first layer produces a mask with values between 0 and 1 that is multiplied with the state. This causes the removal of information from the state and is called the forget gate. The input modulation gate

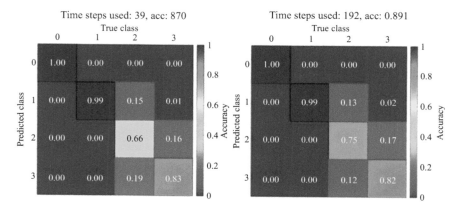

Figure 5.19   Confusion matrices of LSTM-RNN (source: [26])

determines which values of the state are to be updated, while the input gate creates the new candidate values. The results from these two gates are multiplied and added to the state. This step causes the information content to grow. The fourth layer is the output gate. This layer filters the state to produce a relevant output. Because the state of the LSTM is regulated delicately, the LSTM cell can learn very long term relations in the data. Multiple LSTM cells can be stacked in a single RNN.

Using four layers of LSTM cells, a LSTM-RNN was designed and trained to classify the multiple human gaits problem. More details on the exact architecture and training procedure can be found in [26]. Based on the exact same measured data as used in Section 5.4.1.1, besides the original data, spectrograms built with 50% time overlap were used, resulting in spectrograms with 39 and 192 time steps, respectively. The LSTM-RNN was not fed the entire spectrogram at once, instead a section at a time was input until the full spectrogram was processed.

The results of the LSTM-RNN are shown by the confusion matrices in Figure 5.19. The network classifies with near-perfect precision the no-gait and single gait classes, and with high accuracy the two and three gaits classes. The classes with two or more human gaits are sometimes confused with the classes that are off by one of the correct numbers, but it is very seldom off by more than one gait. Also apparent from Figure 5.19 is that the data using 50% time overlap is only slightly harder to classify correctly than the data with 90% time overlap, and showing identical qualitative behaviour. Thus, near-identical performance can be reached with about 25% of the computational cost. Figure 5.20 shows the accuracy of the LSTM-RNN versus the observation time, that is, the classification output of the network while a data sample (consisting of 39 or 192 time steps) has only been partially processed yet. The classification results after only 0.66 s are comparable to the performance of the CNN (discussed in Section 5.4.1.1) after 1.25 s. This shows the possible advantage of a RNN approach over the CNN approach in real-time classification scenarios. Another advantage is that the LSTM-RNN can change its classification output much

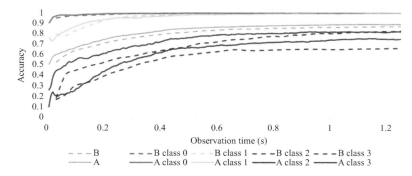

Figure 5.20   Accuracy versus observation time for each class of dataset A (192 time steps) and B (39 time steps) (source: [26])

Figure 5.21   Supervised (left) versus unsupervised (right) training schemes

quicker when new data arrives and for instance alter its prediction from single to two gaits.

5.4.2   Deep unsupervised learning for human gait classification

The main goal of deep neural network (DNN) training consists in obtaining a model which allows to extract relevant information (e.g. target class, specific target features) from the input data. Specifically in the case of supervised training, as explained in the previous subsection and schematically depicted in Figure 5.21 (left), information of interest is already predefined (e.g. target class division) and associated to each input data example used for training. The DNN is then trained to be able to extract this information from input data samples.

In the case of unsupervised training, schematically depicted in Figure 5.21 (right), a DNN is trained using a general cost function, for instance consisting in the minimisation of a reconstruction error, that is not directly associated to a specific operational task (e.g. classification between targets). After training, relevant information captured by the network to optimise the training cost function can be exploited for several different applications (e.g. classification, data processing, data augmentation).

However, mainly due to the difficulty in adapting the inherent linearity of the building blocks of the neural networks in modelling complex data distributions, deep unsupervised generative models showed to be less performant compared to supervised discriminative models, specifically regarding data generation applications [28]. This is apparent especially in the difficulty in dealing with adversarial examples [29], in the presence of which a trained network could make highly varying predictions, causing decrease in performance. In view of this, the concept of adversarial training can be seen as a non-linear regularisation procedure that allows to better deal with non-linear trends in the data distribution. In the next subsections, adapted from [30], the concept of adversarial training applied to DCNNs, using micro-Doppler spectrograms of human gait as input data, is illustrated. Specifically, two main approaches have been followed. At first, a GAN model is built, focusing on the domain where spectrograms are mapped into a compressed semantic representation. Second, an Adversarial Autoencoder (AAE) is designed in order to directly inspect this domain and visualise it in 2D. The features learned by the AAE via the unsupervised learning process and their relation with the pre-defined class-label information are presented.

5.4.2.1 Generative Adversarial Networks

According to the adversarial training concept, two deep neural networks are trained in competition with each other. Specifically in the case of the GAN framework, schematically depicted in Figure 5.22, these networks are known as the generator (G) and the discriminator (D). Given a pre-built dataset consisting of micro-Doppler spectrograms, the generator network, receiving as input a vector z defined in the so-called latent space domain, is trained to produce data samples that highly resemble the training input data samples. On the other hand, the discriminator network is trained to distinguish between examples that are either the synthetic outputs produced by the generator $X' = G(z)$ or the real spectrograms X from the training dataset.

A training data-set has been built from measurements that have been performed using an X-band (around 10 GHz) CW radar. Recordings of gaits of one or more subjects performing different activities, while walking in a radial direction, both inbound and outbound from the radar, have been collected. Micro-Doppler spectrograms were obtained using a sampling rate of 2 kHz, an FFT window of 128 samples and applying a Hamming window, and a time overlap of 90%, resulting in the final spectrogram size of 128×192. Spectrograms were then normalised between 0 and 1 prior to entering the network. The initial target classes' definition and the respective total amount of spectrograms are listed in Table 5.5.

The discriminator (D) and generator (G) networks consist, respectively, in a deep convolutional and a transposed-convolutional network, each including five convolutional or de-convolutional layers. The latent space vector z, which represents the input for the generator network (G), is characterised by a dimensionality of 100 and has values comprised between -1 and 1, which are sampled from a random uniform distribution during the training phase. Further details regarding training hyper-parameters can be found in [30].

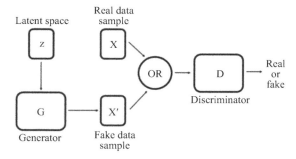

Figure 5.22 GAN architecture (source: [30])

Table 5.5 Classes' definition and number of spectrograms per class (source: [30])

Spectrogram class	Number of spectrograms
Single gait walking inbound	12,072
Single gait walking outbound	11,959
Single gait running inbound	4,041
Single gait running outbound	4,006
Single gait carrying a metal rod walking inbound	3,104
Single gait carrying a metal rod walking outbound	2,810
Double gait walking inbound	2,920
Double gait walking outbound	1,925

After the training process, latent space domain properties were inspected by performing an interpolation between two latent vectors and inspecting the corresponding synthetic spectrograms produced by the generator network. The result is seen in Figure 5.23.

As apparent from Figure 5.23, latent space vectors at close Euclidean distance will correspond to synthesised spectrograms characterised by a high similarity level, revealing that the latent space metric is related to spectrograms' features. This shows how the GAN is internally representing spectrograms characterised by different features, possibly corresponding to different classes, without any prior knowledge about the class definition. Therefore, micro-Doppler feature information can be expected to be structured in a semantically meaningful manner within the latent space domain, thus suggesting a feature-based clustering which would be favourable for instance for classification tasks. In order to be able to access and visualise the latent space domain, the generator is replaced with an autoencoder network, described in the next subsection.

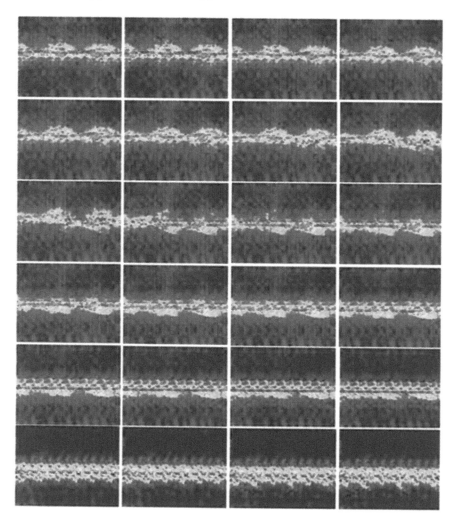

*Figure 5.23   Synthetic spectrograms obtained via linear interpolation in the latent
space. Samples are depicted in order, beginning with the top left with
a walking outbound spectrogram and revealing the feature transition
sequentially until the bottom-right, running outbound spectrogram
(source: [30])*

5.4.2.2   Adversarial Autoencoders

An autoencoder consists of two deep neural networks connected to each other, an
encoder and a decoder. In this specific application, the encoder is a DCNN consisting of five convolutional layers which receives as input a measured spectrogram
and compresses it into a vector of size 100 (i.e. the chosen dimensionality for the

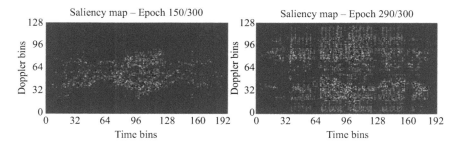

*Figure 5.24   Discriminator saliency maps of the final convolutional layer
(source: [30]).*

latent space domain). Such vector represents the input for the Decoder, a de-CNN which expands the received input into a vector of size 128 × 192, corresponding to the spectrograms' dimensionality, via five de-convolutional layers. The autoencoder network is then adversarially trained with a discriminator (*D*), whose architecture is maintained identical to the GAN case, discussed in the previous subsection. Further details regarding training hyper-parameters for the AAE can be found in [30].

After the training was performed, two main results were investigated during the test phase: first, the saliency maps obtained from the last convolutional layer of the discriminator network and, second, the 2D latent space visualisation.

The saliency maps of the trained discriminator show which pixels affect its decision the most. After a training process consisting of 300 epochs in total, Figure 5.24 reveals that the network works properly for epoch 150, since it focuses on the micro-Doppler signature of the target. Nevertheless, the saliency map corresponding to epoch 290 reveals that the discriminator focuses mostly on the background noise, which is undesirable since it does not contain information associated to the target. To prevent this, the normalised input spectrograms (between 0 and 1) were thresholded at the value of 0.45, meaning that all time-Doppler bins with a smaller value than 0.45 were clipped to 0.45. The thresholding had a significant effect in the latent space visualisations outlined in the following subsection.

Latent space domain analysis has been conducted following two approaches. First, the distribution of the known pre-defined classes in the latent space domain has been inspected. Second, the distribution of the individual data samples in the latent space domain has been analysed in order to highlight the main information extracted by the network from the input data, regardless of the pre-defined class division. Visualisations of the latent space are obtained by using the trained encoder, which receives as input a real spectrogram and outputs the corresponding latent space vector. The two-dimensional representations were obtained using the t-SNE [31] tool for dimensionality reduction. Additionally, t-SNE was also directly applied on the (128 × 192) spectrograms in order to compare how well the latent space vectors encapsulate the information present in the spectrogram related to the pre-defined class division. The result of applying t-SNE on the latent space vectors in the case of thresholded input spectrograms can be seen in Figure 5.25 (left).

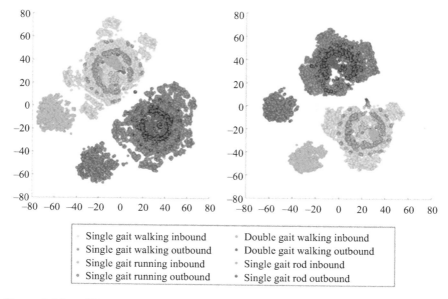

Figure 5.25   *t-SNE visualisation when applied on latent space vectors (left) versus spectrograms (right) in the case of thresholding applied on input spectrograms (source: [30])*

The inbound and outbound information are maintained, as well as the walking and running information. Within the walking clusters, structured islands appear for the double gait and single gait rod classes. These islands, also known as manifolds, represent subdomains within the latent space that can be directly associated to specific micro-Doppler feature information, as will be shown in more detail in the next subsection. On the right plot of Figure 5.25, the result of applying t-SNE directly on the spectrograms shows how the manifolds are still present within the walking clusters.

Latent space visualisation results in the case of not-thresholded input spectrograms are shown in Figure 5.26. On the one hand, structured manifolds are still present in the latent space representations on the left but with less detail than in the previous case. On the other, the right plot of Figure 5.26 shows how the 2D representation obtained by directly applying t-SNE on the spectrograms that have not been thresholded results in the manifold information being lost. Overall, this shows the higher robustness of the latent space representations when input spectrograms including background noise are used during training.

Starting from Figure 5.26 for thresholded input spectrograms, the latent space walking outbound cluster was extracted for further investigation and plotted in a polar coordinate system. By carefully investigating the spectrogram features corresponding to the points in the cluster, a correlation between polar coordinates of each cluster

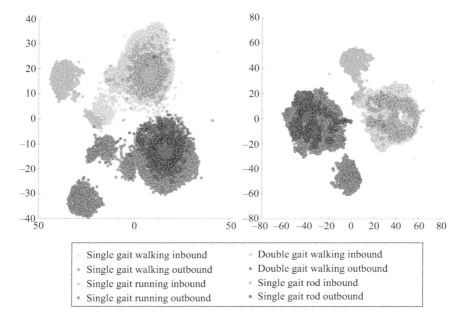

Figure 5.26   t-SNE visualisation when applied on latent space vectors (left) versus spectrograms (right) in the case of no thresholding applied on input spectrograms (source: [30])

point and initial gait phase change and main Doppler velocity shift in the spectrograms was found.

The initial gait phase, as depicted in the numbered spectrograms in Figure 5.27, reveals how the gait phase is dependent on the angular position in the cluster within the latent space. Relating this information to the clusters in Figure 5.25, it can also be explained why the double gait class is more unevenly distributed in an annular arrangement than the single gait class. Indeed, because of the presence of multiple gaits, an exact initial phase gait is more difficult to be identified, especially in the case of partial superposition of the signatures in the spectrogram.

Furthermore, a relation between the radial position within the cluster and the main Doppler velocity observed in the associated spectrogram was found. Specifically, by inspecting cluster points moving outwards in the radial direction, larger Doppler velocities in the associated spectrograms can be observed, as seen in the labelled spectrograms (A–D) in Figure 5.27. This correlation between radial distance from to the cluster centre and main Doppler velocity is highlighted in Figure 5.28, where values of each cluster point (colour coded) correspond to the main Doppler velocity value extracted from the associated spectrogram. Relating this information to the clusters shown in Figure 5.25, the distinct and thinner annular distribution of the single gait rod class can be explained, since the velocity of the upper limits is limited when carrying the metal rod compared to a human gait with arms and hands free.

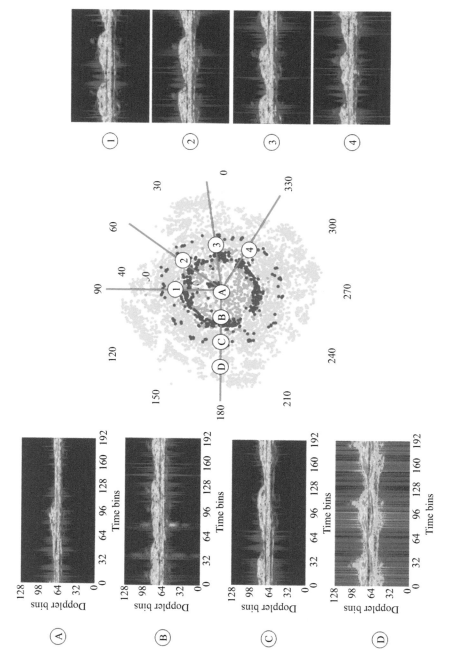

Figure 5.27   Relation between main Doppler velocity, initial gait phase and the position of the cluster point (source: [30])

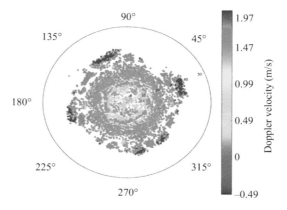

Figure 5.28   Relation between initial gait phase and main Doppler velocity and the position of the cluster point (source: [30])

5.4.3   Conclusion

Data-driven approaches for human-gait classification based on deep neural networks have been illustrated. The specific case of distinguishing multiple human gaits by means of micro-Doppler signature represents an example of the capability of this approach to better deal with challenging classification tasks. For the model-driven approach, these scenarios would be in most of the cases bottlenecked by the accuracy of the available model. On the other hand, in the data-driven approach, the required accuracy in the feature extraction process is achieved by means of an adequate complexity of the neural network employed for the training. Both supervised and unsupervised training schemes have been considered: in the latter case, the neural network shows the capability to extract semantically meaningful features when the training is performed with unlabelled data.

Since the learning process is based on the information included in the training data examples, good test performance can be achieved only if the test data is well represented in the training dataset. Furthermore, differently than in the case of model-driven approach, test input data need to be obtained with same radar parameters (carrier frequency, waveform) as well as pre-processing configuration (in the case of input representations such as spectrograms or cepstrograms). Otherwise, a different neural network should be trained on purpose or, alternatively, multi-modal learning approaches might be considered.

5.5   Discussion

The classification of personnel by ground-based surveillance radar has seen significant developments over the last two decades. This chapter discussed modelling and measuring human walking and running motions, and both model-based and data-driven approaches to human motion classification. When accurate models are available, the model-based approach is fast and effective. In addition, thanks to the separation of radar model and system (gait) models, the model-driven approach offers flexibility for different radar waveform parameters as well as target aspect angles. However, it is infeasible to create all-encompassing models for the wide variety of conceivable human activities, and it seems difficult to extend model-based techniques to new personnel classification problems.

In contrast, the data-driven approach, which is becoming more and more based on deep neural networks, has become favoured by researchers as it enables solving classification problems that are arguably impossible to be addressed using model-based approaches. Even so, the data-driven approach also has serious disadvantages and limitations. The two most important are:

1. *Data hunger*. Large representative datasets with sufficient variety are required to obtain the classification robustness that is required, especially in a military context. For human gait this problem can be considered less severe compared to 'datastarved' situations, like the classification of adversarial vehicles of which little information is available. Still, the more subtle details in micro-Doppler are expected to play a role in classification, the more data is required.
2. *Black box behaviour*. In order to be able to trust any decision of data-driven classification systems, an additional explanation or motivation should be given as output. This allows the human counterpart to validate the outcome and to detect whether weaknesses of the system have been exploited.

What then will the future of radar-based personnel classification be? We can split up the contributors to future improvement in classification into two categories:

1. *Sensor improvement*, that gives us more and clearer signature information
2. *Processing improvement*, that allows us to look deeper into the signatures of human activity

Radar technology and hardware are becoming more advanced and flexible. Introducing technologies such as polarimetry, multi-static configurations [32], and wide band or multi-band radars will contribute to improve classification capabilities as they add more information to the radar signals.

Another actively researched and promising topic is multi-modal operation where data of multiple modalities (e.g. radar, video, audio) are combined within a single system. All these domains carry different statistical characteristics and features, but they are all related to each other as they provide information from the same object.

More information from signatures requires processing that can exploit all these details. Improving the quality of data by signal processing and data transformation techniques is beneficial for all applications. One recent development that merits attention involves spectrogram-enhancement techniques based on bandwidth extrapolation [33–35], as shown in Figure 5.29. Related to this, super-resolution techniques are being developed in the image domain [36], and these could be transferred to the radar domain.

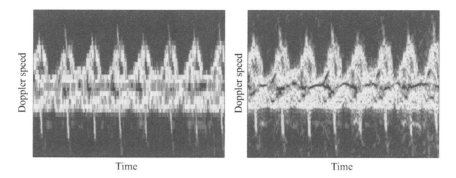

Figure 5.29 Left: X-band spectrogram without enhancement. Right: X-band spectrogram with bandwidth extrapolation enhancement.

Alternatively, deep neural networks could be trained using raw radar signals that have undergone minimal pre-processing. This would make the classification less dependent on radar processing parameters, thereby improving robustness.

In general, we can expect more robust and fine-grained classification of human activities, especially when higher radar frequencies or larger bandwidths are employed. For instance, identification of individuals will be possible based on personal traits in their gait movements, which are embedded in their micro-Doppler signature [10]. This can play an important role in tracking and re-identifying persons in an area under observation. Classification systems will also be able to detect relevant properties of individuals, such as whether he or she is wearing a uniform or carrying weapons. Jointly monitoring both the progression of position and micro-Doppler will allow to identify actions and to estimate intent. It is evident that being able to conclude that someone is trying to prepare an improvised explosive device, for instance, can be of vital importance for military missions.

Though not unique to radar, acquiring high quality data with ground truth labels, especially in uncontrolled environments, will remain a major impediment for developing high performing and robust data-driven classifiers. On the other hand, unsupervised learning techniques may aid in knowledge discovery by processing huge amounts of unlabelled data, and then this knowledge can be codified into the classifier architecture. In addition, advances in data augmentation, generative models and especially multi-modal learning will presumably alleviate but not entirely eliminate the need for acquiring sufficient labelled data.

We conclude this chapter by mentioning the possibility of fusing machine learning with prior expert knowledge which could bring an important improvement to classification algorithms. This effectively combines the model- and data-driven methods into a hybrid approach, where models incorporate the domain knowledge that would structure and guide a flexible learning system that could be efficiently trained for a large variety of realistic scenarios.

References

[1]   P. van Dorp, Human Motion Analysis with Radar, Ph.D. Thesis, Universiteit van Amsterdam, The Netherlands, 2010.

[2]   P. van Dorp and F. C. A. Groen, "Human walking estimation with radar," *IEE Proceedings—Radar Sonar and Navigation*, vol. 150, pp. 356–365, 2003.

[3]   R. O. Duda, P. E. Hart and D. G. Stork, *Pattern Classification*, second edition, New York, NY: John Wiley & Sons, 2001.

[4]   S. Theodoridis and K. Koutroumbas, *Pattern Recognition*, fourth edition, Burlington, MA: Academic Press, 2009.

[5]   V. C. Chen, *The Micro-Doppler Effect in Radar*, Boston, MA: Artech House, 2011.

[6]   R. Boulic, N. Magnenat-Thalmann and D. Thalmann, "A global human walking model with real time kinematic personification," *Visual Computing*, vol. 6, pp. 344–356, 1990.

[7]   A. Ghaleb, L. Vignaud and J. Nicolas, "Micro-Doppler analysis of wheels and pedestrians in ISAR imaging," *Signal Processing*, vol. 2, no. 3, pp. 301–311, 2008.

[8]   N. F. Troje, "Decomposing biological motion: a framework for analysis and synthesis of human gait patterns," *Journal of Vision*, vol. 2, pp. 371–387, 2002.

[9]   N. F. Troje, Retrieving information from human movement patterns, in: *Understanding Events: From Perception to Action*, T. F. Shipley and J. M. Zacks (Eds.), New York, NY: Oxford University Press, pp. 308–334, 2008.

[10]   S. Abdulatif, F. Aziz, K. Armanious, B. Kleiner, B. Yang and U. Schneider, "Person identification and body mass index: a deep learning-based study on micro-Dopplers," *2019 IEEE Radar Conference (RadarConf)*, Boston, MA, USA, 2019, pp. 1–6, doi: 10.1109/RADAR.2019.8835652.

[11]   V. C. Chen, D. Tahmoush and W. J. Miceli, *Radar Micro-Doppler Signature, Processing and Applications*, London: IET, 2014.

[12]   XSense, MVN BIOMECH User Manual, December 2009. [Online]. Available: http://www.xsens.com/en/general/mvn.

[13]   K. D. Trott, "Stationary phase derivation for RCS of an ellipsoid," *IEEE Antennas and Wireless Propagation Letters*, vol. 6, pp. 240–243, 2007.

[14]   C. F. Daudey, *Simulation of Human MicroDopplers Using Captured Human Motion*, Delft: Thales Nederland B.V., 2011.

[15]   C. V. Stewart, Identification of periodically amplitude modulated targets, Doctoral thesis, 1978. [Online]. Available: http://www.dtic.mil/docs/.citations/ADA056516.

[16]   P. van Dorp and F. C. A. Groen, "Feature-based human motion parameter estimation with radar," *IET Radar, Sonar and Navigation*, vol. 2, pp. 135–145, 2008.

[17]   P. van. Dorp and F. C. A. Groen, "Local information from range-speed radar sequences," *IEEE Instrumentation and Measurement, IMTC 2007*, Poland, May 1–3, 2007.

[18] S. Groot, R. I. A. Harmanny, H. Driessen and A. Yarovoy, "Human motion classification using a particle filter approach: multiple model particle filtering applied to the micro-Doppler spectrum," *International Journal of Microwave and Wireless Technologies*, vol. 5, no. 3, pp. 391–399, 2013.

[19] R. I. A. Harmanny and S. R. Groot, "System for characterizing motion of an individual, notably a human individual, and associated method," Patent 11187766.

[20] B. Ristic, S. Arulampalam and N. Gordon, *Beyond the Kalman Filter, Particle Filters for Tracking Applications*, Boston, MA: Artech House, 2004.

[21] R. P. Trommel, R. I. A. Harmanny, L. Cifola and J. N. Driessen, "Multi-target human gait classification using deep convolutional neural networks on micro-Doppler spectrograms," *2016 European Radar Conference (EuRAD)*, London, 2016, pp. 81–84.

[22] Y. LeCun, Y. Bengio and G. Hinton, "Deep learning," *Nature*, vol. 521, pp. 436–444, 2015.

[23] J. Bergstra, O. Breuleux, F. Bastien, *et al.*, "Theano: a CPU and GPU math expression compiler," *Python for Scientific Computing Conference (SciPy)*, Austin, TX, 2010.

[24] S. Dieleman, J. Schlüter and C. Raffel, Lasagne: First Release, 2015.

[25] J. Springenberg, A. Dosovitskiy, T. Brox and M. Riedmiller, "Striving for simplicity: the all convolutional net," https://arxiv.org/abs/1412.6806, 2014.

[26] G. Klarenbeek, R. I. A. Harmanny and L. Cifola, "Multi-target human gait classification using LSTM recurrent neural networks applied to micro-Doppler," *European Radar Conference*, Nuremberg, 2017.

[27] S. Hochreiter and J. Schmidhuber, "Long short-term memory," *Neural Computation*, vol. 9, no. 8, pp. 1735–1780, 1997.

[28] I. J. Goodfellow, J. Pouget-Abadie, M. Mirza, *et al.*, "Generative Adversarial Networks," https://arxiv.org/abs/1406.2661, 2014.

[29] I. J. Goodfellow, J. Shlens and C. Szegedy, "Explaining and harnessing adversarial examples," https://arxiv.org/abs/1412.6572, 2014.

[30] H. G. Doherty, L. Cifola, R. I. A. Harmanny and F. Fioranelli, "Unsupervised learning using generative adversarial networks on micro-Doppler spectrograms," *2019 16th European Radar Conference (EuRAD)*, Paris, France, 2019, pp. 197–200.

[31] L. J. P. van der Maaten and G. E. Hinton, "Visualizing high-dimensional data using t-SNE," *Journal of Machine Learning Research*, vol. 9, pp. 2579–2605, 2008.

[32] F. Fioranelli, M. Ritchie and H. Griffiths, "Analysis of polarimetric multistatic human micro-Doppler classification of armed/unarmed personnel," *2015 IEEE Radar Conference (RadarCon)*, Arlington, VA, 2015, pp. 0432-0437, doi: 10.1109/RADAR.2015.7131038.

[33] J. P. Burg, Maximum Entropy Spectral Analysis, Ph.D. thesis, Stanford University, Stanford, CA, 1975.

[34] J. Liew, Multi-Dimensional Signal Sharpening Based on Bandwidth Extrapolation Techniques, Corvallis, OR: School of Electrical Engineering and Computer Science Oregon State University, 2004.

[35] P. van Dorp, "LFMCW based MIMO imaging processing with keystone transform," *2013 European Microwave Conference*, Nuremberg, 2013, pp. 1779–1782, doi: 10.23919/EuMC.2013.6687023.
v. P. Dorp, "LFMCW based MIMO imaging processing with Keystone Transform," *EuRAD*, 2013.

[36] J. Kim, J. K. Lee and K. M. Lee, "Accurate image super-resolution using very deep convolutional networks," *2016 IEEE Conference on Computer Vision and Pattern Recognition (CVPR)*, Las Vegas, NV, 2016, pp. 1646–1654, doi: 10.1109/CVPR.2016.182.

Chapter 6
Multimodal sensing for assisted living using radar

Aman Shrestha[1], Haobo Li[1], Francesco Fioranelli[2], and Julien Le Kernec[1,3,4]

In nature, information from the various senses is utilised by intelligent beings for identification, recognition and decision. It is easier to identify an object or a threat by mixing the auditory, visual and tactile information, than when only one of the senses are used. Combining separate sources of information, which include different types of data, provides the variety needed for cognition and improved situational awareness. This is the rationale behind sensor fusion [1].

For deployment in real-world healthcare facilities, either in private homes or in hospital and care homes, activity-monitoring systems with a single sensor may not deliver the necessary robust performance requirements. Radar sensing is an emerging approach in the field of assisted living applications [2,3], but limitations in identifying precise movements, especially at unfavourable aspect angles, and a narrow area of detection, mean further co-operative sensing methods are required for a robust system for detection of movements and falls. Sensor fusion is, therefore, one method of mitigating this issue. This chapter will explore the different sensor and fusion topologies, using active radar sensing in conjunction with other sensing technologies as support for assisted living and healthcare applications [2]. Initially, the sensors and their outputs will be described, with specifics of signal processing of the different sensors and the machine learning for classification detailed. The results of applying these methods to the assisted living scenario will then be presented.

This chapter will give insight into activity classification with radar and additional sensing technologies, in particular, wearable inertial and magnetic sensors, focusing on the key information fusion approaches and main improvements using experimental data as validation.

[1]James Watt School of Engineering, University of Glasgow, Glasgow, UK
[2]MS3-Microwave Sensing Signals and Systems, TU Delft, Delft, The Netherlands
[3]ETIS – Signal and Information Processing Lab, University Cergy-Pontoise, Cergy, France
[4]School of Information and Communication, University of Electronic Science and Technology of China, Chengdu, China

6.1　Sensing for assisted living: fundamentals

Assisted living applications utilise the movements produced by the body, or the activities performed to identify potential problems in the physiological state of the person [2]. It relies on tracking anomalies or classifying abrupt movements associated with critical events such as falls. Detecting these abnormalities ensures any degradations occurring within the physiology of a person can be dealt with by a quick and useful response to alleviate symptoms or prevent further injuries.

6.1.1　Radar signal processing: spectrograms

In a human healthcare scenario, the received signal will have micro-Doppler (μD) components from the modulation of the limbs and other body parts with a lower radar cross-section as well as the torso [4]. These modulations are observable only if the pulse compressions or range profiles are sufficiently integrated to obtain a suitable resolution in Doppler. The Doppler shift obtained from the radar backscatter is proportional to the velocity of the target projected on the radial direction to the radar, and this can measure the speed of the target only in the radar line-of-sight or radial direction.

To generate the μD signatures, first, the ranging information is obtained from the raw radar returns. The ranging information is obtained differently depending on the type of radar used, e.g. stretch-processing (or de-ramping) for frequency-modulated continuous wave (CW) (FMCW) radar or matched filtering for pulsed Chirp—linear frequency-modulated pulse—radar. The pulse repetition T period provides coupled ambiguities where $[-cT/2; cT/2]$ gives the range ambiguity where c is the speed of light in m/s, and the Doppler ambiguity is $[-1/2T; 1/2T]$. In other words, without transmitted waveform agility to resolve ambiguities, the radar can only distinguish targets within the ranges defined by the pulse repetition period T. They are coupled, as reducing T will reduce the range ambiguity but increase the Doppler ambiguity and vice versa.

After accumulating range responses every T seconds, a short-time Fourier transform (STFT) (6.1) is performed across a slow time for each range bin:

$$|STFT(m, \omega) = \left| \sum_{N-1}^{n=0} x_m(n)w(n)e^{-j\omega n} \right| \tag{6.1}$$

where $x_m(n)$ indicates the mth range bin in the nth range response and $w(n)$ is the amplitude window function applied across slow time.

This leads to a spectral resolution for the Doppler spectrum of NT^{-1}. The longer one integrates the finer the resolution, and therefore more detail can be seen in the Doppler dimension. On the other hand, the more one integrates, the lower the time resolution for the μD. This means that overlapping is required to have a smooth signature that shows the details of the transitions in the movements being recorded usually set between 70% and 95% signature. One may also add zero-padding to increase resolution by artificially augmenting the size of the STFT but with zero-padding note that the response will be convoluted with a *Sinc* function. Once obtained

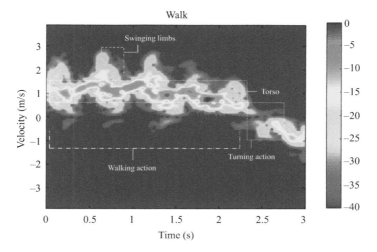

Figure 6.1   μD signature of a person walking towards the radar, turning and walking away normalised to 0 dB. The signatures from the torso and the limbs are highlighted. The dynamic range of the signature shows that the torso movement gives a stronger reflection than the limbs.

the responses for all range bins one can add the different results together to suppress the distance information and only retain the Doppler information against time. The selection of the range of interest for the summation process reduces the noise level and increases accuracy rates for classification.

Figures 6.1 and 6.2 give examples of μD signatures of human activities for walk and a punch-retract action while the subject remains stationary. The difference in the dynamic range of the modulations from the torso (normalised at 0 dB) and the limbs (around −23 dB) are identifiable, and these motions are central behind the idea of using μD signatures for classifying different activities. The dynamic range was clipped at −40 dB.

Figure 6.1 shows an μD signature for walk, where a person is walking towards the radar (positive Doppler), turning (zero-Doppler crossing), and walking away from the radar (negative Doppler). From 0 to 2.5 s, the subject walks towards the radar. The deep red trace represents the mean speed of the subject with the torso as the main reflection contributor and around that red trace yellow–green fluctuations represent the relative Doppler of limbs about the torso main Doppler component. From 1.5 to 2.5 s, the subject decelerates, and the main Doppler component and relative-Doppler components swing range decreases as well. The person reaches a standstill position at approximately 2.5 s, where they turn away from the radar and start walking in the other direction. μD signatures have visually recognisable properties that can help identify different states of the human body. It can show if a person is walking or moving

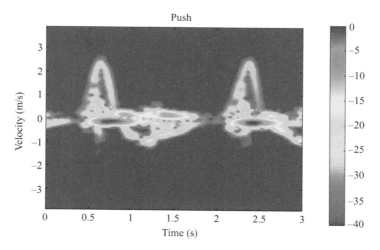

Figure 6.2   Spectrogram of a person punching/retracting while staying stationary normalised to 0 dB

slowly, if they are turning or moving their limbs in a certain manner, and together with sequences of these smaller motions if they are performing specific activities.

Figure 6.2 shows the μD signature of a subject punching/retracting spectrogram while standing still. Initially, at time 0.5 s a sinusoidal-shaped μD signature starts appearing around the torso; this is due to the movement of the arm in the radial direction of the radar, which is accelerating until approximately 1 s, where it is brought to rest causing the μD curve to return to 0 m/s. Then, negative Doppler is observed between 1 and 1.5 s as the arm is slowly retracted towards the main body, after which the same activity is repeated at time 2 s. In this activity, the torso movement is restrained compared to walking, and the shape of the μD curves of the arm are not occluded by the motion of the torso or legs as in the walking action. The lower intensity of the limb motion in the μD signature correlates correctly with the relative size of the arm which has a lower radar cross-section than the legs or torso. It is also explained by the nature of the motion which presents a lower reflective surface compared to the walking action where the arms move along the body.

Therefore, to summarise these simple examples, differences in micromotions of the body parts can be identified through μD signatures [5,6], and the different signatures for different activities shown are perceptibly different. The aspect angle, i.e. the angle between the radar line-of-sight and the trajectory of the target's movement, can also be a significant parameter [4,7], affecting how the μD signatures look like and the accuracy obtainable for classification problems (with examples shown in Figure 6.3).

The generation of μD signatures requires successive Fourier transforms of the windowed signal. A large number of other time-frequency transforms exists, see chapter 12 in [7]. High-resolution quadratic time-frequency distributions (e.g. Cohen's class, extended modified B-distribution) are designed by tuning a 2D (Doppler, Delay) kernel which results in improved time-frequency (TF) representations at the cost of higher computational load and tuning time for increased kernel complexity. Note that some of those techniques result in cross-frequency spurious terms that can only be suppressed by the careful selection and tuning of the kernel [7]. The adaptive TF representation (i.e. wavelet transform) defines a sparse representation of piecewise regular signals (e.g. Haar, Morlet and Daubechies). Wavelets represent the singularities and transients of a signal with fewer coefficients than Fourier transforms. The field of computational harmonic analysis is still very active in wavelet transform research and has grown beyond into sparse representations. The curious reader will find seminal work and MATLAB® tutorials in [8,9] covering the subject of TF signal analysis in breadth and depth. Their implementation in the representation of human radar signatures for classification has the potential to break away from the trade-off between the spectral and temporal resolution of STFT which remains the most common representation.

Recently a method which focuses on the specific spectral region of the cepstral coefficients called hyperbolically warped cepstral coefficients [10], has been suggested as an improved solution. This technique implements a tuned Mel-frequency filter bank on the μD signature to extract cepstral components. This allows the extraction of cepstral features by warping (dilating and compressing) frequency bands similar to the cochlea filter in the human ear and the Mel-frequency filter bank designed for effective speech processing. Using a genetic algorithm for optimisation, a human activity radar μD signature Mel-frequency filter bank is tuned to generate cepstral heat maps that improve classification accuracy over spectrograms.

For the remainder of the chapter, the μD signatures (spectrograms) used for multimodal sensing are generated using STFT.

6.1.2 Wearable sensors: basic information

Wearable sensors devices are a mature technology in the field of health and behaviour monitoring [2,6,11]. Compared to contactless sensing methods based on video-cameras or RF signals, wearable devices require users to wear them during their daily lives. Wearable devices are usually utilised to map the human movements by recording the variation of the physical characteristics (e.g. acceleration, angular speed, and magnetic field strength) related to the human body [12]. The most commonly used wearable sensor is the inertial measurement unit (IMU), which includes an integrated chip with an accelerometer, gyroscope, and magnetometer [13], usually placed on the waist (typically most accurate for classification) or the wrist (most common activity monitor placement).

The accelerometer behaviour can be assimilated to a displaced mass on a string, whereupon movement, it experiences a change in status, and the corresponding acceleration is estimated by the displacement of the string [14]. In the commercial market, piezoresistive, piezoelectric, and capacitive components are used to convert the mechanical displacement into an electric voltage [15]. Piezoresistive materials can effectively measure sudden changes of high acceleration, whereas piezoelectric materials are sensitive to the upper-frequency range and are more temperature tolerant.

On the contrary, capacitor-based accelerometers are sensitive to the lower frequency range.

A gyroscope is utilised to estimate the angular speed and support maintain direction in navigation applications [16,17]; it is typically combined with an accelerometer to construct the inertial navigation system. The main gyroscope frame consists of a gimbal and a rotor, where the spin axis is free to represent any orientations without interference from tilting and rotations. Modern gyroscope sensors are based on micro-electro-mechanical systems (MEMS) technology, which allows packaging multiple gyroscopes for different axes in one chip.

A magnetic sensor or magnetometer can detect weak biomagnetic fields inside the human body. It is categorised into magnetic Hall Effect sensors and magnetoresistance sensors, which include anisotropic magnetoresistance, giant magnetoresistance and tunnel magnetoresistance [18]. Hall sensors are widely used in human activity recognition due to their sensitivity range, whereas magnetoresistance sensors can capture subtle variations of the magnetic field (10^{-6}–10^{-12} T) via an array structure. As the magnetometer is moved when a human body part is performing a movement in a 3D space, different voltages are produced from the conductor according to the amplitude of the motion with respect to the Earth magnetic field [19] and aspect angle. This is known as the Hall-effect, i.e. the external magnetic field can be related to the floating electric current on the conductor, which in turns produces a variable Hall voltage signal related to the human movement.

To summarise, all three wearable sensors, accelerometer, gyroscope, and magnetometer, will produce a voltage signal sampled as a function of time that is related to the movement performed by the human subject. Typically, each sensor will produce one separated signal for each axis in a 3D space (tri-axial sensors), with a total of nine mono-dimensional raw signals to be considered as the starting point information for any activity classification analysis.

6.2 Feature extraction for individual sensors

In cases where the target is the human body with multiple sources of micromotions, the overall movements of the limbs are typically extracted from radar data by generating a spectrogram. The spectrogram matrix and the relevant data within it need to be in a format that can be interpreted automatically by a machine/algorithm to classify different activities without resorting to a human operator [4,5,20].

Feature extraction is part of this process, defined as deriving numerical properties of measured data that are informative and nonredundant and can be highly varied [21]. This also includes a method of dimensionality reduction, as the features contain identifiable information for a certain class but are usually orders of magnitude smaller than the input [4]. An easy example can be that any radar spectrogram is a 2D velocity–time matrix, whereas the extracted features are usually a 1D, shorter vector. Equally, for wearable sensors data, the original signal can contain a lot of samples, but only key metrics are considered as significant features (e.g. moments such as mean, standard deviation, and higher order moments).

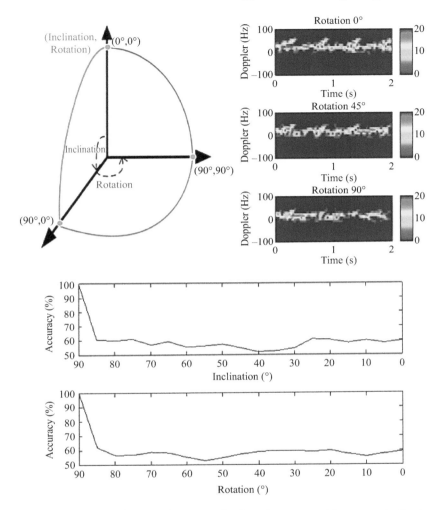

Figure 6.3   Top left: referential system of radar placement with respect to the target at the centre; Top right: μD signatures at different rotation angles; Bottom: classification accuracy of μD signatures for varying aspect angles in rotation and inclination

6.2.1   Features from radar data

Even though the spectrograms include visually perceptible movements, recognisable by an expert human, to translate this for a machine learning application, the properties that define any two distinct movements need to be extracted and converted into numerical quantities.

Two types of feature extraction methods encompass the features currently used in literature: automatic features and handcrafted features. The former uses established dimensionality reduction methods such principal components analysis (PCA) or singular value decomposition (SVD) to reduce the input into vectors with highly varied data. Handcrafted features, on the other hand, use a functional pattern within the input to characterise the signal. In lay terms, automatic feature extraction seeks any significant property that makes the inputs inherently different, whereas with handcrafted features a specific pattern is sought, often relying on expertise from the human operator.

6.2.1.1 Automatic feature extraction for radar

Automatic features are extracted from the decomposition of the data where the maximally variant or the most diverse features are identified and collected for input into classifiers. They include PCA [22], SVD [23], and independent component analysis [24]. None of these methods requires precise derivation of information and modifications or fine-tuning of the algorithm to specifically suit a set of radar data. They are resilient to different levels of signal-to-noise ratio (SNR) and produce salient features in most cases.

PCA of the μD spectrogram finds the salient information and reduces the input dimensionality considerably. In its basic form, PCA converts a set of correlated or covariant features into uncorrelated variables. This reduces the size of the input, but it preserves the variance and the information within the source while containing many identifiable features [25]. The first-order components will have the most variance; therefore, the output will maintain most of the information, compressing the input [22]. The higher order components, however, are usually lacking in variance, only showing the linear relationship between the features. Some information is then lost when the first components are used compared to the source, but this more efficient feature set is often more separable than the source.

SVD decomposes the spectrogram into spectral and temporal content, with the left and the right singular vectors [23,25]. As for the PCA, the first vectors carry the data that contain salient information, but here the extracted subspace has time and frequency information [25]. This means, depending on the specific nature of the data analysed, which may have greater temporal or spectral variety, one subspace can influence the inputs to the classifier model more than the other.

6.2.1.2 Handcrafted features

Handcrafted features, on the other hand, are functional properties of the input, which are commonly identifiable and should theoretically match between different samples of the same class. For μD signatures, some of the prominent handcrafted features are Doppler centroid [26], Doppler bandwidth [26], entropy and grey level of histograms [27], spread-spectrum shape [20], and cadence peaks [20], amongst others.

The Doppler centroid is the centre of energy content in a spectrogram [26]. It is computed for every time sample of the spectrogram (or 'slice') by finding the centre of mass of the Doppler bins from a spectrogram $S(i,j)$ and then multiplying by the corresponding frequency $f(i)$ as expressed in (6.2). This signature is the movement of the largest target area in most human activities and it is a good indicator for the movement of the torso:

$$f_c(j) = \frac{\sum_i f(i) S(i,j)}{\sum_i S(i,j)} \tag{6.2}$$

where f is the index for frequency and j is the index for time, $S(i,j)$ is the energy contained in the pixel (i,j); a spectrogram slice includes all the pixels at time j, and $f(i)$ is the frequency corresponding to the Doppler bin.

In the spectrogram, the Doppler centroid has a broader surrounding energy content. This sideband signature is the bandwidth in (6.3), and it represents the breadth of the movement that the target performs [26]:

$$B_c(j) = \sqrt{\frac{\sum_i \left(f(i) - f_c(j)\right)^2 S(i,j)}{\sum_i S(i,j)}} \tag{6.3}$$

For both the centroid and the bandwidth, the energy values are calculated for the entire spectrogram Doppler slice. This means that these features are affected by the choice in pulse repetition frequency (PRF) and related Doppler ambiguity. From experience, this feature needs to be calculated preferably within a reduced range of interest containing only the minimum and maximum values of expected Doppler for a given application to maximise the information content. This is to avoid adding noise that may degrade the quality of the features. For example, considering an S-band radar with a 100 μs pulse repetition period; the maximum expected Doppler is around 200 Hz, whereas the Doppler ambiguity is ± 5 kHz. One may consider a careful selection of the pulse repetition period, cropping the spectrogram by limiting the range i to the range of expected Doppler spread or limit the inclusion of components with thresholding to avoid the degradation of the informative value of these features.

As the spectrogram is displayed as a figure with a colour map to represent intensity, it can be saved as an image. Thus, classical image-based feature extraction algorithms can be used. Converting the μD signature in Figure 6.1 to greyscale, the intensities of the torso and limbs are translated to different segments of the grey spectrum. A histogram $p(n)$ of the distribution of intensities in the grey scale accurately maps the amplitudes of the movement by those body parts into one snapshot of activities. For example, Figures 6.1 and 6.2 are action snapshots of 3 s, n is the greyscale bin index and $p(n)$ the number of pixels in one snapshot corresponding to that bin, normalised by the number of pixels considered in one snapshot [27].

Image entropy is a measure of the information contained within the image [27]. Considering a greyscale input, the value of entropy (6.4) is a measure of the information contained in the greyscale snapshot, which translates to the total movement of the μD components from the torso and limbs in one feature:

$$E = -\sum_i p(n) \log p(n) \tag{6.4}$$

Table 6.1   A non-exhaustive list of time- and frequency-domain features for wearables sensors [12]

Time domain	Number of features	Frequency domain	Number of features
Norm of XYZ	1	Spectral power	9
Mean	3	Coefficients sum	3
Standard deviation	3	Spectral entropy	3
Autocorrelation (mean STD)	6		
Cross-Correlation (mean STD)	6		
Variance	3		
RMS (root mean square)	3		
MAD (median absolute deviation)	3		
Inter-quadrature range	3		
Range	3		
Minimum	3		
25th percentiles	3		
75th percentiles	3		
Skewness	3		
Kurtosis	3		
Total number of features	49	Total number of features	15

The shift in the grey levels is also an indicator for the different levels of movement within a spectrogram, and calculating the skewness of these shifting levels can provide another useful feature.

6.2.2   Features from wearable sensors

Statistical features in Table 6.1 [13,21,28] can be used to characterise the raw data to infer relevant information and features; they can be partitioned into time and frequency domain. Time-domain features include mean, variance, standard deviation, and other higher order statistical moments as skewness and kurtosis, as well as cross-correlation between data from the different axes of a tri-axial sensor [21,28]. Frequency-domain features are extracted to capture the energy distribution of the signal. For example, one can calculate the magnitude of the power spectral density (PSD) at three different frequency bands, namely 0.5–1, 1–5, and 5–10 Hz (to capture slow, medium and fast changes), and the sum of Fourier transform coefficients. The mean and variance of specific signals are derived by (6.5) and (6.6), where $x(i)$ is the input signal, and N denotes the number of samples:

$$\mu = \frac{1}{N} \sum_{i=1}^{N} x(i) \tag{6.5}$$

$$\sigma^2 = \frac{1}{N} \sum_{i=1}^{N} (x(i) - \mu)^2 \tag{6.6}$$

The correlation function is also significant in classifying activities with signal magnitude changes along two dimensions, such as the activities where the body is turned back and forth. The correlation between X and Y axis, which is equal to the ratio of covariance between two inputs and the product of their standard deviation, is one of the correlation features used.

Frequency features are used to evaluate the change in energy of different activities, whereby activities with fast movement provide a high response in a short frequency band, which can be especially significant for falls (where there are sudden bursts of acceleration). Spectral entropy is calculated by summing the product of the probability density function (PDF) and its log scale; the PDF is obtained by normalising the PSD to 0 and 1.

6.2.3 Classifiers

Typically, the classifiers used in the context of activity recognition with radar utilise pre-existing libraries for a breadth of different implementations based on supervised learning approaches. In literature, simple Bayesian classifiers such as naïve Bayes [20] or K-nearest neighbour (KNN) [2,12] have been used along with more complex support vector machines (SVMs) with different kernel functions, classification trees, and ensemble of trees, and soft-max margin-based artificial neural networks (ANNs) [29]. ANN comprises of an input layer, batches of hidden layers and an output/soft-max layer, whereas a different number of hidden units (neurons) are distributed on each layer. The hidden units between two neighbour layers are fully connected and the output of each neuron is calculated by applying non-linear functions to the sum of its inputs. The key to training the neural network in high efficiency is backpropagation, where the weights of each hidden units are updated at each iteration. In addition, deep learning-based algorithms, in particular, convolutional neural networks and recurrent neural networks are utilised in the field of activity recognition to provide subsequent improvement through intensive computational loads. Instead of using handcrafted features to characterise the data, the deep neural network usually compresses the dimension of the input signals by using combinations of convolutional filtering and pooling.

In the experiments described later on in this chapter, all the training and testing procedures are implemented in MATLAB. The dataset was stochastically divided into two parts, with 70% data for training and 30% data for testing on a per-class basis. The per-class basis was set for stratification in the test set to prevent class imbalance, at the same time ensuring that a range of unknown samples for each class was used for testing the classifier. By using this deterministic approach, the unwanted bias in the results is minimised, which would occur in cases of imbalance between classes in the training and test sets. This process is repeated ten times for each test, and the average results across all the repetitions are presented for conciseness [12,19]. However, for a full analysis, there is an interest in looking at average, mean, minimum, and maximum value out of the repetitions for the classifier to evaluate its robustness. If there is a great variance, the designed classifier may lack the required robustness.

6.3 Multi-sensor fusion

As individual sensors can only measure a certain type of data, there are limitations to the information they can record to describe the movement or activity under observation [1]. For example, in radar human activity recognition, the aspect angle is an open challenge. Indeed, the projection of the subject's speed onto the radial direction reduces as the aspect angle moves from 0° to 90°. As a result, the μD signature gradually shrinks/squeezes as the aspect angle increases [2]. Usually, there is a decrease in the returned signature proportional to the cosine angle relative to the direct line of sight, and this gives a weaker μD signature, potentially less accurate to describe the movement. The feature space derived from the radar might lose in saliency for the classification of activities. As previously shown in Figure 6.3, the Doppler spread in the μD signatures reduces with the angle of the heading of the movement with respect to the radar line-of-sight (aspect angle). This is a scenario where another sensing modality can be introduced to help address potential errors or missing information from the radar spectrogram.

6.3.1 Principles and approaches of sensor fusion

The term information fusion is applied to fusion of any types of data and its sources as discussed in this section. As different sensors have richness of information in different domains, the main reason for sensor fusion is variety, robustness and redundancy, and having information which can be commensurate across sensors.

Depending on the information sources or sensors, existing fusion topologies with specific architecture can be applied in sensor fusion networks. These configurations can be co-operative, competitive, or complementary [1].

Co-operative sensor networks require data from two independent sensors receiving the same type of information from different perspectives. This type of fusion network exploits spatial variety to use commensurate data from different points in space in order to generate new types of data—for example, multi-static radar with different aspect angles and/or interferometric channels.

Competitive fusion networks use the same data with spatial and temporal variation to make the to increase fault-tolerance due to redundancy. In this scenario, the information fused is from the same type of sensors, but the location or timing of the different sensors would vary.

Complementary fusion networks use independent sensors to receive maximally variable data from different domains to get a more comprehensive overall picture. An example would be using an electrocardiogram and electroencephalogram to detect vital signs as they are utilising information from different domains (electric pulses in the chest and voltages in the brain), to come to a decision where the person is identified as alive or not.

In our scenarios of interest, fusion can be performed at different levels with varying amounts of complexity through three methods [1,12,19]: signal-level fusion, feature-level fusion, and decision-level fusion. These are illustrated in Figure 6.4, where the different points in the processing stage where fusion can be applied are

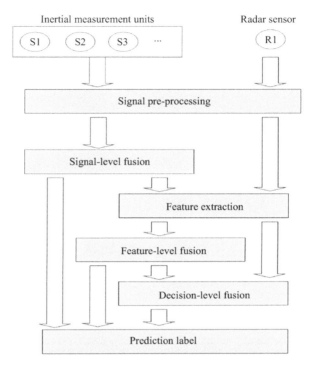

Figure 6.4   Sensor fusion algorithms, topologies, and stages where they are used

highlighted. The sensor is the first step in the chain where the data is pre-processed into digital data. Following this, other processes, including feature extraction and fusion, can be performed.

The point at which fusion occurs defines the level or order of the fusion, where lower level fusion methods utilise the correspondence between the data, whereas higher level methods use confidence metrics for decisions or labels from the classifier. This shows that different levels of fusion could happen in parallel to obtain a final label.

Signal-level fusion is the lowest order of fusion in the context of human activity recognition. It takes place between sensors that record the same quantities from placements on different body parts of the monitored subject. It is a complementary fusion method where commensurate data from the sensors are directly combined through mixing or Kalman filtering. The benefit of signal-level fusion stems from the savings in processing raw data before fusion, which can be computationally intensive. However, signal-level fusion also has drawbacks, as when data from different domains are sampled at different rates or they may have contradictory information, the resulting fused data can contain redundant or erroneous information. This can be addressed by suitable down-sampling, where some samples of the raw data are discarded (with

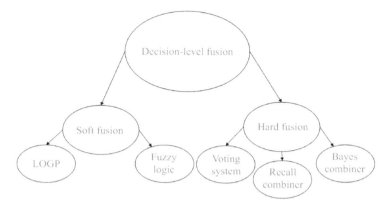

Figure 6.5   Different decision-level fusion methods

loss of information), or up-converting, where data points are resampled increasing redundancies in the data.

Feature-level fusion is a higher order method that combines the features generated after post-processing the signal in the feature space [20]. As features will be generated in a uniform observation or time step, the feature vectors from the different sensors can be concatenated into a single, larger feature vector. This not only has the drawback of increasing the size and dimensionality of the feature space but also brings diverse information together. The main benefit of feature fusion is that since the features extracted are expected to be useful, the new feature subspace will have relevant data and minimal noise. In our context, feature-level fusion can be used for different domains of analysis:

1. The multi-view analysis requires information from complementary, co-operative and competitive sensor networks, which obtain different viewpoints and provide a spatially larger observation of the target scene/activity;
2. The multi-temporal analysis uses complementary networks to detect changes in the time signature through data acquisition at different time steps;
3. The multimodal analysis integrates information from the same event acquired by different sensors to obtain a more complex and dynamic representation of the observation, and this will be the focus of the experiments shown in this chapter.

Decision-level fusion is the highest level of information fusion that considers the confidence levels and/or outputs from the classifiers [12]. In a classification problem, the output of the process chain is the final label provided by the classifier and a measure that expresses the level of confidence in this decision, given as a form of loss or posterior probability. Different types of decision-level fusion methods are listed in Figure 6.5.

If the fusion processing is performed on the confidence metric, then it is called a *soft fusion*. In soft fusion, each sensor data, either as raw or through post-processing, is fed into individual classifiers. Since the data is separated, the classification complexity is reduced, as there is no need for tuning the parameters again for a new feature set. For all the classes of the set, the classifier detects the likelihood of the given class by selecting the label with the lowest 'loss' parameter. When there is insufficient or invariant data, the loss parameter for two classes might be close with no clear winner between them, potentially causing an error in the motion or activity detected. In this case, the 'loss' parameter can be derived from the two different classifiers, scaled logarithmically then added together to correct the classification error. This loss can be corrected in two ways: *pooling* (soft fusion) and voting (often referred to as *hard fusion*).

Logarithmic opinion poll (LOGP) scales the log loss exponentially to a measure of probability in B_n which is a Gaussian mass function [30]. Using a distribution factor which is $1/N$, with N the number of classifiers or sensor sources, $B(\alpha|y)$ gives the highest probable class and in most cases the correct class, which is described in the following equations:

$$B(\alpha|y) = e^{-S_n(\alpha|y)} \tag{6.7}$$

$$B(\alpha|y) = \prod_{n=1}^{N} B(\alpha|y)^d = \prod_{n=1}^{N} e^{-dS_n(\alpha|y)} \tag{6.8}$$

where $n \in [1, N]$ is the classifier index, $\alpha \in [1, C]$ is the class label index, B_n is defined by a Gaussian mass function that converts the negated binary loss returned by the individual classifiers $S_n = [S_{n1} \ S_{n2} \cdots S_{nC}]$ to a posterior probability, and y represents the test samples. The distribution factor d is equal to N^{-1}, where N is the number of classifiers; in this case $N = 2$. Contributions of both classifiers influence the final probability $B(\alpha|y)$, and the test sample will be assigned to the class yielding the highest probability. The Gaussian mass function ensures a higher probability for a smaller error.

Fuzzy logic uses the confidence metric as a fuzzy set, as described in (6.9) [31]. S_{radar} and $S_{inertial}$ are two sets of negated binary loss values including a number of elements equal to the number of classes considered, as generated from the two original classifiers (e.g. SVM classifiers). In other words, S_{Fused} is the new binary loss set which retains the smallest error out of the fused binary losses. The final decision is made by finding the lowest value in this set as in the following equation:

$$S_{radar} = [S_{radar1} \ S_{radar2} \cdots S_{radarC}]$$

$$S_{inertial} = [S_{inertial1} \ S_{inertial2} \cdots S_{inertialC}]$$

$$S_{Fused} = \min\{S_{Radar}, S_{Inertial}\} \tag{6.9}$$

In [32], instead of merging the confidence level of different sensors, *hard fusion* of two sensors (in this case radar and a wrist-worn pressure sensor array) takes place between the prediction labels using a probability combiner for hand gesture recognition. There are several potential combiners in the published literature in this area, including the majority voting system or weighted voting system [33], Recall combiner and Naïve Bayes combiner. A simple voting-based system is not suitable due to the decision clashes in most complex scenarios with many possible classes, whereas Recall combiner is not ideal for binary classes problems since the performance of Recall combiner is typically proportional to the number of classifiers. Naïve Bayes combiner calculates the posterior probability of each class through the prediction label and confusion matrix of the individual sensor. In this case, the probability of certain class after fusion is obtained by the following equation:

$$\log P(C_k|d) \propto \log(P(C_k)) + \sum_{m=1}^{N} \log(p_{m,C_m,k}) \qquad (6.10)$$

where $P(C_k|d)$ is the probability that class C_k is the true class. $P(C_k)$ represents the number of classifiers, with suggested C_k as the prediction label. The classifier belongs to a classifier ensemble whose length is equal to N. $p_{m,C_m,k}$ refers to the confusion matrix element corresponding to classifier m, row C_m, and column k. The final prediction label is the class with the highest posterior probability. Compared with soft fusion, hard fusion requires a lower computational load avoiding selecting and fine tuning the optimal weight function across different classifiers.

6.3.2   Feature selection

Salient features are necessary for having a high degree of accuracy, confidence and reliability in the classification results [31,34]. The key requirement behind a well-performing classifier in most cases is the input to the model, i.e. the set of features used. While there are classifier mechanisms which filter or transform the inputs, starting with a low-sized input set can be helpful, especially as the number of sensory information increases and the feature space can become orders of magnitude larger in dimensionality. This makes the feature space redundant due to 'the curse of dimensionality,' i.e., too many features which do not provide significant information. This eventually results in lower accuracy for a more complex classification algorithm (one could, in this case, mention the old adage *less is more*). To keep the input size reasonable, performing feature selection has been advocated in the literature to reduce the number of features considerably. There are two classes of methods for selecting optimal features [35].

Filter methods, based on information content metrics such as Euclidean distance, entropy, correlation coefficients, find the distance between feature clusters and rank them accordingly enabling a selection of the highest ranked features.

Wrapper methods, which test different combinations in the feature space, find the best performing feature set for a specific classifier. Wrapper methods are resource-intensive and require more iterations and exhaustive search within the feature space to find the optimal input features for the classifier, so when compared to the filter methods they are more complex and slower.

F-score [35] was first introduced to select a feature subset by calculating and ranking the score determined by the distance between data points. The distance between data points belonging to two different groups is as large as possible, while the distance between data points belonging to the same group is as small as possible. It is defined in the following equation:

$$F(X^i) = \frac{\sum_{j=1}^{c} n_j \left(\mu_j^i - \mu^i \right)^2}{\sum_{j=1}^{c} n_j \left(\sigma_j^i \right)^2} \tag{6.11}$$

where $F(X^i)$ is the Fisher score of the ith feature; the parameters n_j and μ_j are the size of the jth class and the mean value, respectively; σ^i and μ^i indicate the standard deviation and mean value of the feature subset regarding to the ith feature.

Relief-F has been modified to fit the multi-class classification problem with the Relief algorithm [35]. The output of Relief-F is a feature weight between -1 and 1, whereas the input of the algorithm should be normalised to [0, 1] by dividing the product of several features and classes. An observation corresponding to a row of feature data set is selected randomly, where the nearest hit belonging to the target class and the nearest miss belonging to the opposite class is generated. The weight of features is updated iteratively by the type of nearest miss, weighted by prior probabilities of each class.

SFS (sequential feature selection) [35] finds the best combinations of features by using a classifier to test all the combinations with progressively increasing (or decreasing) size of the feature space. Forward SFS iteratively adds features according to classifier performance, increasing the input subspace of the classifier. In a backward search, it starts from the original feature subset and reduces the dimension one by one. It is among the most computationally intensive methods.

The implementations of Relief-F algorithm and SFS algorithm are provided as follows.

Relief-F algorithm:

- Set all the weights $W(F)$ to 0
- For $i = 1$: n (n equates to the number of instances)
- Select one observation D_i randomly
- Find k nearest hit h for each Class $C = B_i$
- Find k nearest missing m for each Class $C \neq B_i$
- For $F = 1$: number of features

$$W(F) = W(F) - \sum_{j=1}^{k} \frac{diff(F, B_i, h_j)}{n*k} +$$

$$\sum C \neq class(B_j) \frac{\frac{P(C)}{1 - P(class(B_i))} \sum_{j=1}^{k} diff(F, B_i, m_j)}{n*k}$$

- Update the weight

In the Relief-F algorithm the *diff* () function denotes the difference between the values of features for two observations and k is the class. In this case, $P(C)$ is equal to $1/10$ because the number of observations in each class is equal.

SFS algorithm:

- Initialise with an empty feature set $F(\theta)$
- Find the single feature X which yields the best performance with $F(\theta)$
- The new feature subset selected becomes
  $F(\theta) = F(\theta) + X$
- Update the new feature set iteratively
- Repeat until you find one feature combination which yields improved classification accuracy

---

6.4 Multimodal activity classification for assisted living: some experimental results

In the two key experiments discussed in this section, a combination of non-contact (radar) and wearable sensors was used for monitoring ten daily activities described in this section [12,19].

In one experiment, a single magnetometer embedded within a dedicated IMU was used in conjunction with an FMCW radar. Magnetic sensors have been used in conjunction with accelerometer and gyroscopes for fall detection, although in some works in the literature only accelerometer and gyroscope data have been considered. Therefore, the use of magnetic sensors on their own with a non-contact sensing method such as radar is interesting to consider to explore the achievable performances. This, in turn, can reduce the computational complexity of the processing chain since it has the smallest number of sensors, which potentially reduces battery consumption as well.

In a second experiment, a smartphone with its multiple degrees-of-freedom inertial chip was used together with FMCW radar to assess the effect of jointly combining acceleration and gyroscope data. The FMCW radar and dedicated IMU sensor are shown in Figure 6.6.

6.4.1 Experimental setup

The main hardware components in the experiments were the magnetometer, the IMU and the FMCW radar. The magnetometer was a single Bosch BMM150 Hall sensor sampling at 20 Hz with a ± 1300-μT range and resolution of approximately 0.3 μT. The magnetometer is a subcomponent of the IMU produced by X-IO technologies, containing also an accelerometer and gyroscope sensor.

The FMCW radar produced by Ancortek operated with a centre frequency of 5.8 GHz and has an instantaneous bandwidth of 400 MHz and a PRF of 1 kHz. Doppler ambiguity is linked to the PRF, as a low PRF can cause aliasing within the μD signature appearing wrapped around the opposite Doppler axis; in this case 1 kHz PRF was sufficient to avoid any aliasing in the data. The system was used with one transmit and one receive Yagi antenna, with a gain of 17 dB and transmit power of approximately 100 mW. The antennas have a 48° beamwidth in azimuth; they were 0.5 m apart, therefore operating in a quasi-monostatic configuration. 80 cm height meant that the radar's main lobe would encompass the body of the participants, with the

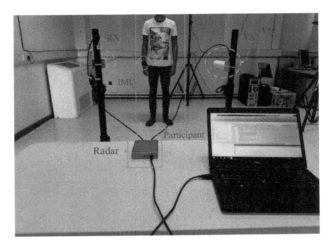

Figure 6.6 *Radar and IMU sensors used in the experiments. The IMU is placed on the wrist of the participant while the Radar system with antennas is located at a distance away from the target. This is intended to be a complementary system where the mass movements of the body are detected by the radar, and the finer arm movements are measured by the IMU.*

increased reflection coming from their torso. A distance of 2 m from the target would replicate real-life conditions where the target would be a short distance away from the radar in an indoor scenario. Variations in the movement were present depending on the person and the activity being performed to replicate realistic movements.

Power requirement is an important parameter for the different sensors. As the wearable sensors were located away from the processor, they required a battery-based supply to operate the sensors and transmit the data. However, the radar system was tethered to the USB port of the data-logging laptop for power. In both experiments, the wearable sensor (either the dedicated sensor or the smartphone) was placed with a Velcro bracelet on the participant's wrist on their dominant hand.

To make the classification challenge harder, sets of similar activities were included in the problem set. The listed activities were selected due to the variety in the movements required and similarity to other activities in the set. Most of these movements, apart from falls, are also commonly performed in daily life and therefore can be good indicators of decreasing ability in terms of personal mobility. The activities are described in Table 6.2, with a pictorial representation provided in Figure 6.7.

The data analysed in this experiment were collected with a group of 20 volunteers aged between 22 and 32 years. In particular, for each volunteer, three repetitions of each of the ten activities were collected simultaneously using the radar and the wearable (or smartphone) sensor, recording 600 datasets per sensor in total [19].

Figure 6.8 shows examples of the data collected, in particular, the radar spectrograms and the corresponding tri-axial raw magnetometer data for four different activities, namely drinking water while standing, taking a phone call, walking back

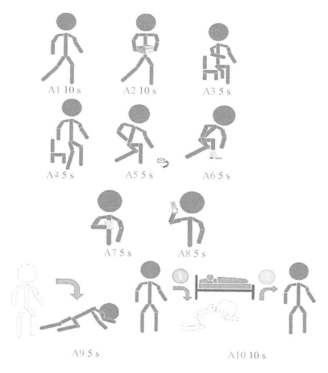

Figure 6.7   Pictorial representation of ten activities performed by the participants

Table 6.2   Description of activities

No.	Description
A1	Walking back and forth
A2	Walking and carrying an object with both hands
A3	Sitting down on a chair
A4	Standing up from a chair
A5	Bending to pick up an object and coming back up
A6	Bending and staying down to tie shoelaces
A7	Drinking a glass of water while standing
A8	Picking up a phone call while standing
A9	Simulating tripping and falling frontally
A10	Bending to check under furniture and coming back up

and forth, and falling. The spectrograms are normalised to the maximum contribution and have been filtered around the DC component to remove the influence of static clutter.

Table 6.3 presents an indicative comparison of performance for different classifiers used with the data discussed in this chapter [12,19]. This list is non-exhaustive,

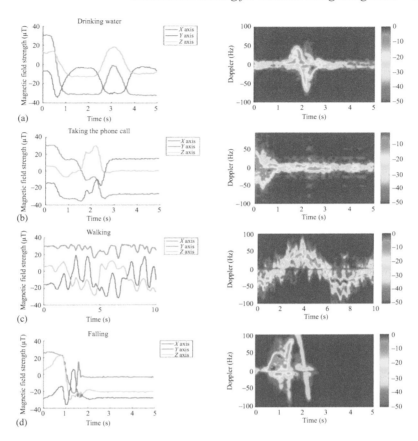

Figure 6.8   Magnetometer and μD spectrogram representations for four activities:
(a) reaching out to drink water from a cup (b) picking up a phone to
receive a call (c) walking back and forth, and (d) frontal fall [19]

Table 6.3   Indicative comparison of classifiers for activity monitoring [12,19]

Classifier	Complexity	Processing time (s)	Accuracy[b] (%)
Linear discriminant	Small	0.37	68–70
K-nearest neighbour	Small	0.48	75–78
Support vector machine	Large	1.4	84–90
Artificial neural network	Large	2.38	86–92
Deep neural network[a]	Very large	>1200	88–95

[a]Another, more complex form of artificial neural network.
[b]Dataset dependent.

and additional classifiers have been used in literature, in particular, a wide range of different neural networks architectures in recent years, some of them discussed in detail in other chapters of this book.

6.4.2   Heterogeneous sensor fusion: classification results for multimodal sensing approach

6.4.2.1   Magnetic sensor and radar results

The signature in Figure 6.8 is normalised to a peak of 0 dB using the strongest signal, and the quality of the radar data in terms of SNR is significantly related to the distance. In an open testing environment, the distance between the radar sensor and the target could be varied from the minimum to maximum detecting range. It is possible that the μD signature is weakened when the target moves too far from the antenna. Hence, the magnetic sensor could be utilised as an 'enhancer' of the radar, as the quality of magnetic data does not depend on where the subject is while performing activities. Additionally, an open challenge for radar is the aspect angle problem: the perceived Doppler shift is proportionally decreased in its intensity with the increase in the cosine angle of the body movement with respect to the radar line of sight. This can generate a weak μD signature. However, the magnetic sensor is not sensitive to the moving direction due to its multiple axes. At the feature fusion level, the 'missing' information in the radar features could be complemented by cascading the feature matrix of the magnetic sensor signal. On the other hand, the magnetic sensor is placed on the wrist of the body. Hence, it would not be capable of quantifying the difference between gaits such as 'walking' or 'limping', or other specific movement patterns for the lower limbs. In that case, radar can be an 'enhancer' towards the magnetic sensor.

Figure 6.9 shows the classification accuracy for the two sensors used and two classifiers (SVM and ANN), as a function of the varying number of optimal features used and selected through the SFS algorithm. There is no monotonic relation between the number of features and the accuracy, as there are certain redundant features which do not improve the classification accuracy when added to the feature set. The accuracy stops improving beyond 30 features for the magnetic sensor and 10 features for the radar in the non-fusion case. The average accuracy with these features is 90% approximately. As adding more features will not bring any improvements to the classification and may reduce the performance, only a subset of all the available features should be used [19].

Feature selection after feature-level fusion combines the feature space of both radar and the magnetic sensors and derives the key features from both. The result of this approach is that the accuracy from feature fusion is shown to outperform the cases of both sensors used individually, reaching accuracy above 96% [19].

Figure 6.10 shows the classification accuracy as a function of the number of neurons used within a single hidden layer ANN (from 1 to 50 neurons), comparing the case of individual sensors vs a fused pool of features. In this case, all the available features were used, with no feature selection, aiming to explore whether the ANN could provide some degree of feature selection capabilities in its inherent processing. Results from using both sensors in conjunction (feature fusion) outperform the use of

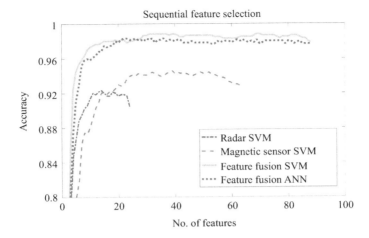

Figure 6.9 Comparison of sensors and classifiers with and without feature fusion and as a function of the number of selected features [19]

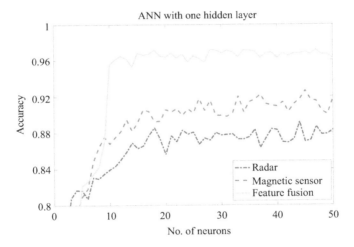

Figure 6.10 Classification accuracy performance of an ANN classifier using selected features set from FMCW radar, magnetic sensor, and feature-level fusion of the two [19]

each sensor individually with an average of 96% with feature fusion, which is similar to the performance of SVM. The 99th percentile of accuracy is reached when over ten neurons are used in the fusion case, which corresponds to the number of classes [19]. For the individual sensors, the number required is comparable to the optimal features from SFS. This appears to suggest that the ANN is implicitly selecting relevant information from the combined space and selecting salient features automatically.

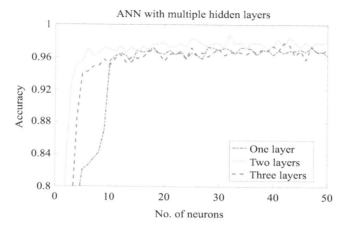

Figure 6.11   Classification accuracy performance of multiple layers of ANN [19]

The investigation of the ANN capabilities is further expanded in Figure 6.11, where the validation accuracy using ANNs with multiple hidden layers (between 1 and 50 neurons) is shown, where the input was the set of all fused features from magnetic sensor and radar. It was found that marginal improvement is achieved from using multiple hidden layers for the ANN after feature fusion is performed. The number of neurons is varied in the last hidden layer and kept constant at 50 for the other layers, but there is a variety of just about 0.8% between using one or two hidden layers with the exception that fewer neurons are required to reach the optimal point.

In real-life applications, the classifier will not have information or data from the test subject, yet it should be able to classify movements and activities from such a person. The performance in this more realistic condition can be tested in a 'leave-one-subject-out' test; in other words a blind test. For the blind test, observations from one specific participant were selected as the testing set, and the remaining participants were used for training. This was done until all the participants were used as the testing set, after which the cumulative accuracy was analysed [19].

Figure 6.12 shows the 'minimum', 'maximum', and 'mean' accuracy obtained from these tests. The 'max' and 'min' variables represent the best and worst case scenario from all the participants, and the 'mean' is the average across all participants. The 'difference' variable compares the stratified test in Figures 6.8 and 6.9 with this approach of 'leave one subject out', with the former deemed to provide higher results on average for the way the classifiers are cross-validated.

The results show that there can be significant variability in accuracy on a subject-by-subject basis, with the extreme case of the individual magnetic sensor, where both ANN and SVM yield an accuracy of approximately 40%. Radar is more robust with both classifiers, as the mean results are 2%–4% lower than the stratified set. The differences are clearer for the magnetic sensor as the accuracy is 12% lower for both classifiers with this sensor. Feature fusion appears to help recover this loss

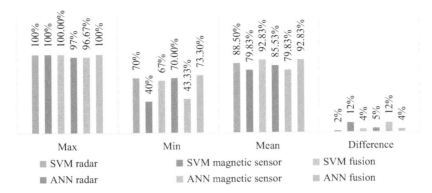

Figure 6.12   *Classifier performance when models are tested with an unknown participant. The difference is the delta from the average stratified test [19]*

[19]. In particular, feature fusion helps the magnetic sensor here, as the 'difference' between the two testing methods is reduced from 12% to 4%. Feature fusion appears to be beneficial for realistic ambient activity monitoring, as the additional degree of freedom from multiple and diverse sensors provides an improvement in performance. As the participants in this test were given the freedom to move in a comfortable way to make the data represent real natural movements, there was a large degree of variability.

In an extreme case, it was noted that one participant who moved slowly and was a general outlier has a completely different signature from the rest of the set. The minimum value for each of the sensor-classifier combinations is related to these outlier cases and shows the challenging issue of activity classification for unknown users, whose data were not available to train the classifier, accounting for the very individual and diverse ways people move. Generating a sufficient amount of training data to be representative of a large cohort of potential users of the technology (e.g. older people to monitor at home) is at this moment an outstanding research challenge. It is unlikely that the huge amount of required experimental data across different people and environments can be realistically collected. Ways to generate synthetic data to complement the experimental ones and validate classification algorithms and systems are likely to be the way forward, using kinematic models (e.g. Boulic), video data (e.g. MOCAP data) converted into synthetic radar data, and specific neural networks such as GANs (Generative Adversarial Networks) to generate synthetic data.

6.4.2.2   IMU and radar results

The accelerometer measures the acceleration of the movements and it is proved to be very sensitive in the field of fall detection, whereas a gyroscope can provide features in terms of body rotation speed and the movement directions. With the help of a magnetic sensor, the fusion with radar has the capability to detect the micro-motions such as gestures and subtle shift of the limb. All the four sensors can be used together; however, there is a trade-off between the classification performance

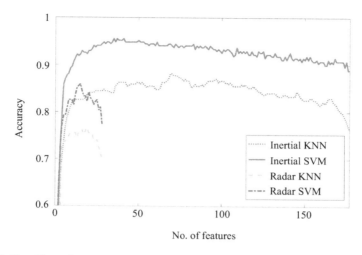

Figure 6.13 *Classification accuracy of individual sensors. The improvements through the sequential feature selection and the classifier used are shown [12].*

Table 6.4 *Comparison of feature selection methods (inertial) [12]*

Method	Accuracy (%)	Time (s)	Features no.
F-Score (SVM)	90.7	1,448	73
F-Score (KNN)	88.2	220.2	76
Relief-*F* (SVM)	91.1	1,210.7	16
Relief-*F* (KNN)	89.3	196.9	58
SFS (SVM)	95.6	14,489.5	35
SFS (KNN)	88.25	903.5	69

and computational intensity in terms of the number of sensors and features. Better fusion schemes have to be implemented to exploit the contribution of each sensor and eliminate the possible redundant or conflicting information between the selected features. For this, decision-level fusion is a good option to combine the posterior output probability of each sensor due to the raised number of classifiers.

Figure 6.13 presents a plot of the classification accuracy for the SFS feature selection method for wearable IMU and a radar sensor. A general summary of the results for all feature selection methods described in the previous sections of this chapter is provided in Tables 6.4 (IMU sensor) and 6.5 (radar). The tables show that the filter methods reduce the number of features used but bring no significant performance improvements, with less than 2% improvement compared to when all available features are used [12].

Table 6.5   Comparison of feature selection methods (radar) [12]

Method	Accuracy (%)	Time (s)	Features no.
F-Score (SVM)	78.8	220.4	17
F-Score (KNN)	74.1	30.6	17
Relief-F (SVM)	74	213.1	20
Relief-F (KNN)	67	24.2	18
SFS (SVM)	85.6	1,316.7	20
SFS (KNN)	79.8	32	19

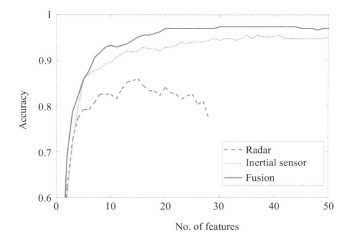

Figure 6.14   Classification accuracy performance of feature fusion with radar and inertial sensor vs individual sensors as a function of selected features

Optimal features suggested by filter methods reduce the required number of features by 60% and 35% for inertial sensor and radar, respectively. Exceptions to this are Relief-F for both sensors. For radar, the Relief-F with KNN increased accuracy by 4% despite being a filter method, and for IMU, Relief-F only reduced the required features set by 8%. SFS reduced the feature subset while bringing an improvement of 5%–7% in accuracy for both sensors with SVM. This is achieved after feature fusion and SFS on the overall, larger features pool. KNN, on the other hand, yielded no performance improvement for the inertial sensor, despite a 9% boost for radar [12]. With feature fusion between the inertial and radar sensors, the highest classification accuracy was 97.4%, as shown in Figure 6.14.

Table 6.6 shows the results of using the different decision-level fusion methods rather than feature fusion. LOGP fusion appeared to be the best performing decision-level method, with an accuracy of 96.7% with only nine misclassification instances over ten iterations of testing. Fuzzy logic performed worse with an accuracy of 94.8% and a higher number of misclassification (16 instances). To improve the

Table 6.6   decision-level fusion results [12]

Method	Average error[a]	Average accuracy (%)
LOGP	9	96.7
Fuzzy logic	14	94.8
Voting system	6	97.8

[a]Average error is the average misclassifications over ten iterations.

decision-level results, a voting system that considered labels from the classifier for each of the sensors and then performed LOGP with the two best-performing sensors was constructed. This system gave the best decision fusion result with an average accuracy of 97.8 and only six misclassifications after ten iterations.

6.5   More cases of multimodal information fusion

6.5.1   Sensor fusion with same sensor: multiple radar sensors

In the previous sections, we described the use of wearable and non-contact radar sensors, i.e. a heterogeneous combination of sensors, to classify activities with higher performance. However, there may be a scope to combine multiple homogenous sensors together. To understand this, a smaller scale experiment with a single 24 GHz CW and 5.8 GHz FMCW radar was performed, and feature-level fusion was applied to identify improvements brought by having two non-contact sensors [35].

Using two radar systems with different centre frequencies is an example of co-operative fusion. As opposed to the heterogeneous fusion case where the IMU measures the finer movements of the arm and the body while being worn or carried, the radar systems are measuring the Doppler shifts in both cases. Given the difference in carrier frequency, hence in the perceived Doppler shift, and the difference in scattering due to the different wavelengths, it is expected that the two sensors may obtain different information to describe the movements and human activities.

The processing was similar to the previous experiments, and the same features were extracted from the spectrograms of both radar sensors. The results from the individual sensors and fused case are shown in Table 6.7, with the classification accuracy provided or each activity. Overall, the FMCW radar performs better than the CW, except for some activities, which are identified better by the CW; feature fusion is still overall the best performing option [35]. In other words, the fused system is utilising the different spectral viewpoints offered by the two different frequencies and using the best result.

However, in certain cases feature fusion may negate correct decisions from either sensor, providing less accuracy than the best sensor used on its own. One such case is for A2 (walking while carrying an object). The accuracy is lower for the fused case at 75.05% compared to FMCW (83.95%). This can happen because the features from

Table 6.7   Comparison of classification results for individual CW and FMCW radar and their feature fusion [35]

Predicted class	CW	FMCW	Fused
A1: Walking	58.82	58.68	61.31
A2: Walking with object	73.89	83.95	75.05
A3: Sitting	70.89	72.89	95
A4: Standing	69.05	80.81	94.22
A5: Picking up an object	83.61	69.94	90.39
A6: Tie up shoes	68.69	74.27	95.14
A7: Drinking from a cup	80.05	78.82	83.14
A8: Taking a call	57.50	72.27	86
A9: Falling forwards	**96.23**	**87.61**	**95.28**
A10: Checking under table	95.67	95.72	96.17

Notes: Falling is a critical activity that we cannot miss nor have other activities misclassified as falls for the practical use of the radar when deployed in the home environment. This is to highlight that fusion increases the overall accuracy, but we need to be careful regarding the individual activity recognition as the performance of falls was higher with CW alone and fell by ∼1% after fusion.

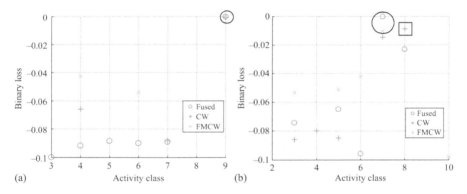

Figure 6.15   Example of how confidence score (binary loss) changes in cases where fusion is applied [35]: (a) A10 and (b) A7 – correct predictions from each radar systems are encircled and incorrect predictions are in a square

CW influence the decision, as only using CW has 73.89% accuracy. This is a case when the worst performing sensor has a detrimental effect on the best performing one, potentially reducing the advantage of fusion and the rationale to perform it. This is a limitation of co-operative and complementary fusion when all sensors are considered useful and equal, which may not be true in all cases, i.e. more complex fusion schemes that rank differently the different sensors may be needed (e.g. weighted fusion schemes or hierarchical classification).

To understand this potential issue further, Figure 6.15 shows how the confidence score, expressed as binary loss, changes with fusion. In Figure 6.15(a), assuming for

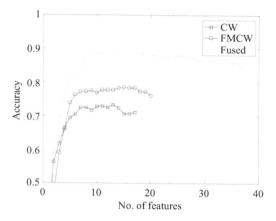

Figure 6.16   Accuracy of the three radar sensor setups over an increasing number of features [35]

example that the target class is activity A9 (fall), we can see that the confidence for A9 is the closest to 0 compared to the other activities for all the sensors. This is the ideal case where the sensors correctly identify the class. In Figure 6.15(b), there are cases of ambiguity in the identified correct class. For example, the FMCW identifies activity A7 (drinking from a cup) while the CW radar selects the alternative, 'confuser' activity A8 (taking a phone call) as the predicted class. Feature fusion helps select the correct class A7, as the incorrect influence from the CW sensor is offset.

There is no universal conclusion that can be presently drawn as to whether fusion is always advantageous for any classification problem or dataset, and how to maximise any advantage of fusion is currently an open problem. The results presented within this chapter aim to highlight cases of clear improvements of using radar sensing with alternative sensors, or multiple diverse radar systems, as well as cases where simple feature fusion schemes do not provide any improvement, but a reduction in performance, hence requiring more complex and well-though fusion schemes.

Figure 6.16 shows the classification accuracy for the case of CW and FMCW radar and their fusion when feature selection via SFS is applied. Twenty-one features from the FMCW radar features are required to achieve an accuracy of 75%, while the CW radar performs more limitedly with 70% accuracy when 19 features are used. With feature fusion, an improvement of classification to 83% occurs. SFS increases this further to almost 90% with only 15 features used. Out of these 15 features selected in the multi-radar fusion scenario, 4 are extracted from the CW radar data, and 11 from the FMCW radar.

6.6   Conclusions

In this chapter, we provided a review of some multimodal sensing approaches and information fusion algorithms, focusing on assisted living scenarios (human activities classification), where radar sensors and related µD signatures are one of the possible sensors and information domain used.

We have provided an overview of the radar signal processing with a brief example of µD spectrograms and the wearable sensors' data and discussed feature extraction for both of these sensor types. Multi-sensor fusion was introduced, and its various principles, approaches, and topologies were presented with methods (soft and hard) operating at the signal, feature, and decision levels. Complementary methods which can assist with the increased feature subspace were discussed, namely Fisher score, Relief-*F* and sequential forward selection for feature selection. Finally, the results from two experiments were presented, where sensor fusion between heterogeneous (radar and wearables) and same sensors (two different radar sensors) was demonstrated.

The objective of this chapter is not to provide an overall comprehensive study of information fusion theory or machine learning, for which valuable textbooks are already available, but to present some of the key aspects and challenges in a simple manner and with support of experimental results accessible to a broad audience of radar engineers. Many outstanding research challenges remain to be addressed at this stage, considering, for example, which sensors are most suitable in which scenarios, what information fusion algorithm has to be used, and when and/or what pre-processing is necessary on each sensor's data (just considering how many different time-frequency distribution and alternative processing rather than spectrograms via STFT have been proposed [36,37]). Adding more sensors can have economic, logistics, and computational implications, but as demonstrated in this chapter, this may improve the overall performance of the system to classify human activities. The usage of sensors that need to be worn or carried, or the presence of cameras may appear as a 'counter-argument' to use radar sensing, whose advantages are contactless capabilities and privacy preservation (i.e. no plain images or videos recorded). However, technological development is pushing towards smart environments and homes and Internet of Things framework, where a plurality of sensors will be in any case available in indoor environments. Hence, the interest to think of radar as 'a sensor in a wider suite of sensors' and to investigate the best approaches to exploit all the available information.

One key issue for the assisted living scenario and in general activity monitoring related to healthcare will be the portability of any classification scheme across different scenarios and different subjects, whether supported by multimodal information fusion or not. How to ensure that a classification algorithm or system trained on a certain subset of data is still performing well in an unknown environment with unknown subjects? Certainly, increasing the number of training data will help, but, especially for experimental radar data, the practical issues for their collection and labelling remains, even if these data can be complemented by synthetic data generated by models, by the transformation of video-based data, or augmentation of experimental data through deep-learning inspired techniques.

Another interesting direction of future research can expand the sensor fusion approaches mentioned in this chapter to the continuous sequence of activities interpreted from deeper, more complex neural networks, rather than simple features

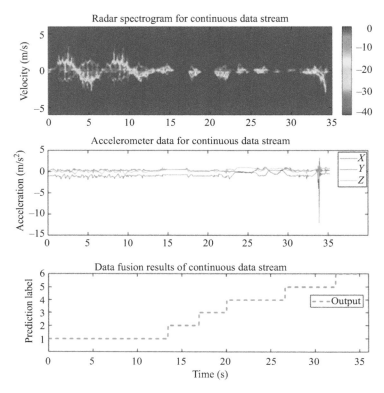

Figure 6.17   Example of continuous activities performed by a subject with data recorded from both radar and wearable accelerometer, with the ideal prediction of activities shown in the bottom figure

extracted from 'snapshots' activities of constrained and fixed duration. Figure 6.17 shows an example of this ongoing challenge, which has continuous data containing different activities and transitions recorded by a 5.8 GHz radar (spectrogram) and a wearable accelerometer. It is a significant challenge to identify automatically critical activities, and isolating the transitions between them correctly, all these are potentially done in real-time as the different streams of data from different sensors are fed to the relevant classifiers.

References

[1]   J. Raol, *Multi-Sensor Data Fusion with MATLAB*. Boca Raton, FL: CRC Press, 2009.
[2]   E. Cippitelli, F. Fioranelli, E. Gambi and S. Spinsante, "Radar and RGB-depth sensors for fall detection: a review," *IEEE Sensors Journal*, vol. 17, no. 12, pp. 3585–3604, 2017.

[3]   F. Erden, S. Velipasalar, A. Z. Alkar and A. E. Cetin, "Sensors in assisted living: a survey of signal and image processing methods," *IEEE Signal Processing Magazine*, vol. 33, no. 2, pp. 36–44, 2016.

[4]   V. C. Chen, D. Tahmoush and W. J. Miceli, "Micro-doppler signatures—review, challenges, and perspectives," in: *Radar Micro-Doppler Signatures: Processing and Applications* (pp. 1–25). London: The Institution of Engineering and Technology, 2014.

[5]   M.G. Amin, Y.D. Zhang, F. Ahmad and K.C.D. Ho, "Radar signal processing for elderly fall detection: the future for in-home monitoring," *IEEE Signal Processing Magazine*, vol. 33, no. 2, pp. 71–80, 2016.

[6]   K. Saho, M. Fujimoto, M. Masugi and L. S. Chou, "Gait classification of young adults, elderly non-fallers, and elderly fallers using micro-Doppler radar signals: simulation study," *IEEE Sensors Journal*, vol. 17, no. 8, pp. 2320–2321, 2017.

[7]   M. Amin, Ed., *Radar for Indoor Monitoring*, 1st edition, Boca Raton, FL: CRC Press, 2017.

[8]   J. M. Giron-Sierra, *Digital Signal Processing with Matlab Examples*, Vol 1–3. Singapore: Springer, 2017.

[9]   S. Mallat, *A Wavelet Tour of Signal Processing – The Sparse Way*, 3rd edition. Burlington, MA: Academic Press, 2009.

[10] B. Erol and S. Z. Gürbüz, "Hyperbolically-warped cepstral coefficients for improved micro-Doppler classification," in: 2016 IEEE Radar Conference (RadarConf), Philadelphia, PA, 2016, pp. 1–6.

[11] J. Favela, J. Kaye, M. Skubic, M. Rantz and M. Tentori, "Living labs for pervasive healthcare research," *IEEE Pervasive Computing*, vol. 14, no. 2, pp. 86–89, 2015.

[12] H. Li, A. Shrestha, H. Heidari, J. L. Kernec and F. Fioranelli, "A multisensory approach for remote health monitoring of older people," *IEEE Journal of Electromagnetics, RF and Microwaves in Medicine and Biology*, vol. 2, no. 2, pp. 102–108, 2018.

[13] H. Heidari, N. Wacker, S. Roy and R. Dahiya, "Towards bendable CMOS magnetic sensors," in: 2015 11th Conference on Ph.D. Research in Microelectronics and Electronics (PRIME), 2015, pp. 314–317.

[14] T. Kose, Y. Terzioglu, K. Azgin and T. Akin, "A single-mass self-resonating closed-loop capacitive MEMS accelerometer," in: 2016 IEEE Sensors, 2016, pp. 1–3.

[15] K. E. Bliley, D. R. Holmes, P. H. Kane, *et al.*, "A miniaturized low power personal motion analysis logger utilizing MEMS accelerometers and low power microcontroller," in: 2005 3rd IEEE/EMBS Special Topic Conference on Microtechnology in Medicine and Biology, 2005, pp. 92–93.

[16] B. Hazarika, N. Afzulpurkar, C. Punyasai and D. Kumar Das, "Design, simulation & modelling of MEMS based comb-drive tunneling effect gyroscope," in: 2012 9th International Conference on Electrical Engineering/Electronics, Computer, Telecommunications and Information Technology, Phetchaburi, 2012, pp. 1–4.

[17]   H. Dong and Y. Gao, "Comparison of compensation mechanism between an NMR gyroscope and an SERF gyroscope," *IEEE Sensors Journal*, vol. 17, no. 13, pp. 4052–4055, 2017.

[18]   V. Nabaei, R. Chandrawati and H. Heidari, "Magnetic biosensors: modelling and simulation," *Biosensors and Bioelectronics*, vol. 103, pp. 69–86, 2018.

[19]   H. Li, A. Shrestha, H. Heidari, J. Le Kernec and F. Fioranelli, "Magnetic and radar sensing for multimodal remote health monitoring," *IEEE Sensors Journal*, vol. 19, no. 20, pp. 8979–8989, 2019.

[20]   F. Fioranelli, M. Ritchie and H. Griffiths, "Centroid features for classification of armed/unarmed multiple personnel using multistatic human micro-Doppler," *IET Radar, Sonar & Navigation*, vol. 10, no. 9, pp. 1702–1710, 2016.

[21]   J. Zhu, R. San-Segundo and J. M. Pardo, "Feature extraction for robust physical activity recognition," *Human-centric Computing and Information Sciences*, vol. 7, no. 1, p. 16, 2017.

[22]   C. Clemente, A. W. Miller and J. J. Soraghan, "Robust principal component analysis for micro-Doppler based automatic target recognition," in: 3rd IMA Conference on Mathematics in Defence, 2013.

[23]   J. J. M. de Wit, R. I. A. Harmanny and P. Molchanov, "Radar micro-Doppler feature extraction using the singular value decomposition," in: 2014 International Radar Conference, Lille, 2014, pp. 1–6.

[24]   V. C. Chen, "Spatial and temporal independent component analysis of micro-Doppler features," in: 2005 IEEE International Radar Conference, Arlington, VA, 2005, pp. 348–353.

[25]   T. Hastie, R. Tibshirani, and J. Friedman, *The Elements of Statistical Learning: Data Mining, Inference, and Prediction*, 2nd edition. New York, NY: Springer, 2009.

[26]   M. Ritchie, F. Fioranelli, H. Borrion and H. Griffiths, "Multistatic micro-Doppler radar feature extraction for classification of unloaded/loaded micro-drones," *IET Radar, Sonar & Navigation*, vol. 11, no. 1, pp. 116–124, 2017.

[27]   X. Shi, F. Zhou, L. Liu, B. Zhao and Z. Zhang, "Textural feature extraction based on time-frequency spectrograms of humans and vehicles," *IET Radar, Sonar & Navigation*, vol. 9, no. 9, pp. 1251–1259, 2015.

[28]   H. S. Al Zubi, S. Gerrard-Longworth, W. Al-Nuaimy, Y. Goulermas and S. Preece, "Human activity classification using a single accelerometer," in: 2014 14th UK Workshop on Computational Intelligence (UKCI), 2014, pp. 1–6.

[29]   J. A. Nanzer, "A review of microwave wireless techniques for human presence detection and classification," *IEEE Transactions on Microwave Theory and Techniques*, vol. 65, no. 5, pp. 1780–1794, 2017.

[30]   C. Chen, R. Jafari and N. Kehtarnavaz, "A real-time human action recognition system using depth and inertial sensor fusion," *IEEE Sensors Journal*, vol. 16, no. 3, pp. 773–781, 2016.

[31]   R. C. King, E. Villeneuve, R. J. White, R. S. Sherratt, W. Holderbaum and W. S. Harwin, "Application of data fusion techniques and technologies for wearable health monitoring," *Medical Engineering & Physics*, vol. 42, pp. 1–12, 2017.

[32] H. Li, X. Liang, A. Shrestha, *et al.*, "Hierarchical sensor fusion for micro-gestures recognition with pressure sensor array and radar," *IEEE Journal of Electromagnetics, RF and Microwaves in Medicine and Biology*, doi: 10.1109/JERM.2019.2949456.

[33] L. Kuncheva and J. Rodríguez, "A weighted voting framework for classifiers ensembles," *Knowledge and Information Systems*, vol. 38, no. 2, pp. 259–275, 2014.

[34] S. Z. Gürbüz, B. Tekeli, M. Yüksel, C. Karabacak, A. C. Gürbüz and M. B. Guldogan, "Importance ranking of features for human micro-Doppler classification with a radar network," in: Proceedings of the 16th International Conference on Information Fusion, Istanbul, 2013, pp. 610–616.

[35] S. Z. Gürbüz, B. Erol, B. Çağlıyan and B. Tekeli, "Operational assessment and adaptive selection of micro-Doppler features," *IET Radar, Sonar & Navigation*, vol. 9, no. 9, pp. 1196–1204, 2015.

[36] A. Shrestha, H. Li, F. Fioranelli and J. Le Kernec, "Activity recognition with co-operative radar systems at C and K band," in: IET International Radar Conference, Nanjing, China, 2018.

[37] S. Z. Gurbuz and M. G. Amin, "Radar-based human-motion recognition with deep learning: promising applications for indoor monitoring," *IEEE Signal Processing Magazine*, vol. 36, no. 4, pp. 16–28, 2019.

[38] J. Le Kernec, F. Francesco, C. Ding, *et al.*, "Radar signal processing for sensing in assisted living: the challenges associated with real-time implementation of emerging algorithms," *IEEE Signal Processing Magazine*, vol. 36, no. 4, pp. 29–41, 2019.

Chapter 7

Micro-Doppler analysis of ballistic targets

Adriano Rosario Persico[1], Carmine Clemente[2], and John Soraghan[2]

Nowadays the challenge of the identification of ballistic missile (BM) warheads in a cloud of decoys and debris is essential in order to optimise the use of counter-missile resources. This chapter is aimed at providing ballistic target (BTs) micro-motion models and examples of signal processing tools to handle this family of targets. The chapter introduces signal models for both warheads and decoys of different sizes and shapes. Additionally, an efficient and robust framework is presented, which exploits the micro-Doppler (μD) information extracted from the time–frequency analysis of the radar echo from the target. Some feature extraction approaches are also presented, including those based on the estimation of statistical indices from the one-dimensional (1D) averaged cadence velocity diagram (ACVD), on the evaluation of pseudo-Zernike (pZ) and Krawtchouk (Kr) image moments and on the use of two-dimensional (2D) Gabor filters, considering the CVD as 2D image. An assessment of the presented algorithms is also reported in this chapter exploiting real radar data realised in a laboratory.

7.1 Introduction

The development of BMs after their advent during World War II marks the beginning of the development of sophisticated defence systems. By contrast, a huge budget annually is invested into research and production of countermeasures in order to minimise the effectiveness of ballistic missile defence (BMD) systems [1]. One of the most common practices is the use of a large number of decoys, or false targets, with the aim to confuse the defence systems. Nowadays different decoy strategies are available, e.g. replica decoys, decoys using signature diversity and decoys using anti-simulation [2]. The lightweight decoys are a very attractive strategy against exo-atmospheric defences. Long-range BMs move on sub-orbital trajectories and their ranges typically depend on the altitude achieved by using one or more boosters. On

[1]Aresys srl, Via Flumendosa, Milano, Italy
[2]Centre for Signal and Image Processing (CESIP), University of Strathclyde, Glasgow, UK

the other hand, the missile warhead sizes and range depend on the weight of the carried payload. Hence, missiles can be equipped with a large number of lightweight decoys without affecting the maximum warhead range [1]. The longer part of a BM flight takes place in the exo-atmosphere and it is commonly known as the mid-course phase. The lightweight decoys are released during the mid-course phase, so both the decoys and, the much heavier, warhead travel on similar trajectories due to the absence of atmospheric drag in the vacuum of space [3]. In addition to intentional decoys, missiles also release incidental debris and deployment hardware, e.g. boosters for missile launch, which can pose an additional source of interference on radar returns. In the absence of reliable target identification, the defence system has to intercept all the detected targets, including decoys, in order to cancel the threat. Therefore, the challenge of BT classification identifying the warhead into a cloud of decoys and debris is of fundamental importance to increase the effectiveness of BMD systems. Defence system efficiency can be critically affected by decoys in two related ways. In fact, in the case in which a decoy is classified as a warhead (*false alarm*), the defence system may run out its limited ammunition of interceptors prematurely. In contrast, the misclassification of a warhead (*leakage*) may lead to catastrophic consequences [2].

The BM flight is generally divided into three phases: boost phase, which comprises the powered flight portion; mid-course phase, as mentioned above, during which the warhead separates from the rest of missile; and the re-entry phase wherein the warhead re-enters the Earth's atmosphere to approach the target. The missile interception in the boost phase would be free from the issue represented by decoys. However, the boost phase does not offer much opportunity to track accurately for intercepting a BM since the launch point will normally be a significant distance from the defence radar system. Moreover, during this phase the missile separates from several boosters, which would result in significant interference. For these reasons, BMD infra-red seekers are largely confined to exo-atmospheric operation due to sensitivity requirements and atmospheric friction effects [4]. A chance to discriminate between warheads and lightweight decoys occurs during the re-entry phase, since decoys would slow down more rapidly due to the atmospheric drag than the warhead. Nevertheless, the warhead interception in this phase may be not useful due to its short duration (few seconds), and because the warhead could have already passed the minimum intercept altitude for an above-the-atmosphere interceptor [2]. Additionally, the re-entry vehicles could be armed with a nuclear or chemical bomb such that the warhead has to be intercepted at a safe altitude to avoid that a nuclear detonation has effects on the Earth's surface, or chemical and biological payloads disperse through the troposphere [4]. Hence, the mid-course phase usually represents the most useful flight part for intercepting missiles, due to its relatively long duration and for the absence of tactical manoeuvring of targets since they are in free-flight motion.

The anti-BM interceptors are usually equipped with an on-board computer to perform control, guidance, target data estimation, mission sequencing and various other critical operations during the entire flight, from pre-launch up to the impact [5]. However, all these operations are made more difficult due to the high velocities of the moving target and the interceptor which demands high data update rates from sensors, in order to provide frequent commands to the control system. For these reasons, fast algorithms with low computational costs are required in order to maximise the interceptor's probability to intercept a warhead.

Warheads and decoys exhibit different micro-motions that, if appropriately exploited, may be used to discriminate one from the other [2]. Specifically, the warheads are typically spin-stabilised to ensure that they do not deviate from the intended ballistic trajectory [3]. However, warheads exhibit precession and nutation motion due to the effect of the Earth's gravity. By contrast, decoys tumble when released by the missiles due to the gravity and the absence of a spinning motion [1,2]. As different micro-motions generate different μD signatures, the use of the latter for warheads and decoys classification is an attractive solution.

In the last decade, a large amount of research has been conducted on the possibility to use μD information to identify different targets in many fields of interest, e.g. human motion classification and air moving target recognition. Most systems use information extracted from the time–frequency distribution (TFD) of radar echoes. In [6,7], the features for human motion classification are empirically estimated from the spectrogram. In [8], a set of features is evaluated by using the singular value decomposition (SVD) on the spectrograms and estimating the standard deviation of the first right singular vector. In [9], the authors propose a method for the extraction of cepstrum- and bicoherence-based features from TFD for aircraft classification. In [10], the features are estimated as the Fourier series coefficients of the spectrogram envelope, whereas in [11] the Mel-frequency cepstral coefficients are employed with the main aim to recognise human falling from other motions which can be used for healthcare applications.

In the literature, there are several papers which present mathematical models of the μD signature from BT with micro-motions, such as in [12] and [13]. In this chapter, in order to understand the μD shifts from warheads and decoys, a high frequency-based signal model for the targets of interest is introduced incorporating the effects of occlusion for all the scattering points. Furthermore, a framework based on the processing of the CVD described in [14] for radar μD classification is presented, in combination with different information extraction techniques. In particular, the classification framework is flexible, allowing for multiple techniques for features extraction from the CVD which require different computational complexity, making the algorithm adaptable to the resources available to the defence system (e.g. ground based or airborne).

Results on laboratory data show the effectiveness of μD analysis for BMD purposes.

7.2 Radar return model at radio frequency

The motion of a rigid body is generally given by a combination of both translations and rotations. In the specific case of BTs, missile warheads exhibit precession and nutation during the flight onto the exo-atmospheric part of their sub-orbits, as shown in Figure 7.1(a). The precession is composed of two micro-motions: the spinning

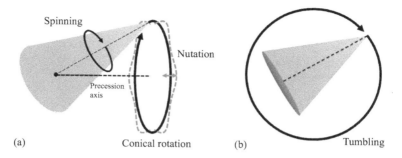

Figure 7.1   BT micro-motions: (a) warhead; (b) decoy

of the target around its symmetry axis, and the conical movement such that the target's symmetry axis rotates conically around the precession axis. The nutation is an oscillation of the symmetry axis, perpendicular with respect to the precession axis. The lightweight decoys instead start to tumble when released by the missile due to the Earth gravity. The tumbling is defined as the rotation of a decoy such that the angular velocity vector is perpendicular to the symmetry axis of the object, as shown in Figure 7.1(b).

In order to describe the effect of target motion with respect to the radar, three coordinates systems are considered: the radar coordinates system $(\hat{U}, \hat{V}, \hat{W})$, centred on the radar; the reference coordinates system $(\hat{X}, \hat{Y}, \hat{Z})$, which is parallel to the previous one and whose origin is the mass centre (MC) of the target; and the local coordinates system $(\hat{x}, \hat{y}, \hat{z})$ such that the axis $\hat{z}$ corresponds with the symmetry axis of the target [12]. The first reference system is used to evaluate the position of the target with respect to the radar, the second for the orientation of the target (and its symmetry axis) with respect to the radar line of sight (LOS), while the third for evaluating the position of the scattering points with respect to the symmetry axis. Figure 7.2 illustrates the three reference systems where $\angle \text{El}_{\text{MC}}$ and $\angle \text{Az}_{\text{MC}}$ are the elevation and azimuth angles of the LOS between the radar and the MC of the target with respect to the radar coordinates system.

Without loss of generality and neglecting the envelope of the transmitted signal, it is assumed that the radar transmits a sinusoidal signal as follows:

$$s_{tx}(t) = \exp\left(j2\pi f_0 t\right) \tag{7.1}$$

where f_0 is the radar carrier frequency. The generic received signal from a target can be written as

$$s_{rx}(t) = \sum_{i=0}^{N_p-1} \sqrt{\sigma_i(t)} \exp\left(j2\pi f_0 \left[t - \frac{2t_i(t)}{c}\right]\right) \tag{7.2}$$

where N_p is the number of scattering points, $t_i(t)$ and $\sigma_i(t)$ are the delay of propagation and the scattering coefficient of the ith scatterer, respectively. An expression of the

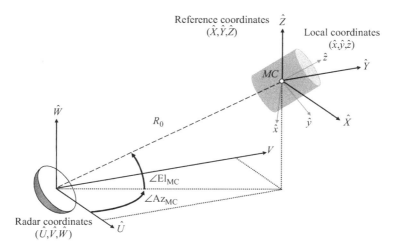

Figure 7.2 Geometry for radar and target with micro-motions

propagation delay for the ith generic point is given by

$$t_i(t) = \frac{2\mathrm{r}_i(t)}{c} \tag{7.3}$$

where $\mathrm{r}_i(t)$ is the distance between the radar and the considered point scatterer. The distance $\mathrm{r}_i(t)$ is the norm of the position vector r_i^{radar}, i.e.

$$\mathrm{r}_i(t) = \|r_i^{\mathrm{radar}}\| = \|r_{\mathrm{MC}}^{\mathrm{radar}} + vt + r_i(t)\| \tag{7.4}$$

where $r_{\mathrm{MC}}^{\mathrm{radar}}$ is the initial position vector of the MC with respect to the system $(\hat{U}, \hat{V}, \hat{W})$, v is the bulk motion velocity of the target and $r_i(t)$ is the position vector of ith scattering point with respect to the $(\hat{X}, \hat{Y}, \hat{Z})$ system.

When the target is at relatively long distance, such that

$$\|r_{\mathrm{MC}}^{\mathrm{radar}}\| \gg \|vt + r_i\| \tag{7.5}$$

the relative distance between the radar and the ith scatterer in (7.4) can be approximated as follows:

$$\mathrm{r}_i(t) \approx \|r_{\mathrm{MC}}^{\mathrm{radar}}\| + \langle v, n \rangle t + \langle r_i(t), n \rangle \tag{7.6}$$

where the operator $\langle \cdot, \cdot \rangle$ defines the scalar product between the two vectors, $r_{\mathrm{MC}}^{\mathrm{radar}}$ and r_i are the position vectors of the MC and the ith scatterer with respect to the radar system, respectively, v is the target bulk velocity vector and n is the direction of LOS, given approximately by

$$n \approx \frac{r_{\mathrm{MC}}^{\mathrm{radar}}}{\|r_{\mathrm{MC}}^{\mathrm{radar}}\|} \tag{7.7}$$

Neglecting the time dependence for conciseness, and considering the coordinate system $(\tilde{x}, \tilde{y}, \tilde{z})$ shown in Figure 7.3, which is centred into MC, and such that $\tilde{x} \equiv \hat{z}_t$ and the vector $r_{\mathrm{MC}}^{\mathrm{radar}}$ belongs to the plane $\tilde{x}\tilde{y}$, the projection along the LOS of the distance between the ith scatterer and the MC is given by

$$\langle r_i, n \rangle = -\tilde{x}_i \cos(\alpha) - \tilde{y}_i \sin(\alpha) \tag{7.8}$$

where $\tilde{x}_i$ and $\tilde{y}_i$ are the $(\tilde{x}, \tilde{y})$-coordinates of the ith scatterer, and the aspect angle α is the angle between the target symmetry axis and the LOS direction. The latter is evaluated as

$$\alpha = \alpha(t) = \cos^{-1}\left(\langle \hat{z}_t, n \rangle\right) \tag{7.9}$$

with $\|\hat{z}_t\| = \|n\| = 1$. Therefore, the received complex signal can be written as

$$s_{rx}(t) = e^{j\frac{4\pi}{\lambda}t} e^{j\phi_{\mathrm{MC}}} e^{j\phi_{bD}(t)} \sum_{i=1}^{N_p} \sqrt{\sigma_i(t)} e^{j\phi_i} \tag{7.10}$$

with ϕ_{MC} and $\phi_{bD}(t)$ the phase rotations due to the initial MC range and bulk motion, given by

$$\phi_{\mathrm{MC}} = -j\frac{4\pi}{\lambda} \|r_{\mathrm{MC}}^{\mathrm{radar}}\| \tag{7.11}$$

$$\phi_{bD}(t) = -j\frac{4\pi}{\lambda} \langle v, n \rangle t \tag{7.12}$$

and where ϕ_i is the phase of the complex coefficient of the ith scatterer, given by

$$\phi_i = j\frac{4\pi}{\lambda} \left(\check{x}_i \cos(\alpha) + \check{y}_i \sin(\alpha)\right) \tag{7.13}$$

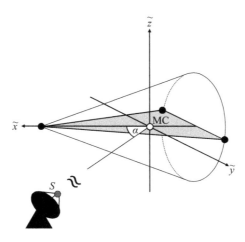

Figure 7.3　Representations of the reference system $(\tilde{x}, \tilde{y}, \tilde{z})$

   The direction of LOS can be expressed in terms of elevation and azimuth angle with respect to the angular velocity vector of warhead conical rotation or decoy tumbling. Let us consider the coordinate system $(\hat{u}, \hat{v}, \hat{w})$ shown in Figure 7.4 such that the axis $\hat{u}$ corresponds to the unit angular velocity vector of conical rotation $\hat{\omega}_c$, and the plane $\hat{u}\hat{v}$ contains the symmetry axis of the object in the initial observation time instant $\hat{z}_0$. The vectors $\hat{z}_t$ and n in the new coordinate system can be written as

$$n = [\cos(\angle\text{El})\cos(\angle\text{Az}), \cos(\angle\text{El})\sin(\angle\text{Az}), \sin(\angle\text{El})]^T \tag{7.14}$$

$$\hat{z}_t = [\cos(\Theta+\Delta\Theta), \sin(\Theta+\Delta\Theta)\sin(\Omega_c t), \sin(\Theta+\Delta\Theta)\cos(\Omega_c t)]^T \tag{7.15}$$

with $\angle\text{El}$ and $\angle\text{Az}$ representing the elevation and azimuth angles relative to the radar position, respectively. Substituting (7.14) and (7.15) into (7.9), the aspect angle can be written as

$$\begin{aligned} \alpha(t) = \cos^{-1}(\,&\cos(\angle\text{El})\cos(\angle\text{Az})\cos(\Theta+\Delta\Theta)\\ &+\cos(\angle\text{El})\sin(\angle\text{Az})\sin(\Theta+\Delta\Theta)\sin(\Omega_c t)\\ &+\sin(\angle\text{El})\sin(\Theta+\Delta\Theta)\cos(\Omega_c t)) \end{aligned} \tag{7.16}$$

It is worth noting that in case of a tumbling decoy, since the angular velocity vector is perpendicular to the symmetry axis of the target, the aspect angle can be calculated by (7.16) by applying the following equivalences:

$$\Theta = \frac{\pi}{2} \tag{7.17}$$

$$\Delta\Theta = 0 \tag{7.18}$$

$$\Omega_c = \Omega_r = \|\omega_r\| \tag{7.19}$$

In this analysis, two possible shapes are considered for the warhead, which are namely cone and cone with triangular fins, while three shapes are considered for the decoy, namely cone, cylinder and sphere. The number of scattering points located on the target depends on the considered shape. In particular, they are generally located in proximity of the edges of the target section obtained by the intersection between the target volume and the incident plane, defined as the plane containing both the symmetry axis and the LOS.

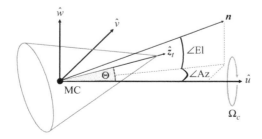

Figure 7.4   Representations of the reference system $(\hat{u}, \hat{v}, \hat{w})$

For simplicity, in this analysis it is assumed that the phase rotations due to target range and bulk motion are compensated, and the singular scattering properties of each scatterer are not taken into account, considering the modulus of scattering coefficients as a binary function whose possible values are $\{0, 1\}$. Specifically, this function depends on the aspect angle $\alpha(t)$, which is 1 when there is an LOS for the scattering points, and 0 otherwise. Finally, it is assumed that the radar resolution allows to distinguish different targets in range and azimuth so that the return from different targets can be processed distinctly. These simplifying approximations allow us to focus the analysis on the μD components that characterise the return from a target.

7.2.1 Cone

For the conical targets, three principal scattering points are considered: the first is in correspondence of the cone tip; the other two points are located on the intersection between the circumference at cone bottom and the incident plane $(\tilde{x}\tilde{y})$, as shown in Figure 7.7(a). The coordinates of the three points into system $(\tilde{x}, \tilde{y}, \tilde{z})$ are

$$
\begin{aligned}
P_1(\tilde{x}, \tilde{y}, \tilde{z}) &= (h_1, & 0, & \quad 0) \\
P_2(\tilde{x}, \tilde{y}, \tilde{z}) &= (-h_2, & R_b, & \quad 0) \\
P_3(\tilde{x}, \tilde{y}, \tilde{z}) &= (-h_2, & -R_b, & \quad 0)
\end{aligned}
\tag{7.20}
$$

where h_1 and h_2 are the distance of the cone tip and centre of the cone base with respect to the MC, and R_b is the base radius.

Let us consider the possible variations of $\alpha(t)$ in the interval $[0, \pi]$. For the cone, $\sqrt{\sigma_i}$ is 0 for P_1 when $\alpha(t) \in [\pi - \gamma, \pi]$ and for P_3 when $\alpha(t) \in [\gamma, \pi/2]$, with γ the semi-angle of the cone; while for P_2 the occlusion never occurs for $\alpha(t) \in [0, \pi]$. Then $\sqrt{\sigma_2} = 1$ with $\alpha(t) \in [0, \pi]$. The values of the coefficients as a function of the aspect angle for the cone scatterers are summarised in Table 7.1.

7.2.1.1 Cone plus fins

For the cone with fins, a scattering point in correspondence with the tip of each triangular fin is considered in addition to the main three scatterers described above.

Table 7.1   *Scattering coefficient magnitude of the three principal scatterers P_1, P_2 and P_3 of the cone, with respect to the aspect angles α*

	$\sqrt{\sigma_1}(\alpha)$	$\sqrt{\sigma_2}(\alpha)$	$\sqrt{\sigma_3}(\alpha)$
$\alpha < \gamma$	1	1	1
$\gamma \le \alpha < \frac{\pi}{2} - \gamma$	1	1	0
$\frac{\pi}{2} - \gamma \le \alpha < \frac{\pi}{2}$	1	1	0
$\frac{\pi}{2} \le \alpha < \pi - \gamma$	1	1	1
$\pi - \gamma \le \alpha \le \pi$	0	1	1

When tips of the fins move on the plane containing the cone base, the coordinates into $(\tilde{x}, \tilde{y}, \tilde{z})$ system are given by

$$P_{\text{fin}_i}(\tilde{x}, \tilde{y}, \tilde{z}) = (-h_2, \quad R_b + H_{\text{fin}} \cos(\varpi_i), \quad R_b + H_{\text{fin}} \sin(\varpi_i)) \tag{7.21}$$

with $i = 1, \ldots, N_{\text{fin}}$, where N_{fin} is the number of fins, H_{fin} the fin height and ϖ_i is the angle between the ith fin and $\tilde{y}$ axis given by

$$\varpi_i = \Omega_s t + \Omega_{s_0} + \frac{2\pi i}{N_{\text{fin}}} \tag{7.22}$$

with Ω_{s_0} representing initial phase of warhead spinning.

The occlusion function of the fin tips depends not only on the aspect angle α, but also on the spinning of the cone as it can cause the fins to be occluded behind the warhead body. In order to evaluate the occlusion function for the fins, the physical optics approximation is considered. This is a valid approximation given the relatively high frequency at which the radar system operates. Since the targets of interest are within the Fraunhofer zone [12], the rays that strike the targets can be considered as parallel. The occlusion of fins occur for $\alpha(t) \in [\gamma_{\text{fin}}, \pi/2]$, where γ_{fin} is the semi-angle of an isosceles triangle whose height is equal to the height of the cone, and the base is equal to the diameter of circumference drawn by rotating fins. Therefore, the coefficient $\sqrt{\sigma_{\text{fin}_i}}$ is 1 when $\alpha(t) \in [0, \gamma_{\text{fin}}]$ and for $\alpha(t) > \pi/2$. The value of the scattering coefficient for $\alpha(t) \in [\gamma_{\text{fin}}, \pi/2]$ is calculated by comparing the $\tilde{z}$-coordinate of P_{fin_i} with a suitable threshold as follows:

$$\sqrt{\sigma_{\text{fin}_i}}(t) = \begin{cases} 1 & \text{if } \tilde{z}_{\text{fin}_i}(t) < \mathscr{X} \\ 0 & \text{if } \tilde{z}_{\text{fin}_i}(t) \geq \mathscr{X} \end{cases} \tag{7.23}$$

where $\mathscr{X} = \mathscr{X}(\alpha)$ is the threshold, which depends on the aspect angle, hence on time. In order to evaluate $\mathscr{X}$, it is necessary to calculate when the straight line joining the radar and tip of the fin becomes tangential to the cone surface, as represented in Figure 7.5. Considering the reference system $(\tilde{x}_0, \tilde{y}_0, \tilde{z}_0)$, obtained moving the origin of system $(\tilde{x}, \tilde{y}, \tilde{z})$ into the centre of the cone bottom as shown in Figure 7.5, the position vectors of the generic fin tip $\overline{OF}$, and of the radar $\overline{OS}$ are

$$\overline{OF} = [0, (R_b + H_{\text{fin}}) \cos(\varpi_i), (R_b + H_{\text{fin}}) \sin(\varpi)]^T$$
$$\overline{OS} = [D' \cos(\alpha), -D' \sin(\alpha), 0]^T \tag{7.24}$$

with ϖ is being the angle of the fin tip with respect to the axis $\tilde{y}_0$, and where

$$D' \simeq D + h_2 \cos(\alpha) \tag{7.25}$$

with $D = \|r_{MC}^{\text{radar}}\|$ representing the distance between the radar and the MC. The conical surface is represented by the function:

$$f(\tilde{x}_0, \tilde{y}_0, \tilde{z}_0) = \tilde{R}^2 - (\tilde{y}_0^2 + \tilde{z}_0^2) = R_b^2 \left(1 - \frac{\tilde{x}_0}{H}\right)^2 - (\tilde{y}_0^2 + \tilde{z}_0^2) \tag{7.26}$$

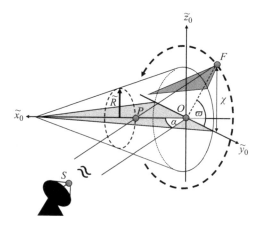

Figure 7.5   Representations of the reference system $(\tilde{x}_0, \tilde{y}_0, \tilde{z}_0)$

where $\tilde{R} = \tilde{R}(\tilde{x}_0)$ is the radius of the generic cone section given by

$$\tilde{R}(\tilde{x}_0) = R_b \left(1 - \frac{\tilde{x}_0}{H} \right) \tag{7.27}$$

where $H = h_1 + h_2$ is the cone height. Considering the generic point of the cone, P, whose position vector is

$$\overline{OP} = \left[H \left(1 - \frac{\tilde{R}}{R_b} \right), \tilde{R} \cos(\xi), \tilde{R} \sin(\xi) \right]^T \tag{7.28}$$

where ξ is the position angle with respect to $\tilde{y}_0$ axis, the lines from P to F and S are, respectively,

$$\overline{PF} = \overline{OP} - \overline{OF} = \begin{bmatrix} H \left(1 - \dfrac{\tilde{R}}{R_b} \right) \\ \tilde{R} \cos(\xi) - (R_b + H_{\text{fin}}) \cos(\varpi) \\ \tilde{R} \sin(\xi) - (R_b + H_{\text{fin}}) \sin(\varpi) \end{bmatrix} \tag{7.29}$$

$$\overline{PS} = \overline{OP} - \overline{OS} = \begin{bmatrix} H \left(1 - \dfrac{\tilde{R}}{R_b} \right) - D' \cos(\alpha) \\ \tilde{R} \cos(\xi) + D' \sin(\alpha) \\ \tilde{R} \sin(\xi) \end{bmatrix} \tag{7.30}$$

In order to evaluate the occlusion threshold, it is necessary to evaluate the angle ϖ and ξ such that $\overline{PF}$ and $\overline{PS}$ are both tangent to the conical surface, as follows:

$$\begin{cases} \left[\dfrac{\partial f}{\partial \tilde{x}_0}, \dfrac{\partial f}{\partial \tilde{y}_0}, \dfrac{\partial f}{\partial \tilde{z}_0} \right]^T \cdot \overline{PF} = 0 \\[3mm] \left[\dfrac{\partial f}{\partial \tilde{x}_0}, \dfrac{\partial f}{\partial \tilde{y}_0}, \dfrac{\partial f}{\partial \tilde{z}_0} \right]^T \cdot \overline{PS} = 0 \end{cases} \tag{7.31}$$

where the components of gradient vector for a generic cone point are evaluated from (7.26) as

$$\frac{\partial f}{\partial \tilde{x}_0} = \frac{-2R_b^2}{H} \left(1 - \frac{\tilde{x}_0}{H} \right) = \frac{-2R_b\tilde{R}}{H}$$

$$\frac{\partial f}{\partial \tilde{y}_0} = -2y_{f_0} = -2\tilde{R}\cos(\xi) \tag{7.32}$$

$$\frac{\partial f}{\partial \tilde{z}_0} = -2z_{f_0} = -2\tilde{R}\sin(\xi)$$

with

$$\tilde{x}_0 = H\left(1 - \frac{\tilde{R}}{R_b} \right); \qquad \tilde{y}_0 = \tilde{R}\cos(\xi); \qquad \tilde{z}_0 = \tilde{R}\sin(\xi) \tag{7.33}$$

From (7.31) and (7.32) follows

$$\begin{cases} (-2\tilde{R}) \quad [R_b - (R_b + H_{\mathrm{fin}})\cos(\xi - \varpi)] = 0 \\[3mm] (-2\tilde{R}) \quad \left[D'\sin(\alpha)\cos(\xi) + R_b - \dfrac{R_b D'\cos(\alpha)}{H} \right] = 0 \end{cases} \tag{7.34}$$

which leads to

$$\begin{cases} \cos(\xi - \varpi) = \dfrac{R_b}{R_b + H_{\mathrm{fin}}} \\[3mm] \cos(\xi) = \left[\dfrac{D\cos(\alpha)R_b}{H} - R_b \right] \dfrac{1}{D\sin(\alpha)} = \left[\dfrac{\tan(\gamma)}{\tan(\alpha)} - \dfrac{R_b}{D\sin(\alpha)} \right] \end{cases} \quad \forall \tilde{R} > 0 \quad (7.35)$$

Finally, the threshold is given by

$$\mathscr{X} = (H_{\mathrm{fin}} + R_b)\cos(\varpi^*) \tag{7.36}$$

where

$$\varpi^* = \cos^{-1}\left[\frac{\tan(\gamma)}{\tan(\alpha)} - \frac{R_b}{D\sin(\alpha)} \right] - \cos^{-1}\left[\frac{R_b}{R_b + H_{\mathrm{fin}}} \right] \tag{7.37}$$

When the radar LOS is perpendicular to the target symmetry axis, a particular assumption for the threshold is taken into consideration when the height of the fin is such that

$$\varpi^*(\alpha, H_{\mathrm{fin}}) > \frac{\pi}{N_{\mathrm{fin}}}, \qquad \text{with } \alpha = \frac{\pi}{2} \tag{7.38}$$

In this case, the $\mathscr{X}$ for $\alpha = \pi/2$ is given by

$$\mathscr{X} = (H_{\text{fin}} + R_b) \cos\left(\frac{\pi}{N_{\text{fin}}}\right) \tag{7.39}$$

Figure 7.6 shows how the threshold varies as a function of aspect angle for the cone dimensions H and R_b of 1 and 0.375 m, respectively, fin height $H_{\text{fin}} = 0.200$ m and at a distance of 150 km.

It is worth noting, that in the evaluation of the occlusion of fins tips, the effect of the fin area is not taken into account for simplicity.

7.2.2 Cylinder

The cylindrical target is represented by four principal scattering points shown in Figure 7.7(b), specifically two for each base, taken by intersecting the circumferences at the bases and the incident plane, such that

$$
\begin{aligned}
P_1(\tilde{x}, \tilde{y}, \tilde{z}) &= (\tfrac{H}{2}, & R_b, & \quad 0) \\
P_2(\tilde{x}, \tilde{y}, \tilde{z}) &= (-\tfrac{H}{2}, & R_b, & \quad 0) \\
P_3(\tilde{x}, \tilde{y}, \tilde{z}) &= (-\tfrac{H}{2}, & -R_b, & \quad 0) \\
P_4(\tilde{x}, \tilde{y}, \tilde{z}) &= (\tfrac{H}{2}, & -R_b, & \quad 0)
\end{aligned}
\tag{7.40}
$$

Table 7.2 presents the values of the amplitude of the coefficients for the cylinder scatterers as a function of the aspect angle. In particular, $\sqrt{\sigma_i} = 0$ for P_1 when $\alpha(t) = \pi$; for P_2 when $\alpha(t) = 0$; for P_3 when $\alpha(t) \in [0, \pi/2]$; and for P_4 when $\alpha(t) \in [\pi/2, \pi]$.

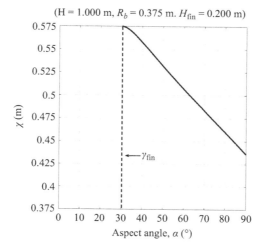

Figure 7.6   Example of threshold values $\tilde{x}$ as a function of aspect angle (α)

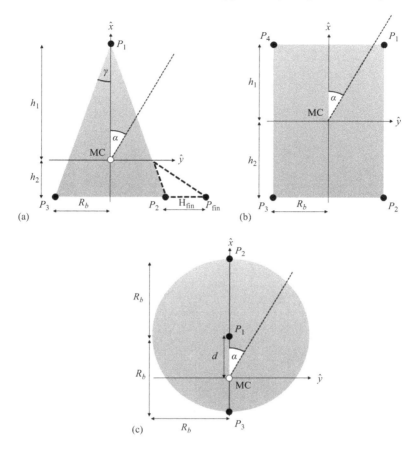

Figure 7.7 Target shape models: (a) conical target; (b) cylindrical target; and (c) spherical target

7.2.3 Sphere

Due to its symmetry among all directions, a tumbling sphere, with rotation centre coinciding with sphere centre, does not lead to variation of relative radar view of its shape, hence no μD information can be extracted. In this analysis, the spherical decoy has a displacement d between the tumbling rotation centre and the sphere centre. Three scattering points are considered, as shown in Figure 7.7(c): two on the spherical surface along the symmetry axis, and one corresponding to the sphere centre. The coordinates of the three points in the coordinate system $(\tilde{x}, \tilde{y}, \tilde{z})$ are

$$
\begin{aligned}
P_1(\tilde{x}, \tilde{y}, \tilde{z}) &= (d & 0, & \ 0) \\
P_2(\tilde{x}, \tilde{y}, \tilde{z}) &= (d + R_b, & 0, & \ 0) \\
P_3(\tilde{x}, \tilde{y}, \tilde{z}) &= (d - R_b, & 0, & \ 0)
\end{aligned}
\tag{7.41}
$$

where R_b, is the sphere radius.

Table 7.2   Scattering coefficient magnitude of the four principal scatterers P_1, P_2, P_3 and P_4 of the cylinder, with respect to the aspect angle α

	$\sqrt{\sigma_1}(\alpha)$	$\sqrt{\sigma_2}(\alpha)$	$\sqrt{\sigma_3}(\alpha)$	$\sqrt{\sigma_4}(\alpha)$
$\alpha = 0$	1	0	0	1
$0 < \alpha < \frac{\pi}{2}$	1	1	0	1
$\alpha = \frac{\pi}{2}$	1	1	0	0
$\frac{\pi}{2} < \alpha < \pi$	1	1	1	0
$\alpha = \pi$	0	1	1	0

For the sphere, the occlusion never occurs for P_1, such that $\sqrt{\sigma_1} = 1$, $\forall \alpha$. For the scatterer P_2, $\sqrt{\sigma_2} = 0$ when $\alpha(t) \in \,]\pi/2, \pi]$, and for P_3 $\sqrt{\sigma_3} = 0$ when $\alpha(t) \in \,]0, \pi/2[$. In Table 7.3, the values of the coefficient amplitudes for the sphere scatterers as a function of the aspect angle are summarised.

7.3 Laboratory experiment

In this section, the experiment conducted for the evaluation of μD effect due to BT micro-motions is described. The radar measurements containing the target motion information are acquired from scaled replicas of potential BTs with a 24 GHz continuous-wave (CW) coherent radar, which transmits in vertical polarisation. The different target micro-motions are emulated by using a robotic manipulator and an additional rotational motor. A representation of configuration for the μD experiment is shown in Figure 7.8. The robotic arm is used for introducing the conical rotation and the nutation for the warhead target, while the additional rotor is used for simulating the warheads spinning and the decoy tumbling. Specifically, the conical rotation is simulated with revolutions of the end effector of robotic arm, while the nutation is introduced by an sinusoidal oscillation of the radius of circumference where the end factor moves, by varying the angles of the robotic arm joints. The additional rotor

Table 7.3   Scattering coefficient magnitude of the three principal scatterers P_1, P_2 and P_3 of the sphere, with respect to the aspect angles α.

	$\sqrt{\sigma_1}(\alpha)$	$\sqrt{\sigma_2}(\alpha)$	$\sqrt{\sigma_3}(\alpha)$
$\alpha = 0$	1	1	0
$0 < \alpha < \frac{\pi}{2}$	1	1	0
$\alpha = \frac{\pi}{2}$	1	1	1
$\frac{\pi}{2} < \alpha < \pi$	1	0	1
$\alpha = \pi$	1	0	1

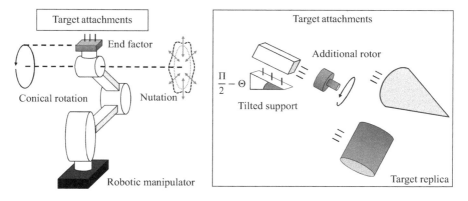

Figure 7.8 Experiment configuration

Figure 7.9 Experiment set-up [15]

is attached to the end factor of the robotic arm by a support which depends on the movement. Specifically, the rotor support is tilted of an angle equal to $\pi/2 - \Theta$ in case of processing warhead, and the length of the support is set such that the target MC results stationary with respect to the conical rotation. The decoy tumbling is simulated by using only the rotor. It has to be underlined that the trajectory of BTs is not taken into account in the experiment considering that the principal movement of the object is compensated. In this way, the classification is based only on the micro-motions of targets of interest.

The aim of the experiment is to simulate an S-Band BMD radar system, with a carrier frequency of 3.3 GHz. The targets are divided in two classes, which are *warhead* and *decoy*. Moreover, both of them are divided in sub-classes, which are associated to a particular type of target. Specifically, the *warhead* class is composed of two sub-classes: cone and cone with triangular fins at the base, which are replicas of warhead without and with fins, respectively. *Decoy* class, in contrast, is divided into three sub-classes: sphere, cone and cylinder. The considered conical warhead has a radius, R_b, of 0.375 m and a height, H, of 1 m, while the fin's height, H_{fin}, is 0.20 m. The sizes of the decoys are usually comparable with the dimensions of the warheads in order to confuse the anti-missile radar system. Therefore, both the cylindrical and conical objects are chosen to have a radius and a height equal to 0.375 and 1 m, respectively, while the sphere diameter is 1 m. However, since the CW radar used in the laboratory has a carrier frequency of 24 GHz, the actual dimensions of the replicas of the targets used to acquire the real data are scaled by a factor of 0.1375. The precession angle chosen for both the types of warhead is 10° while the precession frequency is 0.25 Hz; the angular velocity of spinning and nutation are 1 and 5 Hz, respectively. For the decoys, the tumbling rate is 1 Hz.

The database for experimental data contains for each of five sub-classes, five acquisitions of 20 s for each of the possible nine pairs of azimuth and elevation angles formed using three values for both of them, namely [0°, 45°, 90°]. Hence, the database is composed of 45 acquisitions of 20 s for each sub-class.

Figure 7.10 shows examples of spectrograms (in the dB scale) from a conical warhead and a warhead with fins, obtained by using both simulated and real data for different pairs of (∠El, ∠Az). In this analysis, the spectrogram is defined as the magnitude of the short-time Fourier transform (STFT) of received signal, in order to reduce the differences between the weights of μD components within the target echo due to different scattering properties of each target scatterer [15]. As observed in Figure 7.10(a), the spectrogram from simulated data allows the observations of echo modulations due to the conical rotation and nutation, but not from the spinning, due to the assumption of perfect symmetric conical target. Although the spinning of a target perfectly symmetric with respect to the rotation axis does not introduce any μD shift, from Figure 7.10(c) it is worth noting that some flashes into the spectrogram from the acquired data appear due to the spinning since the target replicas wrapped with a metallic material used in the laboratory experiment ware not perfectly symmetric. From Figure 7.10(b) and (d), it is possible to note that the two spectrograms show the same trend, where the precession leads to a modulation of the maximum Doppler which is due to the fins rotation. It is pointed out that the main differences between the simulated and the real cases for the analysed cases are due to the fact that in the presented signal model the radar cross section (RCS) of each scatterer is not taken into account (hence, even the dependence on signal polarisation), and the initial phase of different micro-motions exhibited by warheads is random in both simulated and acquired signal.

Figure 7.11 represents examples of spectrogram (in the dB scale) from the three different shapes considered for the decoys, from both simulated echo and laboratory acquisitions. Even in these cases, it is worth noting that the spectrograms from both simulated and acquired data show the same trends, in terms of spectrogram periodicity and maximum Doppler shifts from main scatterers of tumbling objects. However, the absence of an RCS and overall electromagnetic (EM) model in the simulated data leads to some differences, as mentioned above. Additional details on the laboratory experiment set-up and target replicas will be described in Section 7.5.

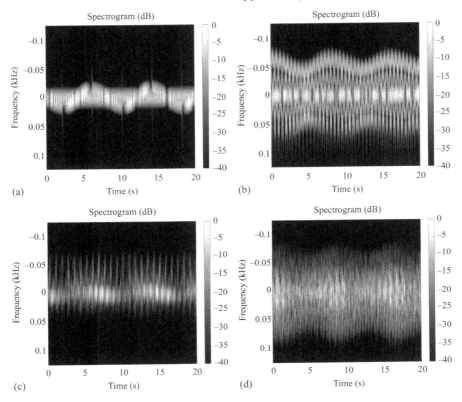

Figure 7.10 *Example of spectrograms obtained by a received signal from a warhead for different pairs of (∠El, ∠Az): (a) simulated echo from conical warhead for (90°,0°); (b) simulated echo from conical warhead with fins for (0°,45°); (c) laboratory acquisition from conical warhead (90°,0°); and (d) laboratory acquisition from conical warhead with fins for (0°,45°)*

7.4 Classification framework

In this section, the algorithm proposed in [14] to extract μD-based features for the target classification is reviewed, introducing the different feature extraction approaches proposed in this chapter. Figure 7.12 shows a block diagram of the classification method, outlining the common steps used for four different feature extraction approaches. The first approach is based on the statistical characteristics of the unit area function obtained by averaging and normalising the CVD; the second and the third approaches are based on the computation of pZ moments and Kr moments of the CVD, respectively; the fourth one is based on filtering the CVD with a bank of 2D Gabor filters. The starting point of the proposed algorithm is the received signal $s_{rx}(n)$, with $n = 0, \ldots, N_s$, containing μD components and comprising of N_s signal

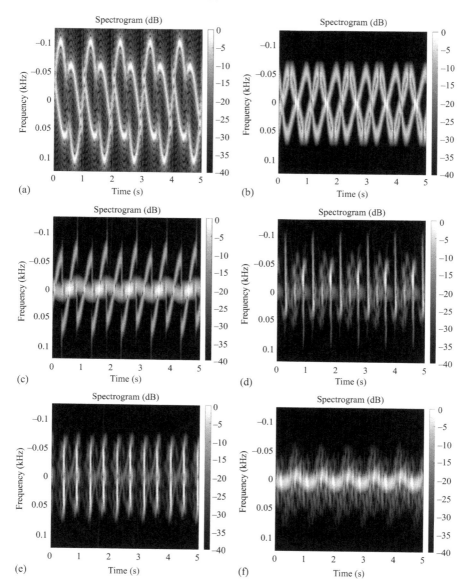

Figure 7.11   *Example of spectrograms obtained by a received signal from a decoy for different pairs of (∠El, ∠Az): (a) simulated echo from conical decoy for (90°,0°); (b) simulated echo from cylindrical decoy for (45°,0°); (c) simulated echo from spherical decoy for (0°,90°); (d) laboratory acquisition from conical decoy for (90°,0°); (e) laboratory acquisition from cylindrical decoy for (45°,0°); and (f) laboratory acquisition from spherical decoy for (0°,90°)*

Figure 7.12   Block diagram of the proposed algorithm

samples. The received signal has to be pre-processed before the evaluation of the μD signature. The first block includes a notch filtering, down-sampling and normalisation. Specifically, the notch filter is applied to filter out the signal component from the stationary parts of the target; the down-sampling is performed according to the maximum μD shift expected, in order to reduce the computational load by reducing the signal samples; the normalisation is performed in order to have an input signal with a maximum amplitude equal to 1, so that the extracted features are not biased by the received signal power. The second step is the spectrogram computation of the pre-processed signal $\tilde{s}_{rx}(n)$:

$$S_{|\text{STFT}|}(v, m) = \left| \sum_{n=0}^{N_s-1} \tilde{s}_{rx}(n)w(n-m) \exp\left(-j2\pi v \frac{n}{N_s}\right) \right| \tag{7.42}$$

with $m = 0, \ldots, N_c - 1$, where v is the normalised frequency and $w(\cdot)$ is the smoothing window.

As depicted in Figure 7.12, the next step consists in the extraction of the CVD, that is defined as the Fourier transform of the spectrogram along each frequency bin:

$$S_{\text{CVD}}(v, \varepsilon) = \left| \sum_{m=0}^{N_c-1} S_{|\text{STFT}|}(v, m) \exp\left(-j2\pi \varepsilon \frac{m}{N_c}\right) \right| \tag{7.43}$$

where ε is known as the cadence frequency. The CVD is chosen because it offers the possibility of using, as discriminants, the cadence of each frequency component within the signal and the maximum Doppler shift, and because the CVD is more robust than the spectrogram since it does not depend on the initial phase of moving objects. Figure 7.13(a) and (b) show the CVDs obtained by processing the spectrograms in Figures 7.10(c) and 7.11(d), and resulting from the laboratory acquisitions of signal scattered by the conical warhead and decoy, respectively. It is worth noting that the zero cadence component in the CVDs is filtered out, by setting the corresponding column in the CVD matrix to zero. Finally, the CVD has to be processed to extract a Q-dimensional feature vector $F = [F_0, F_1, \ldots, F_{Q-1}]$, which can identify each class. Before classification, the vector F is normalised as follows:

$$\tilde{F} = \frac{F - \varsigma_F}{\zeta_F} \tag{7.44}$$

where ς_F and ζ_F are the statistical mean and standard deviation of the vector F, respectively.

The feature extraction block of Figure 7.12 for the four different approaches will be described in the following subsections.

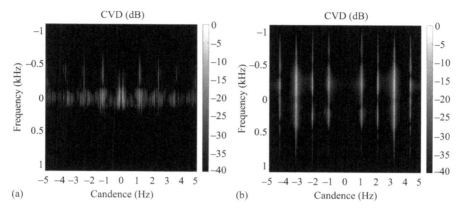

Figure 7.13   *CVDs obtained from processing the spectrograms from the conical warhead in Figure 7.10c (a) and from the conical decoy in Figure 7.11d (b)*

7.4.1   Feature vector extraction

In this analysis, four different feature vectors are considered for performing the target classification.

7.4.1.1   ACVD-based feature vector approach

In the ACVD-based approach, six features are computed from the ACVD. The starting point is the mean of the CVD along each cadence bin; the resulting 1D function is then normalised to have a unit area. From the obtained 1D signature $\Lambda_{\text{avg}}(\varepsilon)$, $\varepsilon = 0, \ldots, N_c - 1$, where N_c is the number of cadence bins, three statistical indices are extracted:

- Standard deviation:

$$F_0 = \sqrt{\frac{1}{N_c - 1} \sum_{\varepsilon=0}^{N_c-1} \left[\Lambda_{\text{avg}}(\varepsilon) - \frac{1}{N_c} \sum_{\varepsilon=0}^{N_c-1} \Lambda_{\text{avg}}(\varepsilon) \right]^2} \qquad (7.45)$$

- Kurtosis:

$$F_1 = \frac{\frac{1}{N_c} \sum_{\varepsilon=0}^{N_c-1} \left[\Lambda_{\text{avg}}(\varepsilon) - \frac{1}{N_c} \sum_{\varepsilon=0}^{N_c-1} \Lambda_{\text{avg}}(\varepsilon) \right]^4}{\left(\sqrt{\frac{1}{N_c-1} \sum_{\varepsilon=0}^{N_c-1} \left[\Lambda_{\text{avg}}(\varepsilon) - \frac{1}{N_c} \sum_{\varepsilon=0}^{N_c-1} \Lambda_{\text{avg}}(\varepsilon) \right]^2} \right)^4} - 3 \qquad (7.46)$$

- Skewness:

$$F_2 = \frac{\frac{1}{N_c} \sum_{\varepsilon=0}^{N_c-1} \left[\Lambda_{\mathrm{avg}}(\varepsilon) - \frac{1}{N_c} \sum_{\varepsilon=0}^{N_c-1} \Lambda_{\mathrm{avg}}(\varepsilon) \right]^3}{\left(\sqrt{\frac{1}{N_c-1} \sum_{\varepsilon=0}^{N_c-1} \left[\Lambda_{\mathrm{avg}}(\varepsilon) - \frac{1}{N_c} \sum_{\varepsilon=0}^{N_c-1} \Lambda_{\mathrm{avg}}(\varepsilon) \right]^2} \right)^3} \tag{7.47}$$

Three other indices, specifically the Peak Sidelobe Level (PSL) ratio and two different definitions of the Integrated Sidelobe Level (ISL) ratio, are computed from the normalised autocorrelation of the sequence $\Lambda_{\mathrm{avg}}(\varepsilon)$, $C_{\Lambda_{\mathrm{avg}}}(m)$, $m = 0, \ldots, M-1$. Specifically,

$$F_3 = \mathrm{PSL} = \max_m \frac{\left| C_{\Lambda_{\mathrm{avg}}}(m) \right|}{\left| C_{\Lambda_{\mathrm{avg}}}(0) \right|} \tag{7.48}$$

while the latter are

$$F_4 = \mathrm{ISL}_1 = \frac{\sum_{m=1}^{M-1} \left| C_{\Lambda_{\mathrm{avg}}}(m) \right|}{\left| C_{\Lambda_{\mathrm{avg}}}(0) \right|} \tag{7.49}$$

and

$$F_5 = \mathrm{ISL}_2 = \frac{\sum_{m=1}^{M-1} \left| C_{\Lambda_{\mathrm{avg}}}(m) \right|^2}{\left| C_{\Lambda_{\mathrm{avg}}}(0) \right|} \tag{7.50}$$

respectively. Hence, the final feature vector extracted is $F^{\mathrm{avg}} = [F_0, \ldots, F_5]$.

7.4.1.2 2D Signature-based feature vectors

For the other three feature extraction methods, the CVD is considered as a 2D image representing the target signature. First, the magnitude of the CVD is normalised to obtain a matrix whose values belong to the set $[0, 1]$ as follows:

$$\Lambda_{\mathrm{CVD}}(\nu, \varepsilon) = \frac{S_{\mathrm{CVD}}(\nu, \varepsilon) - \min_{\nu,\varepsilon} S_{\mathrm{CVD}}(\nu, \varepsilon)}{\max_{\nu,\varepsilon} \left[S_{\mathrm{CVD}}(\nu, \varepsilon) - \min_{\nu,\varepsilon} S_{\mathrm{CVD}}(\nu, \varepsilon) \right]} \tag{7.51}$$

The different feature vectors extracted are based on pZ moments, Kr moments and Gabor filtering, respectively.

pZ Moments-based feature vector approach Introduced in [16], the pZ moments of an image $I(x, y)$ are geometric moments computed as the projection of the image on a basis of 2D polynomials which are defined on the unit circle. Specifically, the pZ moments of order r and repetition l are calculated as follows:

$$\zeta_{r,l} = \frac{r+1}{\pi} \int_0^{2\pi} \int_0^1 W_{r,l}^*(\rho, \theta) I(\rho \cos\theta, \rho \sin\theta) \, \rho \, \mathrm{d}\rho \, \mathrm{d}\theta \tag{7.52}$$

with $r \geq |l|$, and where the polynomial $W_{l,r}$ can be written as

$$W_{l,r}(x, y, \rho) = W_{r,l}(\rho, \theta) = \sum_{h=0}^{r-|l|} \frac{\rho^{r-h}(-1)^h (2r+1-h)!}{h! \, (r+|l|+1-h)! \, (r-|l|-h)!} e^{jl\theta} \tag{7.53}$$

where $x = \rho \cos \theta$ and $y = \rho \sin \theta$, with $\rho \le 1$.

In order to compute the pZ moments, the support of the spectrogram, hence that of the CVD, has to be chosen to be a unit square so that it can be inscribed in the unit circle [14]. A feature vector F^{pZ} is then extracted, whose qth element F_q^{pZ} is the pZ moment $\zeta_{r,l}$ of order r and repetition l calculated from the magnitude of the CVD by (7.52), with $r = l = 0, \dots, \mathcal{K}_1 - 1$ and $q = 0, \dots, (\mathcal{K}_1 + 1)^2 - 1$, where $\mathcal{K}_1$ is the maximum value considered for both the moments order and the repetition.

Kr moments-based feature vector approach The Kr moments of order r of an image $I(x, y)$, introduced in [17], are computed as the projection of the image on a basis of orthogonal polynomials which are associated with the binomial distribution. These are calculated as the product of the classical Kr polynomials, K_r, and a weight factor to overcome the numerical stability problem as follows [17,18]:

$$\overline{K}_r(x, p, \mathcal{N}) = K_r(x, p, \mathcal{N}) \sqrt{\frac{w(x; p, \mathcal{N})}{\rho(n; p, \mathcal{N})}}$$

$$= {}_2F_1\left(-n, -x; -\mathcal{N}; \frac{1}{p}\right) \sqrt{\frac{w'(x; p, \mathcal{N})}{w''(n; p, \mathcal{N})}} \tag{7.54}$$

with

$$w'(x; p, \mathcal{N}) = \binom{\mathcal{N}}{x} p^x (1 - p)^{\mathcal{N} - x}$$

$$w''(x; p, \mathcal{N}) = (-1)^{\mathcal{N}} \left(\frac{1 - p}{p}\right)^n \frac{n!}{(-\mathcal{N})_n} \tag{7.55}$$

where x and n belong to $(0, 1, 2, \dots, \mathcal{N})$, $\mathcal{N} \in \mathbb{N}$, with $\mathbb{N}$, representing the set of natural numbers, p, a real number belonging to the set $(0, 1)$. The classical Kr polynomials are defined through the Gauss hypergeometric function ${}_2F_1$, given by

$$_2F_1(a, b; c; z) = \sum_{k=0}^{\infty} \frac{(a)_k (b)_k}{(c)_k} \frac{z^k}{k!} \tag{7.56}$$

with $(a)_k$ the Pochhammer symbol, given by

$$(a)_k = a(a + 1) \dots (a + k - 1) = \frac{\Gamma(a + k)}{\Gamma(a)} \tag{7.57}$$

Considering a 2D image $I(x, y)$, the Kr moments of order (n, m) are defined as [18]:

$$K_{nm} = \sum_{x=0}^{\mathcal{N}_x - 1} \sum_{y=0}^{\mathcal{N}_y - 1} \overline{K}_n(x, p_1, \mathcal{N}_x - 1) \times \overline{K}_m(y, p_2, \mathcal{N}_y - 1) I(x, y) \tag{7.58}$$

where $\mathcal{N}_x$ and $\mathcal{N}_y$ are the image dimensions along both the axes.

In the Kr moments-based approach, a feature vector F^{Kr} is extracted, whose qth element F_q^{Kr} is the Kr moment K_{rl} of order (r, l) calculated from the magnitude of the CVD by (7.58), with $r = l = 0, \dots, \mathcal{K}_2 - 1$ and $q = 0, \dots, (\mathcal{K}_2 + 1)^2 - 1$, where $\mathcal{K}_2$ is the maximum value considered for the moment orders.

2D Gabor filter-based feature vector approach The 2D Gabor function is the product of a complex exponential representing a sinusoidal plane wave and an elliptical Gaussian in any rotation. The filter response in the continuous domain can be normalised to have a compact closed form [19,20]:

$$\rho(x,y) = \frac{f_{cs}^2}{\pi\, \eta_x \eta_y}\, e^{-\left(\frac{f_{cs}^2}{\eta_x^2}x'^2 + \frac{f_{cs}^2}{\eta_y^2}y'^2\right)}\, e^{j2\pi f_{cs}x'} \tag{7.59}$$

with

$$x' = x\cos(\vartheta) + y\sin(\vartheta), \quad y' = -x\sin(\vartheta) + y\cos(\vartheta) \tag{7.60}$$

where f_{cs} is the central spatial frequency of the filter, ϑ is the anti-clockwise rotation of the Gaussian envelope and the sinusoidal plane wave, η_x is the spatial width of the filter along the plane wave and η_y is the spatial width perpendicular to the wave. The sharpness of the filter is controlled on the major and the minor axes by η_x and η_y.

In the 2D Gabor filter-based approach, the resulting matrix $\Lambda_{\text{CVD}}(\nu, \varepsilon)$ is filtered with a bank of Gabor filters whose impulse responses $\rho_{m,l}$ are obtained for various f_l and ϑ_m, with $l = 0, \ldots, L - 1$ and $m = 0, \ldots, M - 1$, where L and M are the number of selected spatial central frequencies and orientation angles, respectively. The choice of the f_l and ϑ_m depends on the specific application and on the worst case image to represent with the moments. The selection of these parameters has to be conducted in order to get an accurate representation of the image under test. In fact, since by varying ϑ_m the harmonic response of the filter moves on a circumference, whose radius is f_l, it is possible to extract local characteristics in the Fourier domain by choosing a set of values for the two parameters. The value of each pixel of the output image is given by the convolution product of the 2D Gabor function and the input image, $\Lambda_{\text{CVD}}(\nu, \varepsilon)$, as

$$\begin{aligned} \Xi_{l,m}(\nu, \varepsilon; f_{cs_l}, \vartheta_m) &= \rho_{l,m}(\nu, \varepsilon; f_l, \vartheta_m) * \Lambda_{\text{CVD}}(\nu, \varepsilon) \\ &= \int_{-\infty}^{\infty}\int_{-\infty}^{\infty} \rho_{l,m}(\nu - \nu_\tau, \varepsilon - \varepsilon_\tau; f_l, \vartheta_m)\Lambda_{\text{CVD}}(\nu_\tau, \varepsilon_\tau)d\nu_\tau d\varepsilon_\tau \end{aligned} \tag{7.61}$$

with $l = 0, \ldots, L - 1$ and $m = 0, \ldots, M - 1$, where L and M are the number of central frequency and orientation angles, respectively. Finally, the outputs of the filters are processed to extract the feature vector used to classify the targets. In particular, a feature is extracted from the output image of each filter by adding up the values of all pixels, as

$$F_q^{\text{GF}} = \sum_{\nu}^{N_\nu - 1}\sum_{\varepsilon}^{N_\varepsilon - 1} |\Xi_{l,m}(\nu, \varepsilon; f_l, \vartheta_m)| \tag{7.62}$$

where $q = mL + l$, with $l = 0, \ldots, L - 1$ and $m = 0, \ldots, M - 1$, N_ν and N_ε are the dimensions of the image Λ_{CVD} along both axes.

7.4.2   Classifier

The classification performance of the extracted feature vectors are evaluated using the K-nearest neighbour (KNN) classifier, modified in order to account for *unknown* class [15]. In particular, let $\mathcal{T}$ be the training vectors set, for each class v an hypersphere $\mathcal{S}_{\mathbf{CM}_v}(\rho_v)$ is considered, with centre $\mathbf{CM}_v$ and radius ρ_v. In the case in which the tested vector does not belong to any hypersphere, it is declared as unknown. The operation mode of this classifier consists of three phases. In the first phase, the set $\mathcal{N}$ of nearest neighbour training vectors to the tested vector $\mathbf{F}$ is selected from $\mathcal{T}$ as follows:

$$\mathcal{N} = \left\{ \tilde{\mathbf{F}}_1, \dots, \tilde{\mathbf{F}}_k : \forall i = 1, \dots, k, \ \left\| \tilde{\mathbf{F}}_i - \mathbf{F} \right\| < \min_{\tilde{\mathbf{F}} \in \{\mathcal{T} - \tilde{\mathbf{F}}_1, \dots, \tilde{\mathbf{F}}_{i-1}\}} \left\| \tilde{\mathbf{F}} - \mathbf{F} \right\| \right\}$$

(7.63)

The second phase consists into definition of vector ι whose elements represent a label for each vector in $\mathcal{N}$. Each label can assume an integer value in the range $[0, V]$, where V is the number of possible classes. The value 0 is assigned when the tested vector does not belong to any hypersphere of the vectors in $\mathcal{N}$, while the values $[1, V]$ correspond to a specific class. Specifically, $\forall i = 1, \dots, k$, the i-label ι_i is updated as follows:

$$\iota_i = \begin{cases} 0 & \left\| \tilde{\mathbf{F}}_i - \mathbf{F} \right\| > \zeta_v \\ v & \text{otherwise} \end{cases}$$

(7.64)

where v is the value corresponding to the belonging class of $\tilde{\mathbf{F}}_i$. Finally, the $(V + 1)$-dimensional score vector $\mathbf{s}$ is evaluated, whose elements are the occurrences, normalised to k, of the integers $[0, \dots, V]$ in the vector ι. The estimation rule then may be implemented as follows:

$$\hat{v} = \begin{cases} \arg\max_v \mathbf{s} & \text{if} \quad \max(\mathbf{s}) > \frac{1}{2} \\ 0 & \text{otherwise} \end{cases}$$

(7.65)

where 0 is the unknown class.

Assuming that the feature vectors of each class are distributed uniformly around their mean vector, the hypersphere radius ρ_v was chosen equal to $\Upsilon_v\sqrt{12}/2$, where $\Upsilon_v = \text{tr}(\mathbf{C}_v)$ and $\mathbf{C}_v$ is the covariance matrix of the training vectors which belong to the class v. The choice is made according to the statistical properties of uniform distributions. In fact, for 1D uniform variables the sum of mean and the product between the standard deviation and the factor $\sqrt{12}/2$ gives the maximum possible value of the distribution.

The choice of a KNN classifier is justified because it is based on the evaluation of the Euclidean distances between the vector under test and the vectors forming the training set of each class in order to estimate the target class. Hence, the classification performance evaluated with KNN classifier are not polarised by the properties of the classifier, and it depends only on the characteristic of features to occupy multidimensional spaces for each class sufficiently separated. However, in general, other classifiers with similar characteristics could be also selected.

7.5 Performance analysis

In this section, the algorithms are tested on laboratory data acquired from replicas of the targets of interest. In order to analyse the performance of the proposed algorithm, three figures of merit are considered, which are the *Probability of correct Classification* ($\mathscr{P}_C$), the *Probability of correct Recognition* ($\mathscr{P}_R$), and the *Probability of Unknown* ($\mathscr{P}_U$). The meaning of classification is the ability to distinguish between the warhead class and the object class, while recognition means the capability to identify the actual shape of the target within the warhead and the object class. Finally, $\mathscr{P}_U$ is computed as the ratio of the number of analysed objects for which the classifier does not make a decision and the total number of analysed objects.

A Monte Carlo approach is used in order to calculate the mean of the three figures of merit over several cases. Specifically, the means are evaluated over 500 different Monte Carlo runs in which all the available signals are divided randomly into training or testing sets with 70% used for training and 30% for testing. The k value of the classifier has to be chosen greater than 1 in order to consider the unknown class; specifically, it is set to 3.

The performance is shown as a function of the signal-to-noise power ratio (SNR) and observation time. Specifically, assuming that the noise for the acquired signals in a controlled environment is negligible, the analysis was conducted by adding white Gaussian noise to the real data. The laboratory acquisitions of 20 s have been split into segments of 10, 5 and 2 s. In addition, before processing, the received signals are down-sampled by a factor of 10, since the sampling frequency of the 24 GHz radar used in laboratory is 22 kHz, which is significantly higher than the expected maximum μD shift (about 100 Hz). In this way, the computational load is reduced, and the μD profile is clear in the spectrogram. Moreover, for the 2D signature-based approaches, the analysis of the performance is conducted on varying the dimension of the feature vector.

The spectrogram is computed using a Hamming window of 200 samples (90.9 ms), with 75% overlap. The number of points for the DFT computation, N_{bin}, is fixed for the ACVD approach, whereas is adaptively evaluated for the pZ and the Gabor filter methods, in order to obtain a square representation of the spectrogram. Specifically, in these cases N_{bin} is given by

$$N_{\text{bin}} = \left\lceil \frac{N_s - W \ overlap}{W \ (1 - overlap)} \right\rceil \tag{7.66}$$

where N_s is the number of signal samples, where W is the dimension of the Hamming windows in terms of time samples, $\lceil \cdot \rceil$ represents the smaller integer greater than or equal to the argument, and *overlap* is the percentage of overlap expressed in the interval [0, 1]. Finally, it is assumed that the effect of the principal translation motion of the targets is compensated before the signals are processed. In this way, the classification is based only on the micro-motions of targets of interest.

7.5.1 ACVD approach

The performance on the experimental data for the ACVD-based method in terms of $\mathscr{P}_C$ and $\mathscr{P}_R$ is shown in Figure 7.14(a), while Figure 7.14(b) shows the $\mathscr{P}_U$.

It is observed that the performance generally improves as the SNR increases, such that both $\mathscr{P}_C$ and $\mathscr{P}_R$ increase. Moreover, the gap between the two figures of merit decreases as both the observation time and the SNR increase. Observing Figure 7.14, it is pointed out $\mathscr{P}_U$ is almost constant for all analysed cases and it is smaller than 0.1.

7.5.2 pZ Moments approach

In the case of the pZ moments-based approach, the dimension of the feature vector, Q, depends on the polynomial order and repetition which determine the number of pZ moments. The results obtained by using the pZ moments-based feature vectors are shown in Figure 7.15. Observing Figure 7.15(a), (b), (d), (e), (g) and (h), it is worth noting that the performance in terms of correct classification and recognition is very sensitive to high level of noise, with unsatisfactory results for the lower values of moments order and negative value of SNR. However, both $\mathscr{P}_C$ and $\mathscr{P}_R$ generally improves as the moments order, the observation time and the SNR increase, leading to a decrement of $\mathscr{P}_U$, as observed in Figure 7.15(c), (f) and (i). The gap between $\mathscr{P}_C$ and $\mathscr{P}_R$ becomes smaller as the moments order increases, with both the figures of merit reaching a maximum value around 0.90 for the higher value considered for the SNR.

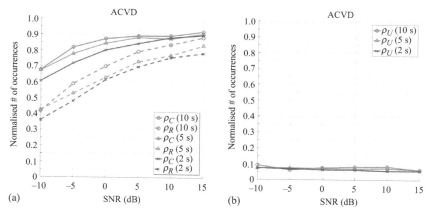

Figure 7.14   *Performance of the ACVD-based feature vector approach for real data on varying the signal duration and the SNR [15]. (a) $\mathscr{P}_C$ and $\mathscr{P}_R$; (b) $\mathscr{P}_U$*

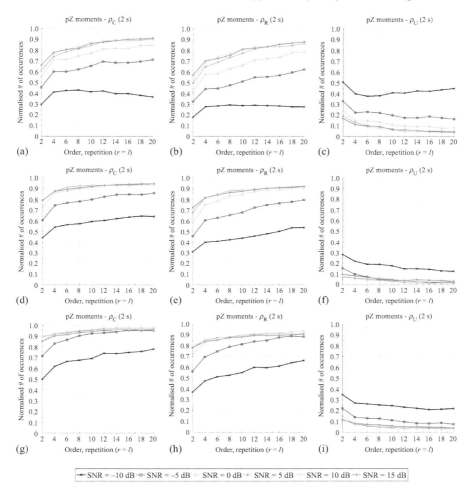

Figure 7.15   Performance of the pZ moments-based feature vector approach for real data in terms of $\mathscr{P}_C$ (first column), $\mathscr{P}_R$ (second column) and $\mathscr{P}_U$ (third column); the analysis is conducted on varying the signal duration (increasing from the top to the bottom), the moments order and the SNR [15]

7.5.3   Kr moments approach

As for pZ moments, even in this case Q, and hence the number of Kr moments, depends on the polynomials order. The graphs in Figure 7.16 show the results obtained by using the Kr moments-based feature vectors. Figure 7.16(a) and (b) show that, for 2 s, long radar observations $\mathscr{P}_C$ and $\mathscr{P}_R$ increase significantly as the SNR increases; while from Figure 7.16(a), (b), (d) and (e), it is pointed out that they are almost constant

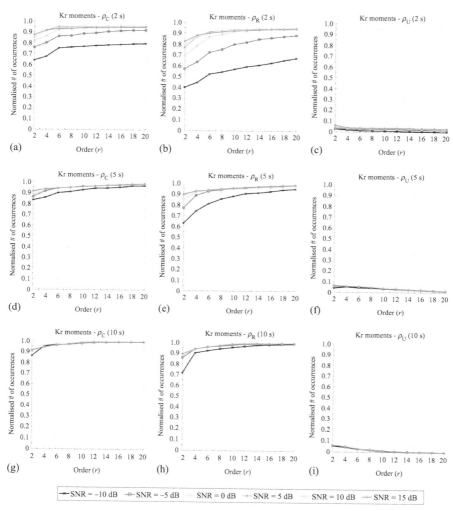

Figure 7.16   *Performance of the Kr moments-based feature vector approach for real data in terms of $\mathscr{P}_C$ (first column), $\mathscr{P}_R$ (second column) and $\mathscr{P}_U$ (third column); the analysis is conducted on varying the signal duration (increasing from the top to the bottom), the moments order and the SNR [15]*

for 5 and 10 s, when the SNR is greater than -10 dB. Moreover, the gap between the two probabilities becomes negligible for SNR greater than -5 dB and moments order greater than 10. Finally, from Figure 7.15(c), (f) and (i), it is worth noting that $\mathscr{P}_U$ decreases as the moments order increases. Nevertheless, the latter is smaller than 0.1 in all the analysed cases.

7.5.4 2D Gabor filter approach

For the 2D Gabor filter approach, the dimension of feature vector corresponds to the number of filters, which depends on the number of values considered for the central frequency and the orientation angle. Recall that Q is given by

$$Q = L \left(\left\lceil \frac{\pi/2}{\vartheta_{step}} \right\rceil + 1 \right) \tag{7.67}$$

where ϑ_{step} is the orientation angular step and L in the number of central frequencies. The latter was fixed at four values; 0.5, 1, 1.5 and 2. The value of ϑ_{step} was set to be an integer in the interval $[3°, 10°]$. In this way, an analysis on varying the density of the considered positions of the harmonic response on each circumference with radius equal to fcs_l is conducted. The values of the orientation angle, ϑ_m, is given by

$$\vartheta_m = m \, \vartheta_{step} \tag{7.68}$$

with $m = 0, \ldots, M - 1$ and where

$$M = \left\lceil \frac{\pi/2}{\vartheta_{step}} \right\rceil . \tag{7.69}$$

From (7.68) and (7.69), it is important to note that the features are extracted moving the harmonic response of the filter considering only the first quadrant, due to the symmetry of the expected image for this application.

The performance of the 2D Gabor filter-based method is shown in Figure 7.17. Observing Figure 7.17(a), (b), (d), (e), (g) and (h), it is clear that both $\mathscr{P}_C$ and $\mathscr{P}_R$ increase as the SNR and observation time increase. In particular, for signal duration of 5 s, both $\mathscr{P}_C$ and $\mathscr{P}_R$ are greater than 0.98 for SNR greater than $-10\,\text{dB}$; for observation time of 10 s, instead, $\mathscr{P}_C$ is greater than 0.99 for the all analysed cases. Finally, the gap between the two probabilities decreases as the SNR increases, being equal for high values of the SNR. Figure 7.17(c), (f) and (i) shows $\mathscr{P}_U$ versus Q, which is clearly smaller than 0.05 in all the analysed case. Therefore, it is worth noting that the performance does not change significantly when varying the feature vector dimension.

7.5.5 Performance in the presence of the booster

The performance of the four algorithms was evaluated also in the case in which the received signal was scattered from an additional object different from warheads and decoys. This analysis is of interest since, during the flight, the missile releases some debris in addition to the decoys, such as the booster used in the boost phase. As in case of decoys, when the booster has been released by the missile, it starts tumbling as shown in Figure 7.18, where the model used for the booster is shown. However, the booster rotation velocity is smaller than that of the decoys, while its dimensions are bigger. It is assumed that the booster has a cylindrical shape, whose diameter and height are 0.75 and 5 m, respectively, with triangular fins, whose base is 0.50 m and height is 1 m; the tumbling velocity is the one-fifteenth of the decoy's.

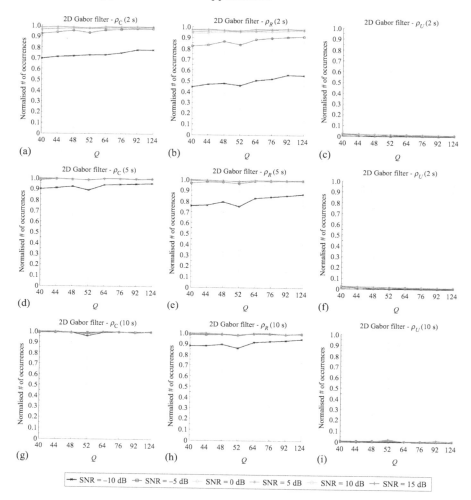

Figure 7.17 *Performance of the Gabor filter-based feature vector approach for real data in terms of $\mathscr{P}_C$ (first column), $\mathscr{P}_R$ (second column) and $\mathscr{P}_U$ (third column); the analysis is conducted on varying the signal duration (increasing from the top to the bottom), the number of features Q and the SNR [15]*

This analysis is conducted by training the classifier with feature vectors belonging to either the warhead class or the decoy class, and then testing it on the booster feature vector. Moreover, the performance is evaluated in terms of $\mathscr{P}_U$, as defined above, and *probability of misclassification (Error) as a Warhead* ($\mathscr{P}_{eW}$), determined by the ratio of the number of times in which the booster is classified as a warhead and the total number of tests. Note, in this specific case, classifying the booster as unknown represents the correct classification as there is no specific *booster* class.

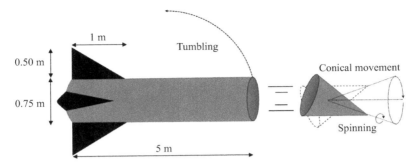

Figure 7.18　Representation of booster: model dimensions and difference of movement respect with warhead

Figure 7.19 shows $\mathscr{P}_U$ and $\mathscr{P}_{eW}$ obtained by the ACVD-based algorithm as the signal duration and the SNR are varied. From Figure 7.19, it is observed that even if $\mathscr{P}_U$ increases and, consequently, $\mathscr{P}_{eW}$ decreases as the signal duration increases, $\mathscr{P}_{eW}$ remains greater than $\mathscr{P}_U$. Moreover, the performance does not change significantly on varying the SNR.

Results obtained by using the pZ moments-based approach are shown in Figure 7.20. From this figure, it is clear that the probability of classifying the booster as unknown increases as the order grows up to 4, independently of the observation length, where the maximum value is reached, and it is above 0.80 for SNR equal to 0 and 5 dB. Considering orders greater than 4, $\mathscr{P}_U$ remains constant for positive

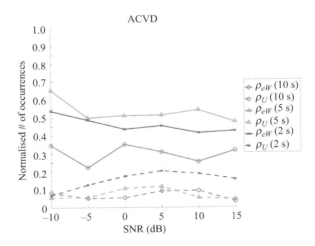

Figure 7.19　Performance of the ACVD-based feature vector approach for real unknown data (booster); the analysis is conducted on varying the signal duration and the SNR [15]

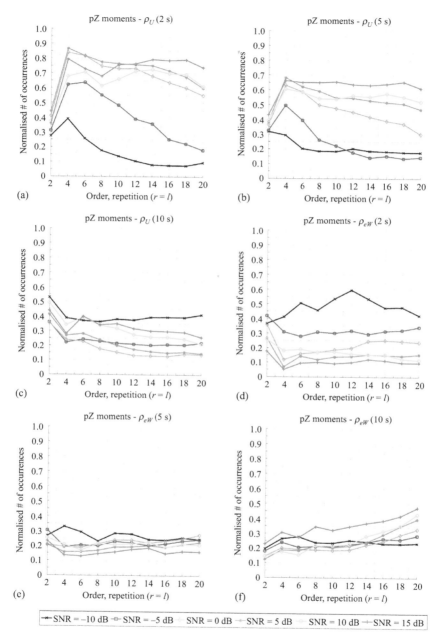

Figure 7.20   Performance of the pZ-based feature vector approach for real
unknown data (booster); the analysis is conducted on varying the
moments order, the signal duration and the SNR [15]

values of SNR, while it significantly decreases for SNR smaller than 0 dB. However, for moments order of about 20, $\mathscr{P}_U$ grows as the SNR increases. It is noticed that $\mathscr{P}_{eW}$ decreases as the observation time increases for negative value of SNR, while it increases for SNR greater than 0 dB. However, the best results are obtained for positive values of the SNR and for signal duration of 2 and 5 s, reaching probabilities of error smaller than 0.20.

Results obtained by using the Kr moments-based approach are shown in Figure 7.21. From Figure 7.21(a), it is noted that for 2 s, long observations $\mathscr{P}_U$ increases as the SNR increases, but reaching a maximum value around 0.50 for the lower values of the moments order. Indeed, $\mathscr{P}_U$ decreases as the order, and hence the feature vector dimension increases. This trend is because vectors composed of a greater number of Kr moments occupy wider spaces for each class such that it is more probable for a test vector being closer to one of them. On the other hand, Figure 7.21(d) shows that $\mathscr{P}_{eW}$ increases as the moments order increases for observation time of 2 s, being greater than $\mathscr{P}_U$ for each value of SNR when r is greater than 8. Observing Figure 7.21(b), (c), (e) and (f), it is worth noting that the performance improves as the observation time increases. From the figures, the values of $\mathscr{P}_U$ and $\mathscr{P}_{eW}$ appear almost constant as the order increases. The best performance in terms of $\mathscr{P}_U$ is obtained for the lowest values of the order, equal to 2, and for the lowest considered SNR, which is -10 dB. Moreover, it is noted that $\mathscr{P}_{eW}$ decreases as the observation time increases, being lower than 0.30 for 10 s long signals and for all the values of moments order and SNR.

Finally, $\mathscr{P}_U$ and $\mathscr{P}_{eW}$ obtained for 2D Gabor filter-based feature vector are shown in Figure 7.22. From the graphs, one can deduce that the performance improves as the signal duration and the SNR increase. In particular, the performance for the signal duration of 2 s is not satisfactory since $\mathscr{P}_{eW}$ is always greater than $\mathscr{P}_U$. However, for observation time of 5 s $\mathscr{P}_U$ becomes greater than $\mathscr{P}_{eW}$ from SNR greater than -10 dB reaching about 0.90 for the highest values of SNR. Finally, for signal duration equal to 10 s, $\mathscr{P}_U$ is constantly greater than 0.90 independent of the values of the SNR and Q; on the other hand, $\mathscr{P}_{eW}$ is smaller than 10^{-2} for values of the SNR greater than 0 dB.

Consequently, it is clear that in case of classification of unknown objects which are not used to train the classifier, such as the booster, the ACVD-based approach does not guarantee satisfactory performance. The pZ moments-based approach is able to give good performance for small signal duration and for high SNR, while the Kr moments approach for longer signal duration and for low SNR. Alternatively the Gabor filter approach provided the optimum results for an observation time of 5 s, for SNR greater than -10 dB, and of 10 s, independent of the noise levels.

7.5.6 Average running time

One of the most common requirements for automatic target recognition (ATR) algorithms in defence applications is the feasibility and reliability in real-time implementation. The four feature extraction methods presented in this chapter have different computational load. In this subsection, the methods are compared in terms of average

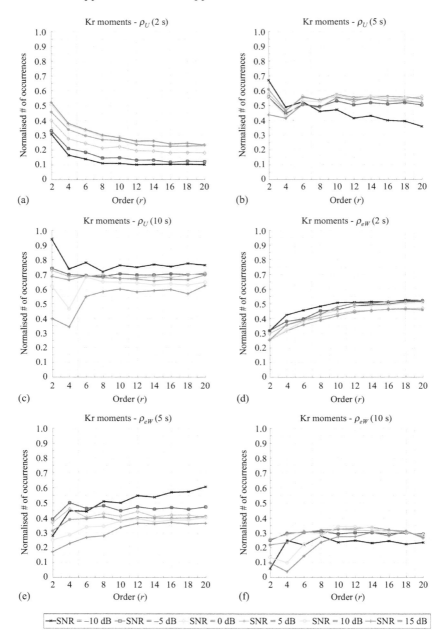

*Figure 7.21   Performance of the pZ-based feature vector approach for real
unknown data (booster); the analysis is conducted on varying the
moments order, the signal duration and the SNR [15].*

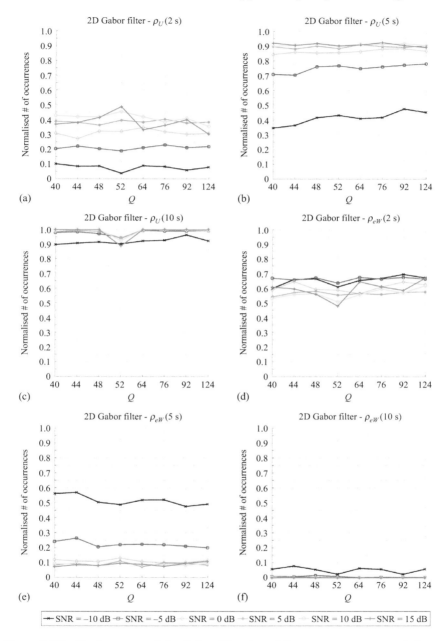

*Figure 7.22   Performance of the Gabor filter-based feature vector approach for
real unknown data (booster); the analysis is conducted on varying the
number of features Q, the signal duration and the SNR [15].*

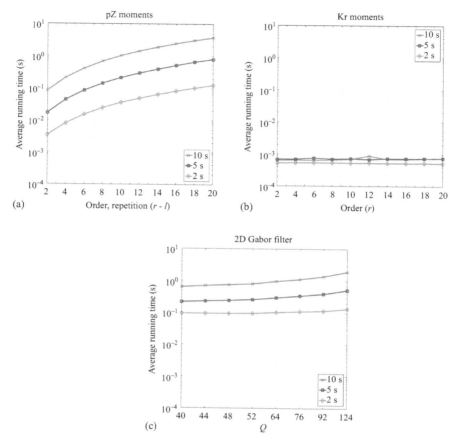

Figure 7.23   *Performance of the pZ moments-, Kr moments- and 2D Gabor filter-based approaches in terms of average running time for the feature vector extraction*

running time needed to extract the feature vectors. The algorithms are implemented in MATLAB® environment, and the average running time is evaluated with a Monte Carlo approach over 1,000 runs. It is worth noting that the time needed for computing the common steps of the different approaches are not taken into account in this analysis, starting the evaluation of the running time after the elaboration of the CVD. Moreover, the pZ and Kr polynomials for the evaluation of the corresponding moments are precomputed and loaded on memory, as well as the 2D Gabor filter bank.

The ACVD-based approach is a very fast method since it requires to average the CVD along the frequency dimension, for extracting six features from a 1D signature. Specifically, the average running time is smaller than 10^{-3} s for all the observation time considered. Figure 7.23 represents the average running time for the extraction of the feature vectors for pZ moments, Kr moments and 2D Gabor filter-based approaches, on varying the signal's duration, hence the CVD dimensions, and the vector dimension. It is observed from Figure 7.23(a) that the computation of pZ moments requires more time increasing the duration of radar observation and the moments order. The highest elaboration time is about 4 s, required for order equal to 20 and for 10 s long signals. Figure 7.23(b) shows that the computation time of Kr moments is almost independent of the considered orders, with a small gap between the evaluation for 2 s long signals and 5 and 10 s, for which the difference is negligible. This trend is because the Kr moments are computed by a matrix product, and the matrix dimensions considered into the analysis do not affect the computational time guaranteed by the hardware used for the evaluation. Finally, for all the analysed cases the average running time is smaller than 10^{-3} s, such as for the ACVD method. The main reason for the faster implementation of the Kr moments with respect to the pZ ones is that they are discretely defined and do not require the inscription of the 2D signature within the unit circle, as the pZ moments do. For the same reason, discretisation error does not exist for Kr moments and the amount of resource required to store the polynomials is reduced thanks to the recurrence relations and the symmetry properties of Kr moments. Figure 7.23(c) shows that the average running time for the 2D Gabor filter-based feature vector increases significantly as the signal duration increases, while it slightly increases by increasing the filter bank dimension within the set of considering values. Comparing Figure 7.23(a) and (c), it is observed that, by fixing the signal duration, for moments order greater than 10 pZ moments require longer running time than the 2D Gabor filtering does for all the considered filter bank dimension. The highest computational time for 2D Gabor filter approach is about 2 s, obtained by using 124 filters in case of 10 s long signals. Table 7.4 summarises the performance of the four approaches in terms of average running time and $\mathscr{P}_C$, on varying the observation time and for SNR equal to 15 dB, when the greater dimension of the feature vector analysed for all the approaches is considered. Specifically, it is observed that while the 2D Gabor filter approach outperforms the other approach in terms of $\mathscr{P}_C$, the Kr moments approach may represent a good compromise between running time and $\mathscr{P}_C$.

Table 7.4 *Performance in terms of average running time and $\mathscr{P}_C$, on varying the observation time, for SNR equal to 15 dB and for the greater dimension of the feature vector analysed in Section 7.5*

	ACVD		pZ		Kr		2D Gabor filters	
	run (s)	$\mathscr{P}_C$	run (s)	$\mathscr{P}_C$	run (s)	$\mathscr{P}_C$	run (s)	$\mathscr{P}_C$
2 s	$\leq 10^{-3}$	0.90	0.15	0.90	5×10^{-3}	0.94	0.15	0.98
5 s	$\leq 10^{-3}$	0.91	0.8	0.94	8×10^{-3}	0.97	0.5	0.99
10 s	$\leq 10^{-3}$	0.92	3	0.95	8×10^{-3}	0.99	1	0.99

7.6   Summary

BTs exhibit characteristic micro-motions that can be exploited to enhance BMD capabilities. The μD effect from warheads and decoys can be characterised using analytical models that can be then used for classification algorithms development. In this chapter, a set of models and a processing framework were introduced with the aim to show the effectiveness of μD for this delicate and demanding challenge. Ballistic threats classification is possible using μD information, and the effectiveness of different classification algorithms would vary depending on the allocated resources and on which platform the algorithms would be implemented. The selection of the best feature vector is outside of the scope of this research work and is still an open research thread that can be further investigated. Therefore, future work would involve a study of the best μD features for BT classification in terms of computational cost and reliability. In addition, the analysis of the framework could be extended by taking into account also target trajectories and possible variations of the micro-motion behaviour. In this way, it would be possible to evaluate the robustness of the framework to the variation of micro-motions within the observation time, and to the accuracy of the main Doppler shift compensation, needed to elaborate the CVD conveniently. Finally, the framework performance could be tested in a multi-target scenario, since the defence systems might not have enough resolution to separate the different targets. The presented mathematical model could be used to investigate and design new model-based classification algorithms or for training convolution neural networks which could classify the BTs starting from I/Q data, CVD or any other equivalent target representation.

References

[1]   Sessler AM, Cornwall JM, Dietz B, *et al.* Countermeasure: A technical evaluation of the operational effectiveness of the planned US national missile defense system. Union of Concerned Scientists MIT Security Studies Program; 2000.

[2]   Weiner SD, and Rocklin SM. Discrimination performance requirements for ballistic missile defense. *The Lincoln Laboratory Journal.* 1994;7(1):63–88.

[3]   Bankman I, Rogala E, and Pavek R. Laser radar in ballistic missile defense. *Johns Hopkins APL Technical Digest.* 2001;22(3):379–393.

[4]   Melvin WL, and Scheer JA. *Principles of Modern Radar. Volume III – Radar Applications.* Raleigh, NC: SciTech Publishing; 2014.

[5]   Rathore HK, Katukuri M, Rao DS, and Murthy BN. Real-time embedded software design for onboard computer of anti-ballistic missile system. In: 2014 International Conference on Computer and Communications Technologies (ICCCT), 2014. pp. 1–5.

[6]   Fioranelli F, Ritchie M, and Griffiths H. Analysis of polarimetric multistatic human micro-Doppler classification of armed/unarmed personnel. In: Radar Conference, 2015 IEEE; 2015. pp. 0432–0437.

[7]   Bjorklund S, Petersson H, Nezirovic A, *et al.* Millimeter-wave radar micro-Doppler signatures of human motion. In: Radar Symposium (IRS), 2011 Proceedings International; 2011. pp. 167–174.

[8]   Fioranelli F, Ritchie M, and Griffiths H. Classification of unarmed/armed personnel using the NetRAD multistatic radar for micro-Doppler and singular value decomposition features. *IEEE Geoscience and Remote Sensing Letters.* 2015;12(9):1933–1937.

[9]   Molchanov P, Egiazarian K, Astola J, *et al.* Classification of aircraft using micro-Doppler bicoherence-based features. *IEEE Transactions on Aerospace and Electronic Systems.* 2014;50(2):1455–1467.

[10]   Hornsteiner C, and Detlefsen J. Extraction of features related to human gait using a continuous-wave radar. In: Microwave Conference (GeMIC), Germany; 2008. pp. 1–3.

[11]   Liu L, Popescu M, Skubic M, *et al.* Automatic fall detection based on Doppler radar motion signature. In: 5th International Conference on Pervasive Computing Technologies for Healthcare (PervasiveHealth), 2011.

[12]   Hongwei G, Lianggui X, Shuliang W, *et al.* Micro-Doppler signature extraction from ballistic target with micro-motions. *IEEE Transactions on Aerospace and Electronic Systems.* 2010;46(4):1969–1982.

[13]   Liu L, McLernon DC, Ghogho M, *et al.* Ballistic missile detection via micro-Doppler frequency estimation from radar return. *Digital Signal Processing.* 2012;22(1):87–95.

[14]   Clemente C, Pallotta L, Proudler I, *et al.* Pseudo-Zernike-based multi-pass automatic target recognition from multi-channel synthetic aperture radar. *IET Radar, Sonar and Navigation.* 2015;9(4):457–466.

[15]   Persico AR, Clemente C, Gaglione D, *et al.* On model, algorithms, and experiment for micro-Doppler-based recognition of ballistic targets. *IEEE Transactions on Aerospace and Electronic Systems.* 2017;53(3):1088–1108.

[16]   Bhatia AB, and Wolf E. On the circle polynomials of Zernike and related orthogonal sets. *Mathematical Proceedings of the Cambridge Philosophical Society.* 1954 1;50:40–48.

[17]   Yap PT, Paramesran R, and Ong SH. Image analysis by Krawtchouk moments. *IEEE Transactions on Signal Processing.* 2003;12(11):1367–1377.

[18]   Kaur B, Joshi G, and Vig R. Analysis of shape recognition capability of Krawtchouk moments. In: International Conference on Computing, Communication Automation (ICCCA), 2015. pp. 1085–1090.

[19]   Kamarainen JK, Kyrki V, and Kalviainen H. Invariance properties of Gabor filter-based features – Overview and applications. *IEEE Transactions on Image Processing.* 2006;15(5):1088–1099.

[20]   Ilonen J, Kamarainen JK, and Kalviainen H. Fast extraction of multi-resolution Gabor features. In: International Conference on Image Analysis and Processing, 2007. pp. 481–486.

Chapter 8

Signatures of small drones and birds as emerging targets

Børge Torvik[1], Daniel Gusland[1], and Karl Erik Olsen[1]

8.1 Introduction

Rapid technological developments in battery technology and embedded computing have made small unmanned aerial vehicles (UAVs), particularly quadcopters, dramatically more affordable and easier to use. Combined with the approved integration in low-altitude airspace, this has led to an exponential increase in the use of small UAVs in private, commercial and military sectors. Modern UAVs are advanced flying sensors, with high-quality communication links, advanced navigation, obstacle avoidance and high-definition imaging capabilities. This has resulted in great advances in remote inspection, agricultural surveillance, search and rescue, battlefield reconnaissance, cinematography and possibly also package delivery.*

Figure 8.1 shows the Minister of Defence in Norway at that time, Anne-Grete Strøm-Erichsen, being delivered the Norwegian Defence Sector's strategy for research and development on a memory stick by the Norwegian nano-drone Black Hornet. The use of UAVs in both commercial and military sectors is projected to continue its increase, ultimately resulting in a densely populated low-altitude airspace. The extensive capabilities of these UAVs come with a large potential for misuse, ranging from minor inconveniences to serious military attacks. A micro-UAV can potentially damage an airplane and therefore pose a threat to air security. In December 2018, hundreds of flights were cancelled at Gatwick Airport due to drone sightings close to the runway. UAVs also pose a threat to privacy with their long operating range and high-resolution cameras.

The carrying capacity of UAVs can also be a factor in causing harm, either by transporting contraband across borders or by carrying weapons and explosives. The growing use of UAVs combined with their significant potential for harm has forced academic and military institutions to research the considerable problem of detecting, tracking and classifying UAVs. As we see further on in this chapter, micro-UAVs can have radar cross section (RCS) and motion patterns similar to that of birds.

[1]Defence Systems, Norwegian Defence Research Establishment (FFI), Kjeller, Norway
*www.engadget.com/2019/04/24/zipline-drone-medicine-vaccine-delivery-ghana/

Figure 8.1   The PD-100 Black Hornet Nano, a nano reconnaissance UAV developed in Norway, was used to deliver the defence sector's strategy for research and development to the Norwegian Minister of Defence, Anne-Grete Strøm-Erichsen, on a USB flash drive during the launch of the strategy. Photo: FFI

Targets of this size and velocity are often filtered out by traditional radar systems, due to low speed and RCS. The ability to detect drones or signature-reduced targets requires sensitive sensors. Because of the potentially high number of detections in such a system, classification is considered to be of critical importance. Despite their inherent similar size, radar recordings of UAVs differ from birds in several ways, such as amplitude fluctuations and their micro-Doppler signature.

This chapter concerns the basis for differentiating between birds and micro-UAVs using micro-Doppler signatures. The chapter starts with a small introduction to the different classes and configurations of UAVs. Further on, a review of the research field, covering RCS investigations, material choice and classification methods is given. Birds and UAVs as radar targets are discussed, and their electromagnetic properties, size and shape are investigated using finite difference time domain (FDTD) predictions. A particular focus is put on moving parts like rotors, propellers and bird wings. These predictions are followed by radar measurements that largely confirm the predictions, and a discussion on how to exploit the target properties for the classification is given. Radar system parameters required to differentiate between birds and UAVs using micro-Doppler are then discussed. Towards the end of the chapter, classification methods are discussed in brief before the chapter is concluded.

8.1.1   Classes and configuration of UAVs

There are many classes, categories and configurations of UAVs. This chapter will focus on CLASS 1 UAVs from the NATO UAS classification guide [1]. The categories

Table 8.1   *UAS classification table for CLASS 1 adapted from NATO*
            UAS Classification Guide [1]

Category	Normal mission radius	Example platform
Small >20 kg	50 km (LOS)	Hermes 90
Mini 2–20 kg	25 km (LOS)	Raven, Scan Eagle
Micro <2 kg	5 km (LOS)	DJI Phantom, Black Widow

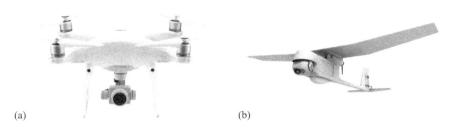

(a)   (b)

Figure 8.2   *Example of a quadcopter and a fixed-wing UAV. (a) DJI Phantom IV*
             quadcopter (Courtesy of DJI). (b) Avinc Raven fixed-wing UAV
             (Courtesy of AeroVironment, Inc.)

'small' and 'mini', shown in Table 8.1, mostly contain military or commercial drones and will in many cases require a licence to operate. Most of the hobbyist UAVs are found in the 'micro' class of UAVs with a mass below 2 kg. In addition to the weight categories introduced in Table 8.1, UAVs also differ in configuration. In this section, we will cover the two most common categories, namely fixed-wing and rotary-wing UAVs (see Figure 8.2). Within these two categories, there are different designs and materials used, such as metal, plastic, reinforced plastic or carbon fibre.

Rotary-wing UAVs use propellers to generate lift to gain or maintain altitude. This makes them less efficient than fixed-wing UAVs, but provides the ability to hover and perform vertical take-off and landing which increases their usability. The two sub-categories considered in this chapter are helicopters and multicopters.

Helicopters use a large primary rotor to provide lift and a smaller rotor, typically a tail rotor, to counteract the torque of the main rotor. Helicopters regulate lift and thrust by regulating both the speed and pitch of their rotors and can be powered by either electric or combustion engine.

Multicopters are often configured using four, six or eight propellers. The rotors are pairwise rotating clockwise and counter-clockwise, yielding a net torque of zero, enabling the multicopter to retain its heading. The propellers of most multicopters have a fixed pitch and the lift is therefore controlled by the propeller speed. Keeping a multicopter stable requires constant control of the individual rotors, which is done by an on-board computer. Because of this constant control, multicopters are typically electrically powered to facilitate the fast regulation of rotor speed.

Fixed-wing UAVs use a propulsion system to generate thrust and fixed wings to generate lift. This makes them very efficient, which in turn yields a long operational range. Fixed-wing UAVs can fly both fast and slow, but are unable to hover. They are typically powered with a propeller connected to an electric or combustion engine, but may also be powered by jet engines.

8.1.2   Literature review

Classification is considered to be of critical importance for UAV detection radar systems. Because of the small physical size and low RCS of UAVs, radars for UAV detection need to be sensitive, resulting in a significant increase in the number of detections. Additionally, UAVs are able to fly low and slow, making simple Doppler filtering suboptimal and contribution from ground clutter unavoidable. UAV classification can be based on several signature properties, such as amplitude fluctuation, polarimetric features, movement patterns from tracks and micro-Doppler information. Classification solely based on RCS and motion pattern classification is questionable, as drones and birds may be associated with similar RCS and motion pattern. Moharejin et al. [2] extract features from radar tracks to classify UAVs in civilian airspace. Torvik et al. [3] use polarimetric features to classify UAVs and birds in the lower frequency bands such as L- and S-bands. These methods are promising, but the most interesting basis for classifying UAVs at higher frequencies is micro-Doppler analysis. To form a good basis for micro-Doppler classification, two major topics need to be addressed, namely the characterisation of the RCS of UAV's moving parts and investigation of classification methods. The former is important to determine the optimal radar parameters for detecting micro-Doppler from UAVs and the latter to find an optimal classification scheme for distinguishing UAVs from birds and possibly even identifying the UAV classes and configurations. A review of radar classification and RCS characteristics is presented by Patel, Fioranelli and Anderson [4].

A significant effort has been made to estimate and measure the RCS of UAVs. Schröder et al. [5] estimate the RCS of a consumer DJI Phantom II at 10 GHz numerically and confirm the simulations using measurements. They conclude that the RCS is in the range of −30 to −20 dBsm. Schröder et al. also explore the major scattering components, such as the battery pack, wires and motors and investigate their contribution to the micro-Doppler signature. Their results indicate that simulating the rotating components is sufficient to model the micro-Doppler signature. Li and Ling [6] conducted an interesting RCS study in two different frequency bands: 3–6 and 12–15 GHz. Several consumer UAVs were put on a turntable, and inverse synthetic aperture radar (ISAR) images were constructed. These images were used to estimate the RCS and reveal the major scattering contributors at different frequencies. The RCS measurements are in the range of −30 to −10 dBsm. They also compare the RCS in the two frequency bands and conclude that the RCS decreased by about 10 dBsm at 3–6 GHz compared to 12–15 GHz. Guay et al. [7] also investigate the RCS of commercial UAVs and conclude that the RCS is approximately −21 dBsm.

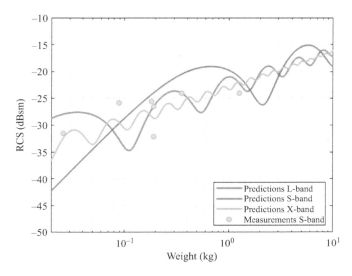

Figure 8.3 *Predicted and measured RCS values of birds. Green dots represent measured values from [8]. Solid lines show rough estimations of expected RCS values of birds modelled as water-filled spheres of similar mass*

They also find that the RCS of UAVs in dynamic flight is considerably larger than the static RCS because of the decreased likelihood of scattering from deep nulls.

The RCS of birds has been investigated for many years. An overview of virtually all measured values published in the literature prior to 1984 is presented in Vaughn's article [8]. This shows measurements in UHF-, S-, C- and X-bands, but without reference to polarisation and bird orientation. Consequently, these data have to be treated as mean values. The underlying data are varied and originate from different sources with birds observed at different aspect angles and with head and wings in different positions. A more thorough discussion and more measurements are found in [9]. The general impression is that RCS measurements of living birds in a controlled environment is challenging. Blacksmith and Mack [10] conclude in despair by referring to the cookbook *Joy of Cooking* for suggestions on how to best make use of agitated birds sabotaging RCS measurements.

Figure 8.3 shows RCS values of birds as function of weight. Green dots show measurements in S-band from [8], whereas solid lines represent RCS predictions of simplified models in L-, S- and X-bands. The predictions assume that birds may be modelled as water-filled spheres of similar weight. This is naturally a rough approximation and the RCS values must only be interpreted as indications. The measured values should also be interpreted as indications of average values only. Factors such as aspect angles, orientation of wings, head and body, polarisation and bird species will affect the RCS. As observed in the figure, there is no measurement of birds heavier than 2 kg available. Typically the weight of a house sparrow is around 25–30 g, of a herring gull or a mallard duck in the range 0.7–1.5 kg, whereas a Canada goose normally reaches a mass of 5–7 kg. Typical RCS values for birds constituting a risk of confusion with drones is therefore found in the range from −30 to −15 dBsm.

The backscattered power from UAV rotors compared to the main fuselage has also been investigated in the literature. This is of critical importance, as the UAVs themselves are small targets and the rotors are even smaller. Schröder *et al.* [5] claim that the micro-Doppler of plastic rotors is nearly invisible to radar, whereas metallic blades cause significant signatures. This is problematic, and according to Farlik *et al.* [11], plastic is the dominating material used for quadcopter propellers. Ritchie *et al.* [12] conducted simulations and measurements for metallic, carbon fibre and plastic propellers. The RCS of the carbon fibre propeller at broadside was found to range from −20 to −5 dBsm in the frequency range from 1 to 10 GHz. It is also noted that the RCS of the propeller is significantly higher at HH-polarisation compared to VV-polarisation. Measurements conducted by the same authors indicate that the RCS of plastic propellers is significantly lower than that of carbon fibre in L- and S-bands, whereas it is more similar in C-band. It can also be noted that the width of the angle for which the maximum RCS of the propeller occurs, also referred to as the rotor flash, gets significantly narrower with increasing frequency. This is further investigated by Harman [13] at frequencies from 5 to 35 GHz. It was found through measurements that the typical broadside 'flash' only occurred at 5 and 10 GHz, whereas the higher frequencies experienced a consistently higher RCS, but a more chaotic flashing nature.

There has also been published research on UAVs as radar targets. Gersone *et al.* [14] perform simulations to investigate the integration efficiency of an L-band staring radar for UAV detection. They conclude that it is possible to achieve a higher integration time than that of conventional rotating radars and still gain signal-to-noise ratio (SNR). This is confirmed by Harman [15] who performed measurements of a number of different micro-UAVs using an ubiquitous radar and analysed their target decorrelation time. He concludes that it is possible to use coherent processing intervals (CPIs) with a 3 dB lobe of duration equal to 100–500 ms, as this can still provide some gain in terms of SNR. He also comments that the micro-Doppler harmonics are still present after a long CPI.

There have been several attempts at distinguishing UAVs from birds using different approaches. Malchanov *et al.* [16] present an approach for classifying small UAVs and birds with an X-band continuous wave (CW) radar with feature extraction and support-vector machine (SVMs). With ten classes, they present a classification accuracy of 95%. Harmanny, Wit and Cabic [17] demonstrate feature extraction using the spectrogram and cepstogram to classify UAVs and birds. Doppler bandwidth and spectrogram periodicity were among the extracted features. Zhang *et al.* [18] use spectrograms and principal component analysis (PCA) to extract features for an SVM classifier. Using measured data, they demonstrate classification accuracies up to 94.7% by fusing dual-band radar sensors. Zhang and Li [19] use cadence-velocity diagram (CVD) to extract the cadence-velocity frequency which is fed to a K-means classifier and achieve an accuracy of 94.2%.

The emergence of methods based on neural network (NNs) and deep learning has been fuelling the development of radar classification methods over the last few years. Several NN implementations have shown impressive results for classification. Regev *et al.* [20] use several multilayer perceptrons (MLPs) that are fed different representations of the data to classify the number of spinning propellers. Kim, Kang and Park [21] use a convolutional neural network (CNN) to classify quadcopters and hexacopters and achieve a solid 99.59% accuracy between the two classes. Brooks *et al.* [22] introduce a simulator for micro-Doppler classification and continue to compare MLPs, recurrent neural network (RNNs) and fully convolutional network (FCNs) for micro-Doppler classification. They conclude that RNNs and FCNs are particularly appropriate for radar classification with a solid classification accuracy close to 100%.

8.2   Electromagnetic predictions of birds and UAVs

Distinguishing between birds and drones of comparable sizes is an important task for many modern surveillance radars. In most cases, man-made targets are targets of interest, whereas birds are generally considered as clutter. Due to the similar RCS and motion pattern between these classes, the false alarm rate caused by birds may, in bird-rich areas, be unacceptably high without any kind of classification routine. Micro-Doppler analysis is a technique that can be used to reduce the number of false alarms. However, insight into target motions and electromagnetic properties is important in order to design an effective system. This section presents theoretical considerations on birds and drones as radar targets, emphasising on the importance of target kinematics, size, shape and material properties.

8.2.1   *Electromagnetic properties, size and shape*

A radar works by transmitting electromagnetic (EM) waves and receiving the signal reflected of a target. The interaction between transmitted radar pulses and target can be described by the Maxwell's equations, which are a set of fundamental laws describing all electromagnetic behaviour [23]. A valid question in this context is whether these types of targets are any different from those typically covered in the classic non-cooperative target recognition (NCTR) or micro-Doppler literature. First of all, birds and drones are of comparable physical size. The spatial resolution available in most long-range operational radars is generally not fine enough to resolve these into a number of resolution cells useful for classification directly. Second, these targets are, especially in the lower frequency bands, often electrically small. By this, we mean that the targets, and particularly the target parts involved in micro-motions essential to the micro-Doppler signature, often are comparable to or smaller in size than the wavelength of the transmitted signal. This is obviously dependent on which frequency band the radar is operating in, as the wavelength in the air for an L-band radar is around 23 cm and for an X-band system is only about 3 cm. This electrical size of rotating/vibrating target parts is important as it affects their RCS and thereby contribution to micro-Doppler analysis. Lastly, these target types consist of a wide range of materials. Whereas targets in the radar literature are often assumed to be large conductive surfaces scattering in the optics region, birds and drones are composed of materials with arbitrary dielectric properties scattering in the resonance or Rayleigh regions.

While drones are built using mostly conductive parts, thermoplastics, fibreglass and carbon fibre-reinforced polymer, birds naturally consists of layers of different biological tissues. These materials constitute a range of different dielectric properties affecting reflectivity and penetration of electromagnetic waves into the target. Reflections occur from differences in impedance between layers of the media the wave is propagating through. Attenuation is highly dependent on the material properties; for more details, see [24]. The attenuation of electromagnetic waves in bird tissues is considered to be similar to human tissues, which is generally high and increases with frequency [25]. The power in the incident wave is significantly attenuated after 1 cm of penetration even in L-band and significant contributions from reflections from internal bird parts are considered unlikely [9]. Potential reflections from the interior of man-made targets are especially challenging for recognition relying on spatial resolution and physical feature extraction from the resulting profiles or images. The potential difference from optical images also makes modelling of such targets difficult, as the interior also may contribute substantially to the target signatures. Methods making use of spatial resolution are not covered in this chapter. However, understanding how material properties may affect the reflectivity of target components responsible for micro-motions is important. This is generally described by the material's permittivity and permeability and the size and shape of the object. The last two quantities must be seen relative to the wavelength and orientation of the electrical field, referred to as the polarisation, of the incoming wave.

How relevant targets and target parts of different shapes and materials respond to carrier frequencies and polarisation in terms of RCS is covered by Torvik in his PhD thesis [9]. The focus of this thesis is on the classification of small and slow targets in long-range air defence radars, typically operating in L- and S-bands. One of the main hypotheses is that polarimetric signatures of electrically small targets or target parts are well defined by their overall shape. Their signatures are thus easier to interpret than the signatures of complex targets in the optics region where many target details may contribute with their individual polarimetric signature. The main drone or bird parts of relevance to micro-Doppler analysis are naturally bird wings, rotors and propellers. The smaller these parts get relative to the wavelength, the better they are represented by dipole models.

8.2.2 RCS predictions

This chapter focuses on micro-Doppler signatures of bird wings and drone rotors of different materials at frequencies ranging from 1 to 10 GHz. Initially, the focus is on the detectability of these targets and their predicted RCS in the specified frequency range. Figure 8.4 shows the three-dimensional (3D) models used for the electromagnetic predictions presented in this section. The RCS of these models is calculated by an FFI implementation of the FDTD electromagnetic prediction code described in [26] and covers aspect angles θ in the XY-plane in the figure ranging from 0° to 360° and 0° to 180° for the rotor and bird wings, respectively. All targets were treated as homogenous

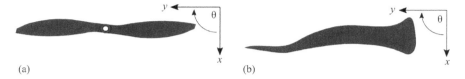

(a)   (b)

Figure 8.4   *3D models used for electromagnetic predictions. (a) DJI Phantom II rotor. Dimensions 22.8 cm × 2.3 cm × 1.0 cm. (b) Muscular parts of gannet wing. The length from tip to wing root is 77.0 cm*

Figure 8.5   Image of a gannet soaring

volumes with dielectric properties as for ABS plastic, bird muscle and carbon fibre. For the two last materials values of human muscle at relevant frequencies from [27] and a perfect electric conductor (PEC) were used, respectively. The rotor shown in Figure 8.4(a) is a model of a Phantom II rotor with a maximum length of 22.8 cm, width of 2.3 cm and thickness of 1.0 cm. The bird wing in Figure 8.4(b) is a model of the muscular and thus reflective parts of a full-size gannet wing. Reflections from bird feathers are not expected as their dielectric properties are close to that of air. The gannet is a large seabird and the maximum length of this wing model is 77.0 cm as shown in Figure 8.5.

Figure 8.6 shows predicted RCS values σ for the DJI Phantom II rotor shown in Figure 8.4. This is representative of rotors for small quadcopters in terms of size and shape. The typical materials used in these blades are carbon fibre and ABS plastic, which have significantly different dielectric properties. This has consequences for the detectability. Carbon fibre may be treated as PEC, whereas plastics are much less reflective. Results are here shown for the two different materials at two orthogonal linear polarisations as a function of aspect angle and four radio frequencies

(1.0, 3.0, 6.0 and 10.0 GHz). The aspect angle is here the azimuth angle θ shown in the xy-plane in Figure 8.4, where $\theta = 0°$ is found along the negative x-axis as the rotor is being illuminated broadside. Figure 8.6(a) and (b) show the RCS prediction results of carbon fibre blades, for simplicity here modelled as PEC, at HH- and VV-polarisation, respectively. The RCSs at these two polarisations are from now on denoted as σ_{hh} and σ_{vv}. In the first case, the propeller is co-oriented with the incident electric field and distinct increases in the response up to towards -10 dBsm are found for all frequencies close to broadside illumination around $\theta = 0°$ and $\theta = 180°$. This is sometimes referred to as the blade flash. However, the width and thus duration of this flash from a rotating blade is seen to be dependent on frequency. The flash width in radians θ_{fw} can be approximated as the beam width of an aperture antenna as a function of the wavelength of the electromagnetic wave in air λ and the length of the rotor L:

$$\theta_{fw} \approx \frac{\lambda}{L} \tag{8.1}$$

This approximation is naturally better at frequencies where the rotor is small compared to the wavelength and the details of the blade do not significantly affect its RCS. For scattering in L-band around 1.0 GHz, this approximation should be good. Inserting a carrier frequency of 1.0 GHz and $L = 0.228$ m into (8.1) gives an approximate opening angle of 75° which agrees well with the blue curve in Figure 8.6(a). The same goes for the flash width at 3.0 GHz in red, where the main lobe is observed with an opening angle close to the theoretical 25° around boresight. At higher frequencies, the blade shape contributes more to its signature. The widths of the flashes are close to the theoretical values; however, the response at other angles reveals that we are dealing with a target deviating from the shape of a rectangular box. The maximum RCS values found in these flashes are relatively constant around -15 dBsm over all frequencies at this polarisation.

In case of VV-polarisation shown in Figure 8.6(b), the E-field is perpendicular to the rotor length. This results in a weaker response, especially at the lower frequencies where the extent of the blade in the direction of the E-field is a small fraction of the wavelength. The differences between frequencies are much larger at this polarisation. At 1 GHz, shown by the blue line, the maximum RCS value is around -55 dBsm, which must be considered to be very low. The maximum value increases significantly with frequency until, at 10 GHz, it is comparable with the value found for HH-polarisation. The angles θ where the maximum RCS is found is also changing as a function of frequency. For the lowest frequency, the maximum is not found at broadside, but at 45° to the side. The maximum response is observed to gradually approach broadside with increasing frequency.

The two lower panels of Figure 8.6 show the results for the same rotor now modelled by ABS plastic material with a permittivity of $\epsilon_r = 2.2$ instead of the conductive carbon fibre. The RCS values predicted for HH-polarisation, σ_{hh}, has now dropped significantly compared to what was observed in the carbon fibre case. The RCS at low frequencies is especially low. At 1 GHz σ_{hh} is found at -65 dBsm, whereas at 10 GHz it increases to around somewhat above -30 dBsm during the flashes. This

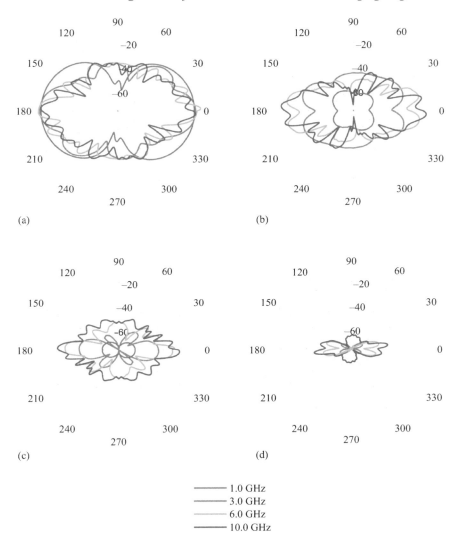

Figure 8.6 *Electromagnetic prediction of RCS (dBsm) of DJI Phantom II 22.8 cm*
rotor. FDTD prediction code used in all simulations. (a) σ_{hh} carbon
fibre rotor. (b) σ_{vv} carbon fibre rotor. (c) σ_{hh} plastic rotor. (d) σ_{vv} plastic
rotor

indicates that the values of σ_{hh} for plastic blades are generally found to be significantly lower than those of σ_{hh} for carbon fibre blades, and that the difference increases
considerably at lower frequencies. In Figure 8.6(d), showing σ_{vv} for plastic blades,
the predicted values are so low that detection of single rotor flashes at long distances
is highly questionable.

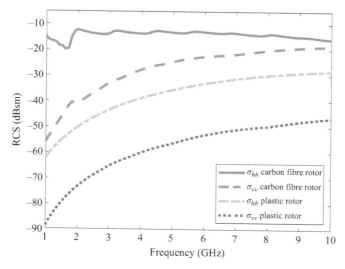

Figure 8.7   Maximum RCS of DJI Phantom II plastic and carbon rotor for linear
polarisations, independent of azimuth angle θ as a function of
frequency

Figure 8.7 shows the maximum RCS of the DJI Phantom II rotor observed in any direction in the xy-plane. σ_{hh} of the rotor is dramatically larger than σ_{vv} for both carbon fibre and plastic in the lower frequency bands. At higher frequencies, the difference is still almost 20 dB for plastic rotors. For carbon fibre rotors, the difference drastically decreases with frequency, and at 10 GHz, there is very little difference between the polarisations.

Figure 8.8 shows similar RCS predictions for the muscular parts of a gannet wing shown in Figure 8.4. This length is somewhat shorter than the physical length of a gannet wing, as the wing tip is assumed to consist of mostly feathers and non-reflective material, not muscles. The size of the gannet makes this wing relevant for the largest birds and those with RCS comparable to drones. These predictions were only performed for azimuth angles θ in the interval 0°–180° as illumination from the side where the wing is attached to the body makes no sense. Results for σ_{hh} in dBsm are shown in Figure 8.8(a). The values at 1.2 GHz in blue fluctuate slowly between −30 and −15 dBsm. No distinct increase in any specific direction is found. The shape of the wing seems to matter even at the lowest frequency. More fluctuation is observed at the two highest frequencies of 3.3 and 10.0 GHz. However, there is no significant difference in the mean values between the frequency bands at this polarisation. This is very different for σ_{vv} shown in Figure 8.8(b). Here the RCS at 1.2 GHz, shown in blue, is significantly lower than that at 3.3 GHz, while the highest frequency is associated with an even higher RCS. This low RCS at the lowest frequency is connected to the low extent of the bird wing in the direction of the incoming electrical field at VV-polarisation.

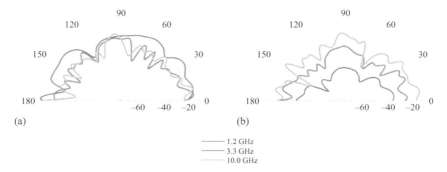

Figure 8.8 *Electromagnetic prediction of RCS (dBsm) of muscular parts of a 77-cm long gannet wing. Azimuth angles covered in the range 0°–180°, and dielectric properties as for human muscle are used. (a) σ_{hh} wing. (b) σ_{vv} wing*

8.3 Target analysis using radar measurements

Most UAVs are highly agile and have low mass, which makes them dependent on wind conditions. This amplifies their natural erratic behaviour and affects their properties as radar targets. It is therefore interesting to investigate their target properties in flight, such as body RCS, rotor RCS and their corresponding decorrelation time. This is particularly interesting because a long CPI increases the velocity resolution which can potentially benefit classification. These investigations were conducted using real radar data provided through the NATO STO task group SET-245. The data contain several UAVs and birds of opportunity with different flight paths. It should be noted that there is a finite combination of flight paths within the data set and the results presented are averaged to give an impression of the most likely outcome in a real-world scenario.

8.3.1 Body RCS

The RCS of the UAV body does not directly affect the UAVs micro-Doppler signature. The difference in the RCS of the UAV in relation to the to the rotor, propeller or wings does, however, indicate whether or not micro-Doppler classification is achievable in a realistic scenario. There have been several thorough investigations of the RCS of commercial UAVs. As mentioned in the introduction, Schröder *et al.* [5] estimate numerically the RCS of a DJI Phantom II at 10 GHz to be in the range of −30 to −20 dBsm. Li and Ling [6] measured the same type of UAV and concluded that the RCS is in the range of −30 to −10 dBsm and that it is generally higher during dynamic flight. We estimated the RCS of several UAVs by using measurements of several UAVs in flight. Table 8.2 shows the measured RCS of the UAV body in their basic classes. These results are the mean of several different multicopters with plastic and carbon blades, fixed wings and several birds. The estimated RCS is calculated

*Table 8.2   Estimated body RCS for horizontal and vertical polarisation for L-, S-,
          C- and X-bands in dBsm*

Frequency band	Horizontal polarisation				Vertical polarisation			
	L	S	C	X	L	S	C	X
Multicopter plastic	−14.3	−12.6	−12.4	−8.4	−11.2	−10.6	−8.5	−10.9
Multicopter carbon	−9.3	−9.6	−8.5	−9.7	−9.4	−9.9	−11.9	−10.7
Fixed wing	−7.3	−8.7	−8.5	−8.3	−9.7	−9.7	−8.4	−8.2
Birds	−21.6	−22.6	−23.4	−24.3	−24.2	−22.6	−24.0	−24.3

from the mean of the body throughout the whole data set. It can also be seen that the birds in these measurements were generally significantly smaller in physical size than the included UAVs. The same is true for the included multicopters, which were mostly different configurations of quadcopters, compared to their physically larger fixed-wing counterparts. These experimental mean values are slightly higher than the predictions, possibly because of the limited variation of aspect angles in the data set. The RCS of UAVs is highly angle dependant and the data set contains mostly approaching and receding UAVs. Another interesting observation is that multicopters with plastic blades appear to have a frequency-dependent RCS. This is not the case for their carbon fibre counterparts, which is likely caused by the blade return interfering with the body.

8.3.2   Rotor RCS

Using the same measurements of UAVs in flight, the RCS of the rotors, propellers and wings was estimated. To estimate the rotor RCS, the data are split into short time intervals, the peaks are identified and the mean of the peaks is calculated. These peaks are assumed to originate from the rotors. The mean of the body at the same time interval is also calculated and the results are subtracted from each other. This gives an estimate of the maximum RCS of the rotor blade corresponding to the rotor flash. This is repeated throughout the whole data set and the mean is presented in Table 8.3. The multicopters are separated by the material of their rotor blades, as carbon fibre and plastic blades exhibit severely different properties. The table shows that the carbon rotors have a higher RCS than their plastic counterparts. The RCS of the carbon rotors also appears to decrease slightly with frequency, which agrees well with the RCS simulations of a DJI Phantom II rotor blade in Figure 8.7. The absolute value of the RCS is also in agreement with the simulations to within a few dB. The same is true for the plastic rotors, which are in good agreement with the simulations in X- and C-bands. The measurements deviate from the simulations in L- and S-bands, which might be because the rotors blades are made of composite materials or because the flash is very low. It is therefore likely that other contributors affect the measurements at these frequencies. Despite this deviation, the RCS of plastic rotors still exhibit a

*Table 8.3   Estimated RCS of the moving parts of the targets at HH-polarisation in
L-, S-, C- and X-bands in dBsm*

Frequency band	Horizontal polarisation				Vertical polarisation			
	L	S	C	X	L	S	C	X
Multicopter plastic	−38.7	−35.4	−35.5	−27.8	−38.7	−34.5	−33.8	−31.3
Multicopter carbon	−16.1	−17.7	−19.7	−20.7	−21.2	−24.6	−26.4	−26.0
Fixed wing	−32.1	−30.6	−25.8	−23.2	−31.4	−28.8	−24.7	−22.5
Birds	−35.1	−35.8	−37.3	−37.2	−38.6	−36.7	−36.3	−37.2

clear increase with frequency. The RCS of the fixed-wing rotors also increases with frequency, although not to the same extent as the plastic rotors of the multicopters. They also have generally higher RCS than the multicopter plastic rotors. This might be because they are larger than the multicopter rotors, or that they are made from composite materials.

Table 8.4 shows the relative power of the micro-Doppler response. This is estimated by subtracting the rotor RCS in Table 8.3 from the body RCS in Table 8.2. This measure of relative power can in turn yield a good estimate of the SNR necessary to capture the micro-Doppler features and hence an estimate of when classification based on micro-Doppler features can be achieved. From the table, it is again evident that multicopters fitted with carbon fibre blades are associated with prominent micro-Doppler signatures. They also appear to be more visible in the lower frequency bands. Plastic rotors are a different story. In L-band, there is close to 18 dB difference in the relative power of the carbon rotors compared to plastic rotors for multicopters, and this difference decreases to approximately 8 dB in X-band. The results for fixed-wing UAVs are similar to multicopters with plastic rotors, which can be explained by larger main fuselages and rotors which, when combined, yield the same relative power. It should be noted that these results are the mean of many measurements and observation angles. There will be many examples of these numbers being both higher and lower, yet a mean value gives valuable information on the feasibility of micro-Doppler classification in a real-world scenario. As for the birds, the relative power is comparable to that of rotors. A reasonable SNR requirement for robust micro-Doppler classification depends on the processing technique and the sophistication of the classification algorithm. However, it can be assumed that for spectrogram-based classification, the micro-Doppler features should have more power than the noise floor for a given detection. In this case, we would have to look at the lowest relative power for the micro-Doppler features which are at approximately 20 dB in X-band and 26 dB in L-band. Given that the response at that SNR level would be comparable to the noise floor, it is reasonable to assume that a body SNR of approximately 25–35 dB should be sufficient for robust micro-Doppler classification, given that the rotors are not occluded or out of plane with the incident wave.

Table 8.4   Power of the micro-Doppler response relative to the originating body in dB at HH-polarisation

Frequency band	Horizontal polarisation				Vertical polarisation			
	L	S	C	X	L	S	C	X
Multicopter plastic	−24.4	−22.8	−23.0	−19.4	−27.5	−23.9	−25.3	−20.4
Multicopter carbon	−6.8	−8.0	−11.2	−11.0	−11.8	−14.7	−14.5	−15.3
Fixed wing	−24.8	−21.9	−17.3	−14.9	−21.7	−19.1	−16.4	−14.4
Birds	−13.5	−13.2	−13.9	−13.0	−14.4	−14.1	−12.3	−12.9

8.3.3   Maximum CPI

Achieving a sufficient SNR at long distances for small targets such as birds and UAVs puts strenuous requirements on the radar system. In addition, it might be beneficial to have a high Doppler/velocity resolution for classification purposes. It is therefore interesting to see how long CPIs can be used to gain SNR for these targets in dynamic flight. To investigate this, a random starting point in time was selected for each of the targets. After that, N pulses were added to construct a Doppler spectrum. The top value of this spectrum originates from the UAV body and the noise level is estimated from the edge of the Doppler spectrum. This was repeated for CPIs from 10 ms to 1 s. The maximum integration time is assumed to be the point when there is no longer an SNR gain from adding more pulses caused by target decorrelation. This operation is repeated 1,000 times with random starting points, representing different positions and flight paths. The mean value of this investigation is presented in Table 8.5. There are several interesting observations to be made. First, all the integration times are very long and considerably longer than the observation times used by most scanning radar systems. This indicates that scanning radars could use longer time on target to gain SNR and that radars that in general use longer CPIs, such as ubiquitous radars, might be beneficial for these targets. In addition to the fact that the maximum CPIs are very long, they also appear to have a slightly decreasing trend with frequency. The results for multicopters in X-band are in agreement with the results presented by Harman [15] when taking into account that he refers to the maximum integration interval as the half power width of the CPI. The values in Table 8.5 are mean values and the decorrelation is highly dependent on the change in target aspect angle.

Having investigated the maximum CPI of the body of UAVs and birds, it is interesting to investigate the maximum CPI of the micro-Doppler features from rotating parts as well. The rotation of rotor blades is a short event, with a rotation rate of 1,500 to 9,000 revolutions per minute (RPM) for a typical quadcopter. The event of one revolution only lasts a few milliseconds. When we apply a CPI longer than this event, the data in each CPI contain several rotations of each rotor. The resulting spectrum is dominated by the modulation peaks originating from the rotor RPM. The Doppler signature originating within each rotation is no longer visible, but the modulation

Table 8.5   Estimated maximum integration time for birds, fixed-wing and multicopter UAVs in different radar frequency bands

UAV type	Frequency band			
	L	S	C	X
Multicopter	752	794	750	690
Fixed wing	643	602	538	471
Birds	780	740	645	590

peaks form 'skirts' around the UAVs as shown in the spectrograms in Figure 8.12. Luckily, the rotor RPM is relatively stable over short-time intervals and it is therefore possible to achieve close to coherent integration. To investigate how long CPIs can be used to capture the micro-Doppler features of plastic rotors, a data set of a DJI Phantom II with plastic blades was used. Again a random starting point was selected and an initial CPI of 10 ms was used. The UAV body is detected and removed and six of the maximum micro-Doppler features are selected. The mean of these features is here called the mean micro-Doppler feature level. The CPI is doubled until a CPI of 1 s is achieved. The noise floor will increase 3 dB for each doubling of the CPI and for fully coherent integration a 6 dB signal power increase is expected for each CPI doubling, yielding a 3 dB SNR increase. This process was repeated 1,000 times and the mean for L-, S-, C- and X-bands is presented in Figure 8.9. The y-axis in the figure denotes the delta signal power, with theoretical fully coherent gain represented by the 6 dB line. The x-axis is plotted logarithmically. The plot shows that the integration is fairly frequency independent and a similar gain is achieved in all the measured frequency bands. It is also worth noting that the integration never reaches a full coherent integration gain of 6 dB. This can be explained by the fact that this is a feature caused by moving propellers, hence the signal is not present at all times. Even though it is not as high as full coherent gain, it is fairly close and note that throughout the tested CPI of 1 s, the average gain is never below the 3 dB line that would cause an SNR reduction.

8.3.4   Micro-Doppler analysis

As discussed throughout this chapter, there are several target properties relevant to the micro-Doppler analysis. So far material properties, size and shape of UAV rotors, propellers and bird wings have been central. Figure 8.10 shows the signature of a small passerine bird flying away from the radar collected simultaneously at HH-polarisation in L-, S-, C-, and X-bands. Figure 8.10(d) shows the typical flying pattern with periodic flapping and gliding for such a small bird. The returns from the bird body are observed as the solid response around 6 m/s, whereas the micro-motion of the wings are seen at ±2 m/s at both sides of the body response during the flapping periods. A weak ground clutter return is found around 0 m/s. The main purpose of

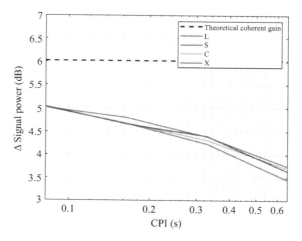

Figure 8.9 Averaged micro-Doppler integration gain for multicopter with plastic rotors in L-, S-, C- and X-bands

this figure is to show the effect of velocity resolution for a fixed CPI of 100 ms in the different frequency bands. In L-band shown in Figure 8.10(a), the signature of flapping wings is hardly visible. The wings of a passerine bird are small compared to the wavelength in L-band making the RCS small at the same time as the velocity resolution of 1.2 m/s using a CPI of 100 ms in L-band is insufficient to resolve the wing movement. In S-band shown in Figure 8.10(b) is the velocity resolution just fine enough to give an indication of wing movements. Figure 8.10(c) shows the same situation in C-band. Here the velocity resolution has increased to 0.3 m/s and the wing movements are clearly observed. This figure underlines the challenges of using micro-Doppler analysis for classification of birds in L- and S-bands frequently used for long-range air defence radars.

Joint time–frequency representations of target signatures can be a good tool to visualise target characteristics. Such representations rely on distinguishing between target scatterers in time and radial velocity and naturally a certain resolution in velocity and time is required. In short-time Fourier transform (STFT)-based methods, these two measures are interdependent, meaning that reducing one increases the other and vice versa. To resolve short events in time, such as rotor flashes, the time resolution would need to be increased resulting in lower velocity resolution Δv. The relation can be expressed as

$$\Delta v = \frac{\lambda \, \Delta f_d}{2} = \frac{\lambda}{2 \, \tau_d} \tag{8.2}$$

Here λ and τ_d denote the wavelength in the air and dwell time, respectively. In the calculation of a STFT-based spectrogram, τ_d equals the window length over which each Doppler-spectrum is calculated. In case of a flapping bird, it is visually explanatory to resolve each flapping period. The spread in the Doppler-spectrum is

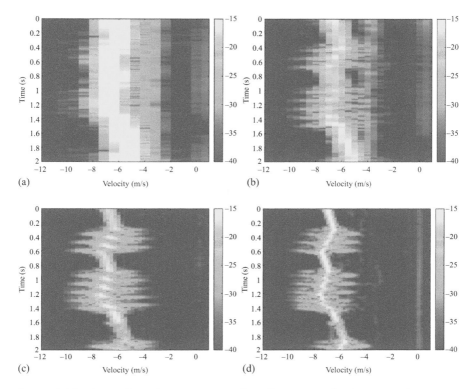

Figure 8.10 *Micro-Doppler signatures of small passerine bird measured*
simultaneously at HH-polarisation in four different frequency bands.
Data provided through NATO STO SET-245. (a) L-band; (b) S-band;
(c) C-band; (d) X-band

dependent on the aspect angle, wing beat frequency (WBF) and wing length. The WBF is dependent on the bird species and generally decreases with increasing bird size. However, the wing length normally increases with decreasing WBF leading to a partial compensation of the wing tip velocity. Figure 8.11 shows rough estimations of how many Doppler/velocity-resolution cells of different bird species can be expected to occupy during normal flapping activity as a function of azimuth angle. The azimuth angles covered correspond to the birds being seen from the front at 0° to broadside at 90°. In this estimation, a contribution is expected to come from about 60% of the bird wing, as these are the most muscular parts, the coherent interval is adapted to 35% of the actual wing beat interval and the elevation angle up to the target is set to 5°. Actual wing lengths and wing beat frequency data are collected from [28].

The number of occupied Doppler resolution cells is small in L-band independent of bird species. This corresponds well with the findings in Figure 8.10. The bird species of highest relevance to that data is the chaffinch or the starling. Even if the CPI is adapted to the WBF, only a couple of Doppler resolution cells are expected to

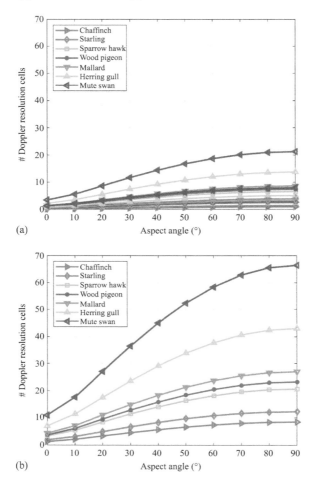

Figure 8.11   Spread in Doppler cells for different bird species in S- and X-bands from [9]. Contribution from 60% of the bird wing length and 120° wing opening angle. Coherent integration interval is kept at 35% of the wing beat interval and elevation angle is 5°. (a) L-band; (b) X-band

be covered at the optimal aspect angle. In X-band, shown in Figure 8.11 the situation is much better, which is also supported by observations in Figure 8.10.

Multicopter drones produce a significantly different micro-Doppler signature than birds. Figure 8.12 shows spectrograms of simultaneous measurements of a DJI Phantom II drone with plastic rotors at HH-polarisation in different frequency bands. Since a CPI of 40 ms including several periodic flashes is selected here, the Doppler-signatures are observed as harmonic lines in the spectrum in this figure. The L-band

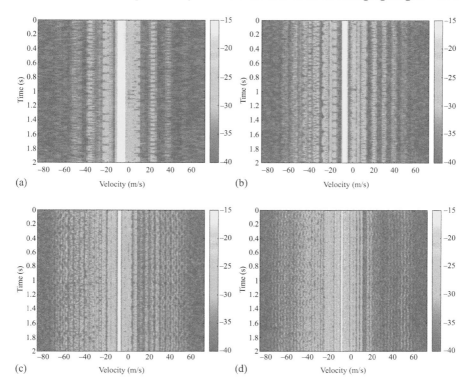

Figure 8.12 *Time–frequency analysis of DJI Phantom II drone with plastic rotors measured at HH-polarisation in four different frequency bands (dBsm). Data provided through NATO STO SET-245: (a) L-band; (b) S-band; (c) C-band; (d) X-band*

data presented in Figure 8.12 show the drone body as the solid line moving away from the radar at 7 m/s. The width of the line spectrum around it is found to be around ±70 m/s. Increasing the carrier frequency through Figure 8.12(a)–(d) shows a significant broadening of this spectrum not only in terms of velocity resolution cells, but also the response of the rotors. This is due to the significantly increasing RCS as a function of frequency observed in Figure 8.7. The response of the blades is significantly lower in terms of RCS in L-band compared to, for example, X-band.

8.3.5 Micro-Doppler analysis on polarimetric data

The discussion in Section 8.2.2 and especially the findings in Figures 8.6 and 8.8 indicate that a rotor blade or a bird wing illuminated in the Rayleigh region is well described as a dipole scatterer. Looking at the response in terms of magnitude or phase of such scatterers at different polarisations may add information for classification in lower frequency bands. Its polarimetric signature may, for example, be used to estimate the apparent orientation along the line of sight. The benefit of introducing polarimetric features to the classification of small and slow-moving targets in long-range air defence radar systems are thoroughly covered in [9].

A scattering matrix can be defined based on the magnitude of the backscattered amplitude S_{rt}, with subscripts t and r denoting transmitted and received polarisation, respectively, and the phase θ:

$$\mathbf{S} = \begin{bmatrix} |S_{hh}| \, e^{j\theta_{hh}} & |S_{hv}| \, e^{j\theta_{hv}} \\ |S_{vh}| \, e^{j\theta_{vh}} & |S_{vv}| \, e^{j\theta_{vv}} \end{bmatrix} \tag{8.3}$$

In the monostatic case, $|S_{hv}| \, e^{j\theta_{hv}} = |S_{vh}| \, e^{j\theta_{vh}}$. According to [29], the monostatic scattering matrix for a dipole is dependent on its orientation ϕ relative to the illuminating electric field about the radar line of sight as

$$\mathbf{S}_{\text{dipole}} = \begin{bmatrix} \cos^2(\phi) & \frac{1}{2}\sin(2\phi) \\ \frac{1}{2}\sin(2\phi) & \sin^2(\phi) \end{bmatrix} \tag{8.4}$$

The RCS already introduced in this chapter is defined as $\sigma_{rt} = S_{rt}^2$. The apparent orientation of such a Rayleigh scatterer may now be estimated by looking at the differential RCS now defined as

$$\sigma_{dr} = 10 \log_{10} \left(\frac{\sigma_{hh}}{\sigma_{vv}} \right) \tag{8.5}$$

Figure 8.13 shows the spectrogram of differential RCS σ_{dr} for a flapping crow with $\tau_d = 150$ ms measured in L- and S-bands simultaneously. This parameter is believed to hold information about the orientation about the radar line of sight of a dipole-like scatterer like the crow wing scattering in the Rayleigh region. A clear periodicity is found in both Figure 8.13(a) and (b) showing L- and S-bands, respectively. Strong yellow corresponds to high σ_{dr} and indicates horizontal orientation of the wing.

The same idea is presented in [9] for drone rotors that are small compared to the wavelength. Figure 8.14 shows spectrograms of a DJI Phantom II drone with carbon fibre rotors gathered simultaneously in L- and S-bands. These spectrograms are formed with a significantly shorter CPI of $\tau_d = 3$ ms in order to resolve individual blade flashes. This short dwell time results in poor velocity resolution, especially in L-band found in Figure 8.14(a). Still, signatures of the blades are resolved in both bands and high values of bright yellow corresponding to high values of σ_{dr} indicate the horizontal orientation of the rotor blades. The differential RCS is only one of many polarimetric variables suggested for classification in [9].

8.4 Radar system considerations

The required detection range and classification range against low-slow-small (LSS)-targets such as UAVs and birds depend on the operational task of the radar. The

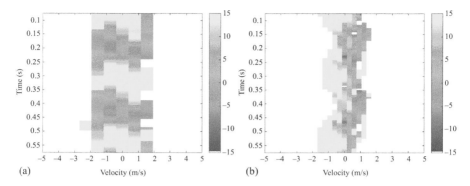

Figure 8.13   *Spectrogram of differential RCS σ_{dr} for flapping crow with dwell time $\tau_d = 150$ ms from [9]. (a) L-band: σ_{dr} @ 1.3 GHz. (b) S-band: σ_{dr} @ 3.25 GHz*

selection of system properties is dependent on the primary use of the radar. In systems where the classification of such targets is important, high priority may be given to requirements important to micro-Doppler. In other systems, such capacity may be considered of secondary importance and long-range capability may be prioritised.

8.4.1   Carrier frequency and polarisation

One system parameter influenced by this trade-off is the carrier frequency. As discussed earlier in this chapter, higher frequencies like X-band may be beneficial for micro-Doppler analysis for several reasons. For short-range radars primarily designed to perform classification of birds and micro drones, such frequencies are a natural choice. For long-range radar systems lower frequencies, like L- and S-bands, are typically selected. The problem of non-cooperative target classification in these bands is covered in the thesis of Torvik [9]. In these bands, less target information is available and the use of polarimetry is suggested to add more information to the classification process. As already covered in Section 8.2.1, the choice of polarisation is of importance for extracting useful signatures of some target parts relevant to micro-Doppler analysis. If one linear polarisation is to be used, horizontal is the best choice due to the typically near horizontal orientation for multicopter rotors.

8.4.2   Bandwidth

LSS targets are physically small, and using range resolution for classification is considered unlikely without several gigahertz of bandwidth. Separation of individual targets is, on the other hand, important to isolate signatures from different targets. Normally resolution in range, angle and velocity contribute to consistent tracking of targets. For micro-Doppler analysis, it is clearly beneficial to separate targets in angle or range as some targets are associated with a wide signature in velocity. For this reason, bandwidths providing a range resolution capable of separating individual

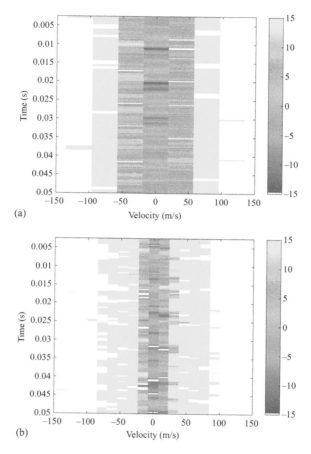

Figure 8.14   Spectrogram of σ_{dr} Raven UAV with dwell time $\tau_d = 3$ ms from [9]. (a) L-band: σ_{dr} @ 1.3 GHz. (b) S-band: σ_{dr} @ 3.25 GHz

targets as much as possible without resolving them into separate bins are relevant. In practice, this means a couple of metres. This number is naturally worth considering in the context of drone swarms and bird flocks.

8.4.3   Waveform and coherency

The physically small targets of interest in this chapter are also small in terms of RCS as covered in Section 8.3. The desirable detection distance is important for the system design. In cases where the distance is large, a pulsed radar system is normally the solution. At shorter distances, frequency-modulated continuous wave (FMCW) radars may be considered. In any case, micro-Doppler analysis based on coherent processing techniques, and naturally, a coherent radar is required.

8.4.4 Pulse repetition frequency

Successful micro-Doppler analysis of drone and bird signatures may put requirements on the radar system's pulse repetition frequency (PRF). For traditional feature-based classification schemes, the minimum PRF is determined by the maximum potential Doppler frequency component of the signature in order to avoid ambiguities in the spectrum. If the goal is to distinguish between UAVs and birds, this will likely not be a problem. However, if it is desirable to distinguish different types of UAVs, an ambiguous Doppler spectrum can pose a problem. The maximum frequency is dependent on the rotation rate of the propeller, the length of its blades, the drone's radial velocity relative to the radar, the carrier frequency of the radar system and the orientation of the plane of rotor rotation compared with the radar line of sight. The maximum Doppler shift, $f_{d\,max}$, is experienced as the rotors are illuminated in the plane of rotation and can be described as

$$f_{d\,max} = \frac{2v_{max}}{\lambda} = \frac{2\omega L}{\lambda} = \frac{\pi f_{RPM} L}{15\lambda} \tag{8.6}$$

Here, the maximum velocity v_{max} is the total relative velocity between the rotor blade tip and the radar including translational movement of drone and radar. Generally, this velocity is kept well below the speed of sound in air. The angular velocity ω (rad/s) and f_{RPM} (rev./min) are measures of the rotational rate of the rotor. The length L (m) is the total length of the blade such that the distance from the tip to the centre of rotation is $\frac{L}{2}$.

Figure 8.15 shows the minimum PRF required to sample the Doppler spectrum unambiguously for two different frequencies as a function of rotor rotation rate in RPM and blade length in metres. Figure 8.15(a) shows the case in L-band at 1.2 GHz. The rotation rate interval from 1,500 to 9,000 RPM are considered relevant for micro drones. PRF values for blade tip velocities relative to the radar above the speed of sound are not calculated. For a drone operating at 5,000 RPM and with a blade length of 22 cm, relevant for DJI Phantom drone, a PRF around 1 kHz is sufficient in L-band. For X-band, as shown in Figure 8.15(b), the same situation leads to a PRF-requirement of up to 8 kHz. This value corresponds well with observations in Figure 8.12(d). These PRF requirements make sense if classification is to be based on features like the Doppler bandwidth. However, the impact on other classification methods such as CNNs may not be equally dramatic.

Birds, however, require a significantly lower PRF to sample the Doppler spectrum unambiguously. Figure 8.16 shows the minimum PRF for different bird species as a function of carrier frequency and aspect angle. Small birds normally have short wings, but higher wing beat frequencies than larger birds as already discussed in Section 8.3.4. These two properties partially compensate for each other. However, the figure shows that the larger birds are still associated with larger Doppler bandwidths. L-band requirements are shown in Figure 8.16(a) and indicate that a PRF less than 250 Hz is required to sample bird Doppler spectra unambiguously. Figure 8.16(b) shows that this requirement is increased to almost 700 Hz in X-band. It is obvious

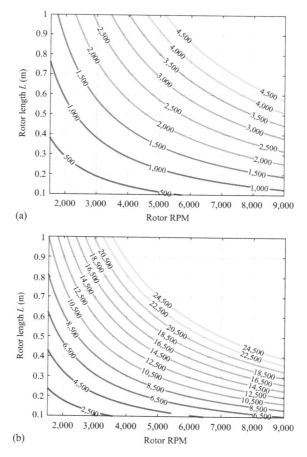

Figure 8.15   Minimum PRF (Hz) to sample Doppler spectrum from drone unambiguously as a function of RPM and blade length. Maximum radial velocity of drone and blade tip velocity is kept below 330 m/s. (a) L-band; (b) X-band

that drones put much stronger requirements on PRF for unambiguous sampling than birds.

Another factor putting requirements on PRF is the wish to detect every blade flash. This may be more relevant for helicopters than for multicopter platforms and is described for recognition of helicopters in [30]. Some of the same considerations can be done for multicopters. Pulsed radars then need to put energy at the drone frequently enough to ensure to illuminate the blades during every blade flash. As was shown in Section 8.2.2, the shape of the flashes vary with the rotor's electrical size and not necessarily according to (8.1). However, if this equation is used, the minimum PRF can be calculated as a function of frequency, blade length and rotation rate as shown

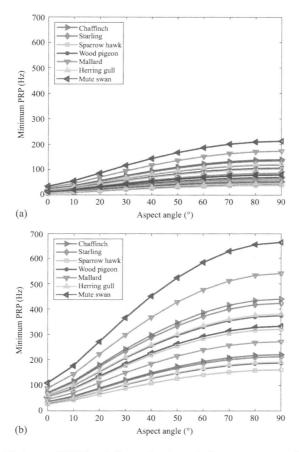

Figure 8.16 *Minimum PRF for different birds with flapping wings. Contribution from 60% of the bird wing length and 120° wing opening angle. Elevation angle is 5°. (a) L-band; (b) X-band*

in Figure 8.17. CW radars naturally illuminate the targets all the time; however, slow sweep rates in FMCW radars may result in different parts of the frequency spectrum interacting with the rotor blades from sweep to sweep.

8.4.5 Pencil-beam or ubiquitous radar

Dwell time and revisit time for each target are important for target detection, classification and tracking. Radar systems may implement different strategies for target illumination ranging from 2D mechanically rotating systems to 3D pencil-beam systems and ubiquitous radars. The choice of strategy has consequences for dwell and revisit time, which again influence the time available for pulse integration, coherent processing providing velocity resolution and update times between detections. Target

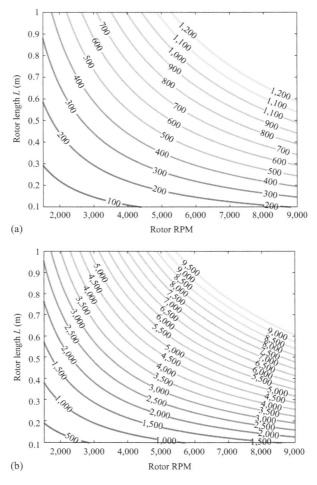

Figure 8.17   Minimum PRF (Hz) to capture single rotor blade flash as a function of RPM and blade length. Maximum radial velocity of drone and blade tip velocity is kept below 330 m/s. (a) L-band; (b) X-band

tracking is beneficial for micro-Doppler analysis and important for an operative radar system. Tracking agile targets, such as UAVs and birds, requires a high update rate. Mechanically rotating antennas generally struggle to provide short enough revisit times in this context. Electronically steered arrays offer significantly more frequent updates and the possibility to prioritise difficult targets. The dwell time on each target is in mechanically rotating 2D systems depending on the antenna beamwidth and its rotation rate. Generally, 3D systems allow for less dwell time in search modes, but electronically steering again provides the possibility to prioritise more dwell time on specific targets. Ubiquitous radars on the other hand floodlight the relevant scene continuously and perform all beam-forming digitally based on the ideal digitisation of the output of each transceiver/receiver module (TRM) behind each antenna element. Data from all parts of the scene are available at all times, which leaves the task of the resource management to prioritising the use of signal processing power.

LSS targets are potentially very slow movers. Velocity resolution may, therefore, be important to separate the target from surface clutter. The combination of dwell time and the carrier frequency is in this context important. The velocity resolution Δv was described in (8.2) and is also important to micro-Doppler analysis as it provides the potential to resolve different scattering centres. Being able to adaptively control the dwell time τ_d according to the motion of the target and the nature of the signature is beneficial. Ubiquitous radar provides long observation times, but will inherently result in less gain in the antenna pattern compared to a pencil beam. The potential of retrieving this by integration over longer dwell times is covered in Section 8.3. The consequences of dwell time for classification is covered in Section 8.5.

8.5 Classification methods

As discussed earlier in this chapter, there are several discriminatory features that separate birds from UAVs. As previously stated, the velocity of the two are comparable and therefore not a good feature for discrimination. The RCS and the RCS distribution is usually different. However, this is not a robust feature, as it is dependent on drone configuration, observation angle and bird species. There are promising methods based on for example track history and movement pattern, but these will not be discussed in this section as this chapter focuses on classification techniques based on micro-Doppler signatures. Micro-Doppler signatures of the two mentioned classes appear to contain robust discriminatory features. Given that a radar system is capable of detecting these features, there are several properties to investigate. With regards to classification, there are roughly two approaches; the first is to understand the physical phenomena and use that knowledge to describe discriminatory features to be used in a conventional classifier. Alternatively, one can find a good way of representing the data and feed it to a machine learning algorithm that can learn to recognise the discriminatory features of the different classes. In addition to that, one can also choose whether to include one or several observations of the target, based on tracker input. This section will start by exploring features available for a conventional classifier.

Naturally, micro-Doppler classification methods are based on spectra, spectrograms or features extracted from either of these. In its essence, the target signature collected by the radar is a time series of complex numbers, but viewing it in the frequency domain as either a spectrum or spectrogram may highlight micro-motions and separate the target signature from clutter, which is beneficial for classification. Many examples of birds and UAVs as radar targets have been presented in this chapter and some features might be evident. It is clear that the best features to employ will also be dependent on the radar system's carrier frequency, PRF and the SNR of the target. In the lower frequency bands, features based on polarimetric data might be beneficial for robust target classification, but for higher-frequency systems polarimetry does not necessarily increase the classification performance. If the radar system is capable of unambiguously sampling the Doppler bandwidth of the targets, the spectrum width of the target is a discriminatory feature that can both potentially differentiate birds from UAVs and possibly UAVs from each other. The typical Doppler bandwidth of birds is significantly lower than that of UAVs. Processing the data as a spectrogram adds the possibility to investigate the time dependence of frequency components in the target signature. This makes it possible to look at features such as spectrogram periodicity and the frequency of a given periodicity, which again is substantially different for birds and UAVs, and potentially also for different types of UAVs. Extracting periodicity in the signal from the target can also be achieved using the cepstrogram. Harmanny *et al.* [17] show that cepstrograms can be quite robust and can help identify the presence of rotors and even the number of rotors present.

Finding effective hand-crafted features may be challenging as this has been proved in the field of image classification. In radar data that might be easier, given that the target is separated in both range and velocity, but there are still events that can confuse a classifier. A radar designed to detect UAVs typically has to operate in complex environments with clutter and other confusers. There might be other targets such as bicycles, cars or air-conditioning fans that at times can resemble the micro-Doppler signatures of UAVs. Because of this, it might be beneficial to find a suitable data representation and use machine learning to find the discriminatory features, which has proved to be a great success in image classification. Many researchers have used several NN methods for UAV classification with great success, starting with standard MLPs [20], continuing with CNNs [21] and RNNs [22]. These methods have proved very successful, but unfortunately it is harder to understand what the model learns and uses as basis for classification. Understanding what the network learns is currently receiving a lot of attention in the machine learning community. Most researchers feed a raw spectrum of the magnitude of the spectrogram to an MLP, but very few utilise the complex nature of the signal.

Moving to CNNs is a sensible step, as CNNs are specialised to exploit spatial correlations in the data by using convolutional filters, essentially learning spatially connected features. Using CNNs instead of MLP is particularly beneficial when working with long CPIs or spectrograms as they are able to decompose the data while preserving the spatial locality.

With both NNs and CNNs, data representation is challenging because different observation times lead to different input sizes, not easily handled by any of these methods. Brooks [22] proposes to use FCN which can naturally handle inputs of varying size because it does not have fully connected layers at the end of the network. It should also be noted that RNNs are a natural choice for this kind of sequential learning. Particularly long short-term memory (LSTM) RNNs are popular because of their ability to learn both short- and long-term sequence dependencies which in case of a spectrogram include long- and short-term events in time. The RNN would view the spectrogram as a sequence of 1D vectors over time, which it essentially is. Klarenbeek *et al.* utilised an LSTM RNN for multi-target human gait classification and showed a better accuracy with RNN compared with CNN.

The importance of a tracker for micro-Doppler classification should also be emphasised. As already observed earlier in the chapter, it is perfectly achievable to perform micro-Doppler classification on one spectrum alone, given that the reflected energy from the rotors is present in that time step. Micro-Doppler signatures are naturally elusive due to the agility of the UAVs, long detection distances and low RCS. Using a tracker makes it possible to link consecutive detections to one target which will increase the data foundation for classification. This can be used in the classifier to increase the confidence of the classification with each added measurement.

8.6 Conclusion

Rapid development in battery technology and embedded computing has made UAVs affordable, easy to fly and readily available for the general public. Beyond recreational use, modern drone technology has resulted in advances in remote inspection, agricultural surveillance, search and rescue, battlefield reconnaissance, cinematography, package delivery and much more. The increased use of drones in civilian, commercial and military has led to a growing need for systems able to maintain safety and security. Radar is an appropriate sensor for surveillance of air, sea and ground independent of light and weather conditions.

As small UAVs potentially have similar RCS and motion pattern as that of birds, there is a need to distinguish between the two classes. One approach in this context is NCTR based on micro-Doppler signatures. The usefulness of such signatures are dependent on both target and radar system characteristics. Micro-Doppler methods utilise target motions such as movement of the wings for birds and rotating propellers for UAVs. Material properties and size relative to the wavelength of such parts affect the micro-Doppler signatures. Whereas bird wings consist of biological material, drone rotors and propellers are available in different materials, such as plastic and carbon fibre. Such factors impact the RCS of the components and affect the range on which classification is achievable. The frequency and polarisation of the electromagnetic wave are also of interest. Bird wings, rotors and propellers tend to be small compared to the wavelength during operation in lower frequency bands like L- and S-bands. In fact, such target parts tend to fall in the Rayleigh scattering region and can be modelled as dipole scatterers. Scattering in this region, as well as in the resonance region, makes traditional micro-Doppler processing challenging. However, there are indications that polarimetric measurements can help out under these conditions.

If the primary task of the radar system is to detect and classify drones and birds, higher carrier frequencies are beneficial. In X-band and above main components of such targets scatter in the optical region. Here target details contribute more to the signature, and distinct polarimetric dependencies observed in lower frequency bands are less prominent. Thus, the benefit from using polarimetric features is less in these bands. However, the RCS of small rotating components is normally higher and the velocity resolution is significantly finer for fixed observation times. These factors contribute to micro-Doppler signatures of birds and drones often being useful for classification in these bands.

The range to the target where such classification can be performed is of great interest. Results in this chapter point in the direction of potentially using long coherent integration times to ensure target detection. The same results also show that similar integration of micro-Doppler signatures is possible to increase the classification ranges. This can potentially pave the way for new radar architectures such as ubiquitous radar systems that can simplify hardware design and increase processing flexibility.

Traditional classification methods have demonstrated good results with regard to separating birds from UAVs. Complex environments can make it difficult to find effective features and the desire to separate different types of UAVs has motivated the usage of machine learning approaches. Classification using NN has shown exceptional results in micro-Doppler classification and is believed to be able to handle complex environments. The current state of the art appears to be centred around NNs, CNNs and RNNs combined with either spectra or spectrograms. Most researchers use the absolute value of the spectra or spectrograms and research on how to best represent sequential radar data needs to be conducted.

Detecting, tracking and classifying commercial drones is a challenge for radar systems and much more research is needed to counter the threat of future UAVs. The general usage of UAVs in the commercial sector is likely to increase, making the job of finding the UAVs with ill intentions harder. One should expect that the UAVs of tomorrow can feature reduced RCSs or will mimic the movement patterns of birds. Alternative propulsion systems, even gliders, might be utilised to reduce the impact of the propellers and confuse classification. Future UAVs will likely appear in swarms and might even mimic the behaviour of bird flocks.

Acronyms

CVD cadence-velocity diagram
NN neural network
FCN fully convolutional network
CNN convolutional neural network
RNN recurrent neural network
MLP multilayer perceptron
PCA principal component analysis
CW continuous wave
SVD singular-value decomposition
SVM support-vector machine
CPI coherent processing interval
FDTD finite difference time domain
FMCW frequency-modulated continuous wave
LSS low-slow-small
PEC perfect electric conductor
PRF pulse repetition frequency
RCS radar cross section

RPM revolutions per minute
STFT short-time Fourier transform
SNR signal-to-noise ratio
TRM transceiver/receiver module
UAV unmanned aerial vehicle
WBF wing beat frequency
NCTR non-cooperative target recognition
ISAR inverse synthetic aperture radar

References

[1]   Ploeger FW. Strategic Concept of Employment for Unmanned Aircraft Systems in NATO. Joint Air Power Competence Centre; January. 2010. pp. 1–28.

[2]   Mohajerin N, Histon J, Dizaji R, and Waslander SL. Feature extraction and radar track classification for detecting UAVs in civilian airspace. In: 2014 IEEE Radar Conference. Cincinnati, OH; 2014. pp. 0674–0679.

[3]   Torvik B, Olsen KE, and Griffiths HD. Classification of birds and UAVs based on radar polarimetry. *IEEE Geoscience and Remote Sensing Letters*. 2016;13(9):1305–1309.

[4]   Patel JS, Fioranelli F, and Anderson D. Review of radar classification and RCS characterisation techniques for small UAVs or drones. *IET Radar, Sonar & Navigation*. 2018;12(9):911–919.

[5]   Schroder A, Aulenbacher U, Renker M, *et al.* Numerical RCS and micro-Doppler investigations of a consumer UAV. SPIE Security + Defence, Edinburgh, United Kingdom; 2016.

[6]   Li CJ, and Ling H. An investigation on the radar signatures of small consumer drones. *IEEE Antennas and Wireless Propagation Letters*. 2017;16:649–652.

[7]   Guay R, Drolet G, and Bray JR. Measurement and modelling of the dynamic radar cross-section of an unmanned aerial vehicle. 2017;11:1155–1160.

[8]   Vaughn CR. Birds and insects as radar targets: A review. *Proceedings of the IEEE*. 1985;73(2):205–227.

[9]   Torvik B. Investigation of non-cooperative target recognition of small and slow moving air targets in modern air defence surveillance radar. PhD thesis, University College London; 2016.

[10] Blacksmith J P, and Mack RB. On measuring the radar cross sections of ducks and chickens. *Proceedings of the IEEE*. 1965;53(8):1125.

[11] J Farlik, M Kratky, Casar J, and Stary V. Radar cross section and detection of small unmanned aerial vehicles. In: 2016 17th International Conference on Mechatronics – Mechatronika (ME), Prague; 2016. pp. 5–7.

[12] Ritchie M, Fioranelli F, Griffiths H, and Torvik B. Micro-drone RCS analysis. In: 2015 IEEE Radar Conference, Johannesburg; 2015. pp. 452–456.

[13] Harman S. Characteristics of the radar signature of multi-rotor UAVs. In: Proceedings of the 13th European Radar Conference Characteristics; 2016. pp. 93–96.

[14]   Gersone F, Balleri A, Baker CJ, and Jahangir M. Simulations of L-band staring radar moving target integration efficiency. In: 2018 IEEE Conference on Antenna Measurements & Applications (CAMA), Vasteras; 2018. pp. 1–4.

[15]   Harman S. Analysis of the radar return of micro-UAVs in flight. In: 2017 IEEE Radar Conference (RadarConf), Seattle, WA; 2017. pp. 1159–1164

[16]   Molchanov P, Egiazarian K, Astola J, *et al.* Classification of small UAVs and birds by micro-Doppler signatures. *International Journal of Microwave and Wireless Technologies.* 2013;6(3-4):435–444.

[17]   Harmanny RIA, de Wit JJM, and Prémel Cabic G. Radar micro-Doppler feature extraction using the spectrogram and the cepstrogram. In: 2014 11th European Radar Conference, Rome; 2014. pp. 165–168.

[18]   Zhang P, Yang L, Chen G, *et al.* Classification of drones based on micro-Doppler signatures with dual-band radar sensors. In: 2017 Progress in Electromagnetics Research Symposium – Fall (PIERS – FALL), Singapore; 2017. pp. 638–643.

[19]   Zhang W, and Li G. Detection of multiple micro-drones via cadence velocity diagram analysis. *Electronics Letters.* 2018;54(7):441–443.

[20]   Regev N, Yoffe I, and Wulich D. Classification of single and multi propelled miniature drones using multilayer perceptron artificial neural network. In: International Conference on Radar Systems (Radar 2017), Belfast; 2017. pp. 1–5.

[21]   Kim BK, Kang HS, and Park SO. Drone classification using convolutional neural networks with merged Doppler images. *IEEE Geoscience and Remote Sensing Letters.* 2017;14(1):38–42.

[22]   Brooks DA, Schwander O, Barbaresco F, Schneider J, and Cord M. Temporal deep learning for drone micro-Doppler classification. In: 2018 19th International Radar Symposium (IRS), Bonn; 2018. pp. 1–10.

[23]   Knott EF, Shaeffer JF, and Tuley MT. *Radar Cross Section.* London: Artech House; 1993.

[24]   Pozar DM. *Microwave Engineering.* Hoboken, NJ: Wiley; 1997.

[25]   Gabriel C, Gabriel S, and Corthout E. The dielectric properties of biological tissues: I. Literature survey. *Physics in Medicine and Biology.* 1996;41:2231.

[26]   Taflove A, and Hagness SC. Computational Electrodynamics – The Finite-Difference Time-Domain Method. 3rd ed. Boston, MA: Artech House; 2005.

[27]   Gabriel C. *Dielectric Properties of Body Tissues.* Available from: http://niremf.ifac.cnr.it/tissprop/.

[28]   Houghton EW, and Blackwell F. Use of bird activity modulation waveforms in radar identification. Bird Strike Committee Europe Vojens, Denmark; 1972.

[29]   Lee JS, and Pottier E. *Polarimetric Radar Imaging: From Basics to Applications.* Boca Raton, FL: CRC Press; 2009.

[30]   Tait P. *Introduction to Radar Target Recognition.* IEE Radar, Sonar, Navigation, and Avionics Series. London: Institution of Engineering and Technology; 2005.

Chapter 9

Hardware development and applications of portable FMCW radars

Zhengyu Peng[1], Changzhi Li[2], Roberto Gómez-García[3], and José-María Muñoz-Ferreras[3]

The Doppler effect is the change in frequency or wavelength of an electromagnetic wave reflected from a target that is moving relative to the radar. In addition to the Doppler effect induced by the bulk motion of a radar target, the vibrations of the target and the motions of other mechanical structures in the target undergo micro-motion dynamics, which induce Doppler modulations on the returned signal, referred to as the micro-Doppler effect [1]. Micro-Doppler has been used in various applications, such as identifying specific types of vehicles and categorising activities of humans [2]. To obtain micro-Doppler signatures, different types of radar systems can be used. Among these kinds of radars, such as ultra-wideband (UWB) radar, Doppler radar, and frequency-modulated continuous-wave (FMCW) radar, FMCW radar is one of the most popular types in micro-Doppler detection. Compared with UWB radar, FMCW radar features simple architecture and easy signal processing [3–6]. Different from Doppler radar, which can only detect relative displacement of targets, FMCW radar is able to obtain both the absolute range and Doppler information of a target. This chapter introduces the fundamental theory, system design, prototyping, and signal processing of FMCW radars for micro-Doppler detection. At the beginning of the chapter, the basic theory of an FMCW radar will be introduced. Then, the design principles of each part of an FMCW radar, including the signal synthesiser, link-budget, basic antenna design, and receiver sensitivity will be discussed in detail. Novel hardware techniques for radar system integration and miniaturisation will also be covered.

Besides the hardware of FMCW radar, range-Doppler and micro-Doppler effects in radar signal processing will be discussed in detail in the following section. In discussing the range-Doppler and micro-Doppler effects, portable radar prototypes and realistic radar baseband signals will be used to give readers a better insight of radar signal processing, range-Doppler imaging and micro-Doppler effects. Finally,

[1]Aptiv Corporation, Kokomo, IN, USA
[2]Department of Electrical and Computer Engineering, Texas Tech University, Lubbock, TX, USA
[3]Department of Signal Theory and Communications, University of Alcalá, Alcalá de Henares, Madrid, Spain

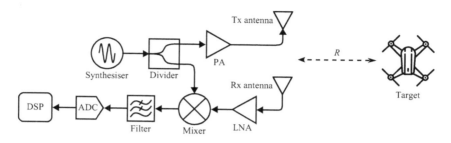

Figure 9.1   *Top-level architecture of an FMCW radar system*

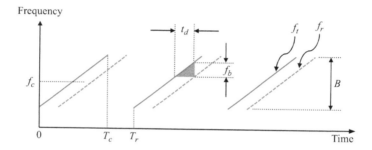

Figure 9.2   *Waveform of an FMCW radar*

emerging applications and trends for range-Doppler and micro-Doppler effects with FMCW radar will be briefly introduced.

9.1   FMCW radar fundamentals

The top-level architecture of an FMCW radar is shown in Figure 9.1. A typical FMCW radar consists of a transmitter and receiver. In the transmitter, a signal synthesiser generates a modulated continuous wave signal, which is also named as the chirp signal. Figure 9.2 plots the frequency versus time of chirp sequences. In a linear chirp, which is the most popular modulation type in FMCW radar, the instantaneous frequency $f(t)$ of a chirp varies linearly with time:

$$f(t) = f_c + kt \tag{9.1}$$

where f_c is the centre frequency, $-T_c/2 \leqslant t < T_c/2$, T_c is the duration of one chirp, and k is the slope of the chirp, which can be expressed as

$$k = \frac{B}{T_c} \tag{9.2}$$

where B is the bandwidth. The time-domain expression for the chirp can be written as follows [4]:

$$s_T(t) = A_T \exp\left(2\pi j \left(f_c t + 0.5kt^2\right) + j\phi_T\right) \tag{9.3}$$

where ϕ_T is the initial phase, and A_T is the amplitude. A power divider is used to distribute the chirp signal to the transmitter antenna and the mixer in the receiver. A power amplifier (PA) can be used before the transmitter antenna to increase the transmitting power level. This chirp signal $s_T(t)$ is transmitted into free space, and then reflected by a target. The reflected signal is collected by the receiver antenna of the FMCW radar. In the receiver, the received signal is processed by low-noise amplifiers (LNAs) and gain blocks. The received signal $s_R(t)$ can be written as

$$s_R(t) = A_R \exp\left(2\pi j \left(f_c(t - t_d) + 0.5k(t - t_d)^2\right) + j\phi_R\right) \tag{9.4}$$

where A_R is the amplitude of the received signal and ϕ_R is the phase that is introduced by the receiver channel. The round-trip delay

$$t_d = \frac{2R(\tau)}{c} \tag{9.5}$$

where $R(\tau)$ is the change of the target range versus "slow-time" τ.

The received signal is mixed with the replica of the transmitted chirp signal through a mixer. The baseband signal can be written as follows [4]:

$$s_b(t) = s_T(t)s_R^*(t) = A \exp\left(4\pi jkt_d t + 2\pi jf_c t_d + 2\pi jkt_d^2 + j\phi\right) \tag{9.6}$$

where $s_R^*(t)$ is the conjugation of the received signal. A is the amplitude of the baseband signal, and ϕ is the total phase change introduced by the transceiver path.

Among the four terms in (9.6), $2\pi kt_d^2$ is an unwanted phase term known as residual video phase (RVP). It is usually negligible, or otherwise can be compensated. ϕ is a constant phase which is introduced by the hardware system. This constant phase can be easily calibrated. After removing the negligible RVP and the constant phase, (9.6) can be rewritten as follows [4]:

$$s_b(t, \tau) = A \exp\left(j\frac{4\pi kR(\tau)}{c}t + j\frac{4\pi f_c R(\tau)}{c}\right) \tag{9.7}$$

For $-T_c/2 \leqslant t < T_c/2$, the frequency domain expression of the baseband signal $s_b(t, \tau)$ can be written as

$$S_b(f, \tau) = AT_c \exp\left(j\frac{4\pi f_c R(\tau)}{c}\right)\mathrm{sinc}\left(T_c\left(f - \frac{2kR(\tau)}{c}\right)\right) \tag{9.8}$$

There are two main parts in (9.8). The sinc function determines the peak location of the spectrum of the baseband signal. The peak of the sinc function is at

$$f_p = \frac{2kR(\tau)}{c} \tag{9.9}$$

It can be clearly seen that the peak location is directly related to the range of the target. Thus, with an FMCW radar, the range of a target can be easily obtained by finding the peak location of the baseband spectrum and performing a conversion from frequency to range by

$$R(\tau) = \frac{f_p c}{2k} = \frac{f_p c}{2BT_c} \tag{9.10}$$

The other important term in (9.8) is the "slow-time" phase:

$$\phi(\tau) = \frac{4\pi f_c R(\tau)}{c} \tag{9.11}$$

Let $f = f_p = 2\gamma R(\tau)/c$ in (9.8) since we are only interested in the position of the target. Equation (9.8) can be re-written as

$$S_b(\tau) = AT_c \exp\left(j\frac{4\pi f_c R(\tau)}{c}\right) \tag{9.12}$$

It can be seen that the phase of $S_b(\tau)$ is a function of the target's range evolution $R(\tau)$. Consider a simple case when the target is moving away from the radar at the initial range R_0 with a constant speed v. Thus,

$$R(\tau) = R_0 + v\tau \tag{9.13}$$

And (9.12) can be written as

$$S_b(\tau) = AT_c \exp\left(j\frac{4\pi f_c R_0}{c}\right) \exp\left(j\frac{4\pi f_c v\tau}{c}\right) \tag{9.14}$$

Another Fourier transform can be applied to $S_b(\tau)$ along the "slow-time" τ:

$$\mathscr{S}_b(l) = AT_c \exp\left(j\frac{4\pi f_c R_0}{c}\right) \text{sinc}\left(T_s\left(l - \frac{2f_c v}{c}\right)\right) \tag{9.15}$$

where l is the "slow-time" frequency, and T_s is the length of the total chirp sequence used for Fourier transform.

The peak of (9.15) is at

$$l = \frac{2f_c v}{c} \tag{9.16}$$

which is directly related to the speed of the target. Thus, with the Fourier transform along the "slow-time", the Doppler properties and velocity of the targets can be extracted. This method is called "range-Doppler" processing.

The performance of an FMCW radar can be characterised by three basic parameters. They are range resolution, range precision, and range accuracy. Range resolution (ΔR) defines the FMCW radar's capability of distinguishing two closely located targets in range. Figure 9.3 shows the case when two targets exist in the radar's field of view (FOV), and these two targets are very close to each other. If the two targets are too close, they cannot be separated in the range spectrum. The minimal distance between two targets resolvable by an FMCW radar is determined by the bandwidth of the chirp:

$$\Delta R = \frac{c}{2B} \tag{9.17}$$

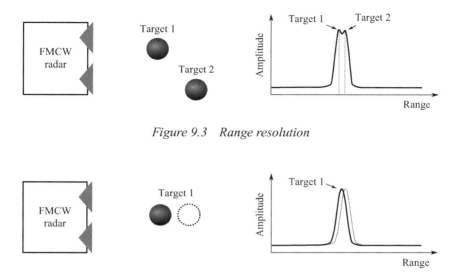

Figure 9.3 Range resolution

Figure 9.4 Detection precision

The range precision of an FMCW radar defines the minimal displacement a radar can differentiate for a single target, as shown in Figure 9.4. Signal-to-noise ratio (*SNR*) is the main factor that affects the range precision.

The range accuracy is another parameter which describes the error between the detected range and the ground truth. The range accuracy is related to systematic bias and *SNR*. Systematic bias includes path delays of the transmission lines in the radar transceiver and delays introduced by transceiver components. These delays usually cause a constant offset in detection, which can be minimised by careful calibration.

For speed detection with range-Doppler processing, a high-speed target will produce ambiguity in Doppler. If the moving direction of the target is known, the maximum unambiguous speed v_{ua} is

$$v_{ua} = \frac{c}{2T_r f_c} \tag{9.18}$$

where T_r is the chirp repetition period, shown in Figure 9.2. On the other hand, if the direction of the target is unknown,

$$v_{ua} = \pm \frac{c}{4T_r f_c} \tag{9.19}$$

where the sign indicates the directions of the motion.

It should be noted that FMCW radar suffers from range-Doppler coupling, which degrades the accuracy of range and Doppler detection. However range-Doppler coupling for short-range radars with slow moving targets is small and can be neglected [7].

9.2   Radar transceiver

The basic architecture of an FMCW radar transceiver is illustrated in Figure 9.1. On the transmitter channel, a chirp generator generates the frequency-modulated chirp signal repeatedly. This chirp signal is amplified by PA and then transmitted through the Tx antenna. A replica of the chirp signal is fed into the receiver channel as a local oscillator (LO) signal. On the receiver channel, the reflected chirp signal is processed by LNA and gain blocks. Then, a mixer is used to mix the received chirp signal with the replica of the transmitted chirp signal. This procedure is also known as deramping. The output of the mixer is the beat signal. Usually, a high-pass filter is necessary to remove the low-frequency mixer bias. An analog-to-digital converter (ADC) is then used to convert the beat signal into the digital domain for further processing.

9.2.1   Transmitter

The chirp generator is the key component in an FMCW radar transceiver. A chirp generator can be realised with various techniques, such as using a fractional-N phase-locked loop (PLL), a digital direct synthesiser (DDS) with an integer-PLL, or simply a free-running voltage-controlled oscillator (VCO) with sawtooth control voltages. Due to its good linearity and high integration capability, fractional-N PLL-based chirp generators have been widely used in commercial FMCW radar chips, especially in automotive radars working at 24 and 77 GHz. DDS-based chirp generators require a high-performance digital synthesiser, which is usually bulky and not optimal for compact applications. A free-running VCO is the simplest, but suffers from poor linearity.

Linearity is one of the most important characteristics for a chirp generator. The linearity of a chirp signal can be described by the frequency deviation Δf_{FMCW}, which is the frequency difference between the real chirp and the ideal chirp, as shown in Figure 9.5 [8].

$$\Delta f_{\mathrm{FMCW}}(t) = f_{\mathrm{ideal}}(t) - f_{\mathrm{real}}(t) \tag{9.20}$$

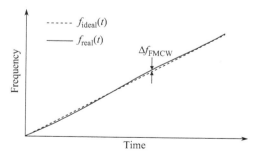

Figure 9.5   Chirp linearity

The frequency deviation can be further characterised as random deviations and periodic deviations based on their properties. Random deviations in the frequency ramp are caused by noise effects. These deviations cannot be classified directly as nonlinearities. The random deviations can be termed as random frequency ramp variations A_n (Hz). Periodic deviations in the frequency ramp are mainly caused by frequency spurs. These deviations may result from switching in digital circuits or transient response of the PLL. Periodic deviations can be characterised by a certain frequency deviation f_e (Hz) and deviation amplitude A_e (Hz). Thus, the real frequency ramp $f_{real}(t)$ can be described as

$$f_{real}(t) = f_{ideal}(t) + e(t) \tag{9.21}$$

with

$$e(t) = \underbrace{2\pi A_e \sin{(2\pi f_e t)}}_{\text{periodic}} + \underbrace{A_n(t)}_{\text{random noise}} \tag{9.22}$$

The impact of these two types of disturbances on the baseband beat signal is determined. The random noise in the chirp mainly affects the noise level of the baseband beat signal, which degrades the SNR of the range spectrum. On the other hand, the periodic deviations in the chirp could introduce some spurs in the range spectrum, which may trigger false detections [8].

9.2.2   Receiver

As shown in Figure 9.1, a typical receiver of an FMCW radar usually consists of receiver antennas, LNAs, a mixer, and a baseband circuitry. The LNAs are used to increase the signal power as well as maintain a relative low-noise floor of the receiver chain. The mixer is used as a multiplier which multiplies the received signal with a replica of the transmitted signal. The output of the mixer contains both the low-frequency beat signal and the high-frequency multiplication products. A filter is used in the baseband circuitry to remove the high-frequency multiplication products, as well as the DC bias. After the filter, an ADC is used to convert the analogue signal into the digital domain for further processing.

For the receiver chain, one of the most important properties is the coherence in processing, which means the phase and frequency used in the receiver should be the same as in the transmitter. For the radio frequency (RF) path, an FMCW radar is coherent in nature since the received signal is mixed with a replica of the transmitted signal. In order to preserve the slow-time phase history $\phi(\tau)$ in (9.11), it is usually required to share the clock across the whole system in a PLL-based FMCW radar to make sure the ADC starts at the same time during each chirp period (Figure 9.6). For an FMCW radar system with a free-running VCO, the sawtooth voltage sequence, which is used to control the frequency of the VCO, can be used to align the beat signal and maintain the coherence between chirp and beat signal [4,9].

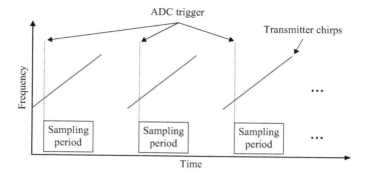

Figure 9.6   Receiver timing to maintain coherence

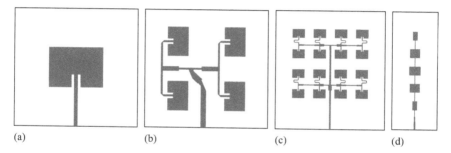

(a)   (b)   (c)   (d)

*Figure 9.7   Patch antennas examples [3,4,11]. (a) Single patch antenna. (b) 2 × 2
patch array. (c) 4 × 4 patch array. (d) 1 × 5 series-fed patch array*

9.3   Antenna and antenna array

The properties of the antennas in an FMCW radar determine several key specifications. For example, the beamwidth of the antenna determines the angular coverage,
i.e. the FOV, of the radar. The gain of the antenna impacts the detection range. Typically, antennas with higher gain increase the detection range of an FMCW radar;
however, increasing the gain narrows the beamwidth and reduces the FOV. In addition, increasing the carrier frequency of the radar helps to reduce the size of the
antenna. However, a higher frequency has a higher path loss and a shorter detection
range with the same transmitting power. It is essential to balance these characteristics
of the antennas based on different requirements for various applications [10].

Patch antennas have been widely used in portable FMCW radar systems due to
their low profile and low cost. Figure 9.7 shows four examples of patch antennas used
in portable FMCW radar systems. Figure 9.7(a) shows a single rectangular patch
antenna which has large FOV and low antenna gain. Figure 9.7(b) is a 2 × 2 patch
array, which has been used in [4]. The total size of the antenna in Figure 9.7(b) is

65 mm × 65 mm. Its half-power beamwidths in the E-plane and H-plane are 40.3° and 44.4°, respectively. The directivity is 11.8 dBi. Figure 9.7(c) is a 4 × 4 patch array, which has been used in [3]. The centre frequency of this antenna is 24 GHz, and its size is 45 mm × 50 mm. The half-power beamwidth in the E-plane is 16.5°, and on the H-plane, the beamwidth is 18.3°. The antenna directivity can reach to 19.8 dBi. Figure 9.7(d) is a 1 × 5 series-fed patch array, which features a wide beamwidth in the H-plane and a narrow beamwidth in the E-plane.

9.3.1 Beamforming

All the antennas shown in Figure 9.7 have fixed antenna beams. With fixed beam antennas, a mechanical steering system is necessary for an FMCW radar to perform a two-dimensional (2D) scan. However, a mechanical steering system increases the size, weight, and cost, while degrading the reliability. The beamforming technology features lightweight and high steering speed without any mechanical moving parts. It is optimal for applications with portable FMCW radars.

The beamforming technology utilises an array of antennas. By controlling the phase and amplitude of the narrow-band signal in each antenna channel, the array can form different radiation patterns for various applications. The radiation pattern of an array can be expressed as the multiplication of the embedded radiation pattern of each element and the array factor. Consider a simple case, where there is an N-element linear array. The array factor can be written as

$$AF = \sum_{n=1}^{N} w_n e^{jk\phi_n(\theta)} \tag{9.23}$$

$$\phi_n(\theta) = x_n \sin \theta \tag{9.24}$$

where N is the number of elements, k is the wave number, which equals $2\pi/\lambda$, λ is the wavelength, x_n is the location of the nth element, and θ is the broadside angle. w_n is the complex weight of the nth element, which corresponds to the phase and amplitude of the excitation.

For a half-wavelength-spaced array, AF can be simplified to

$$AF = \sum_{n=1}^{N} w_n e^{j(n-1)\pi \sin \theta} \tag{9.25}$$

For a 16-element linear array with an uniformed excitation, where

$$w = [w_1, w_2, \ldots, w_{16}] = [1, 1, \ldots, 1] \tag{9.26}$$

the corresponding radiation pattern has a main lobe pointing to 0° (broadside) with 12.8 dB sidelobe level, as shown in Figure 9.8.

The direction of the main lobe can be steered to a different angle by adding phase components into w. For example, let

$$w = [1, e^{-j\pi \sin(-40°)}, e^{-j2\pi \sin(-40°)}, \ldots, e^{-j15\pi \sin(-40°)}] \tag{9.27}$$

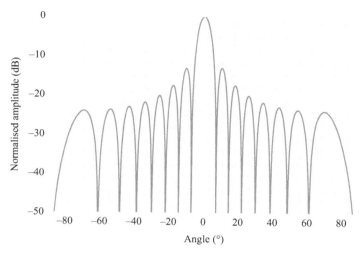

Figure 9.8   λ/2 spaced 16-element array factor with uniform excitation (broadside direction)

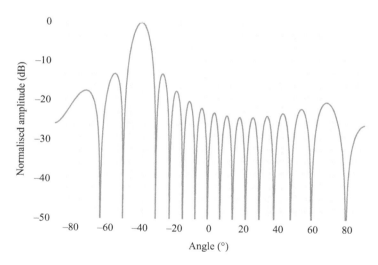

Figure 9.9   λ/2 spaced 16-element array factor with the main lobe steered to −40°

The main lobe of the radiation pattern will be directed to −40°, as shown in Figure 9.9. The sidelobe level is 12.8 dB.

In addition, to further changing the amplitude of the weights, this will offer more flexibility in controlling the array radiation patterns. A simple example is applying the popular Chebyshev window to the weight *w*, as shown in Table 9.1. The corresponding

Table 9.1   Weighting for a Chebyshev window and a −40° pointing

w_1	0.1138	w_2	$0.1964\,e^{-j\pi\,\sin(-40°)}$
w_3	$0.3319\,e^{-j2\pi\,\sin(-40°)}$	w_4	$0.4926\,e^{-j3\pi\,\sin(-40°)}$
w_5	$0.6613\,e^{-j4\pi\,\sin(-40°)}$	w_6	$0.8163\,e^{-j5\pi\,\sin(-40°)}$
w_7	$0.9353^{-j6\pi\,\sin(-40°)}$	w_8	$e^{-j7\pi\,\sin(-40°)}$
w_9	$e^{-j8\pi\,\sin(-40°)}$	w_{10}	$0.9353^{-j9\pi\,\sin(-40°)}$
w_{11}	$0.8163\,e^{-j10\pi\,\sin(-40°)}$	w_{12}	$0.6613\,e^{-j11\pi\,\sin(-40°)}$
w_{13}	$0.4926\,e^{-j12\pi\,\sin(-40°)}$	w_{14}	$0.3319\,e^{-j13\pi\,\sin(-40°)}$
w_{15}	$0.1964\,e^{-j14\pi\,\sin(-40°)}$	w_{16}	$0.1138\,e^{-j15\pi\,\sin(-40°)}$

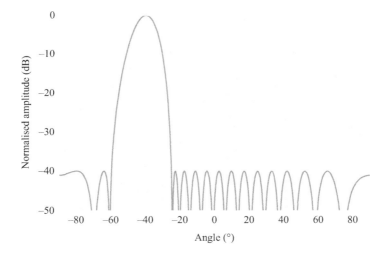

Figure 9.10   λ/2 spaced 16-element array factor with the main lobe steered to −40° and Chebyshev window

figure is illustrated in Figure 9.10. It can be seen that a radiation pattern pointing to −40° with −40 dB sidelobe can be obtained. Other windows can also be used to achieve radiation patterns with different sidelobe properties [12].

It is not always possible to have an array with λ/2 spacing, due to the limitations of the antenna size and mutual coupling. In addition, increasing the spacing of an array can help to obtain a larger aperture and improve the angular resolution. However, having a large spacing between antenna elements introduces grating lobes. Figure 9.11 illustrates an example of a 16-element linear array with λ spacing between elements. The same weighting values shown in Table 9.1 have been applied to the 16 antenna elements. In the radiation pattern, it can be seen that besides the desired main lobe at −40°, a grating lobe with the same amplitude as the main lobe appears at around 20°.

In radar applications, it will not be possible for the radar to distinguish if the target is from the direction of the main lobe or the grating lobe. Thus, with this array, the FOV of the radar has to be limited within the angles around the broadside direction without the grating lobes to avoid target aliasing.

Typically, beamforming can be realised either in the RF domain or in the digital domain. For the beamforming in the RF domain, which is also recognised as the RF beamforming, a phase shifter is required for each antenna channel to control the phase of the RF signal. The block diagram of an RF beamforming receiver architecture is illustrated in Figure 9.12. As shown in the figure, the RF beamforming receiver

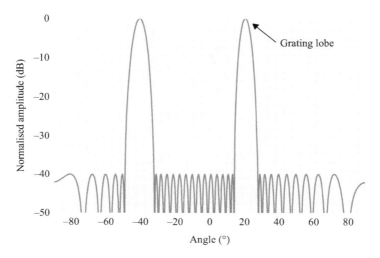

Figure 9.11 λ *spaced 16-element array factor with the main lobe steered to* $-40°$ *and Chebyshev window*

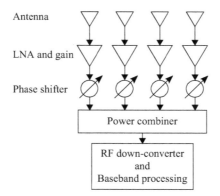

Figure 9.12 *RF beamforming architecture*

consists of an antenna array. Each antenna element in the array is connected with a dedicated LNA and RF gain modules, as well as some necessary filters. Then, a phase shifter is connected to each channel to control the phase of the RF signal. After the phase shifter, all the channels are combined together and down-converted to baseband for further processing. Compared with the digital beamforming architecture, the RF beamforming architecture needs relatively less computational resources for baseband processing and lower power consumption. However, due to the accuracy limitation of RF phase shifters, RF beamforming architecture suffers from beamforming accuracy. In addition, the costs of phase shifters are very high at frequencies above K-band. New RF beamforming techniques are emerging for high frequencies. Researchers have been using delay-line-based techniques [13,14], such as the Butler matrix [15], for low-cost applications. However, delay-line-based solutions have few steering angles due to the limitations in size and signal routing. Researchers have also proposed an LO phase-shifting solution which utilises a complex multi-phase VCO and multi-phase select switches. In this method, the number of steering angles is limited by the number of phase paths and switches of the multi-phase VCO. The design of a multi-phase VCO is also very challenging.

For a highly integrated RF beamforming system, the vector modulator technique is favourable. Figure 9.13 illustrates the conceptual block diagram of a vector modulator. For an RF input signal $x = \sin(2\pi f_c t)$, it is first divided into an in-phase (I) signal and a quadrature (Q) signal by a quadrature power divider. f_c is the carrier frequency, and t is time. Then, each of the I and Q signals is multiplied with a constant value. After multiplication, these two signals are added together. The output signal y can be expressed as

$$y = A_I \sin(2\pi f_c t) + A_Q \cos(2\pi f_c t) = \mathrm{Re}\{Ae^{j\phi}e^{-j2\pi f_c t}\} \tag{9.28}$$

$$A = \sqrt{A_I^2 + A_Q^2} \tag{9.29}$$

$$\phi = \arctan\frac{A_Q}{A_I} \tag{9.30}$$

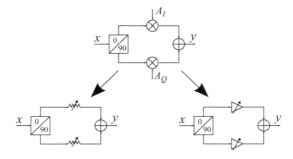

Figure 9.13   Conceptual representation of a vector modulator

By comparing the output signal y with the input signal x, it can be clearly seen that the output signal has a phase shift of ϕ and an amplitude change of A. In circuit realisation, the multiplication parts in the vector modulator can be realised by variable attenuators or variable gain amplifiers (VGAs), as illustrated in Figure 9.13. It should be noted that in order to achieve 360° phase control, the variable attenuators or the VGAs must have 180° phase-shift capability. Based on this concept, integrated phased-array chips working at frequencies higher than 24 GHz have been developed by researchers. A printed circuit board realisation of a 24 GHz RF beamforming radar system has also been developed based on vector modulator with RF attenuators. Two-dimensional target mapping result in short-range localisation application has been reported.

Digital beamforming techniques have been widely used in modern communication and radar systems, thanks to the advance of high-speed ADCs and digital-to-analog converters (DACs). In a digital beamforming architecture, the phase shift and amplitude control are realised in the digital domain. High accuracy in both phase shift and amplitude control can be achieved in the digital domain. Figure 9.14 shows an example of a four-channel digital beamforming receiver. In this receiver, the RF signal received by each antenna channel is processed by a dedicated RF front end, which includes LNAs, gain blocks, filters, and a frequency down-converter. The down-converted signal is then sampled by an ADC into the digital signal. The phase and amplitude of each channel are controlled in the digital domain by using basic digital operators. The digital beamforming architecture has high beamforming accuracy and flexibility. However, due to the massive use of ADCs or DACs, the power consumption is usually high. Researchers have been working on solutions to reduce the required number of ADCs or DACs in a digital beamforming system. Techniques such as sparse array, compressive sensing, path sharing and multiple-input and multiple-output (MIMO) have been investigated to effectively reduce the hardware intensity of digital beamforming systems.

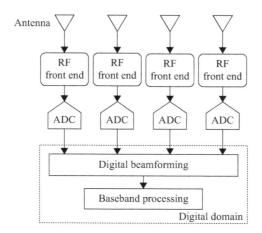

Figure 9.14 Digital beamforming architecture

9.3.2   *Two-way pattern and MIMO*

For a conventional radar system, which includes separated transmitter antenna array and receiver antenna array, the effective radiation pattern is the combination of the transmitter pattern and receiver pattern. The combined pattern is called two-way pattern, which is the multiplication of the transmitter pattern and the receiver pattern. Since an antenna or an antenna array has nulls and lobes in its radiation pattern, it is possible to use the nulls in the transmitter pattern to cancel the unwanted lobes in the receiver pattern, and vice versa. One of the appealing cases is using the two-way pattern to remove the grating lobes.

Figure 9.15 shows an example of the linear array configuration with four transmitter antenna elements and four receiver antenna elements. The distance between the four transmitter antenna elements is 2λ, and the spacing between the receiver antenna elements is $\lambda/2$. The radiation patterns of the transmitter array and the receiver array when the main lobes are steered to $-20°$ are illustrated in Figure 9.16.

Figure 9.15   An example of array configuration with four transmitter antennas and four receiver antennas

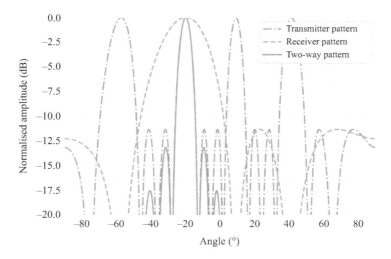

Figure 9.16   Radiation patterns for the linear arrays in Figure 9.15 and two-way radar radiation pattern

Since the transmitter array is a sparse array, three grating lobes can be observed at $-58°$, $9°$, and $42°$. The half-power beamwidth of the main lobe in the transmitter pattern is about $8°$. On the other hand, the pattern of the receiver has one wide main lobe without any grating lobes. It can be also observed that the directions of the three grating lobes on the transmitter patterns are located exactly at the directions of three nulls on the receiver patterns. Thus, the signals transmitted through the grating lobes are rejected by the receiver array. The effective pattern, namely the two-way pattern, of the transceiver array can be obtained by multiplying the transmitter pattern with the receiver pattern, as shown in Figure 9.16. This two-way pattern has a main lobe with $8°$ half-power beamwidth, and it does not have grating lobes.

As shown in Figure 9.16, the transmitter array and the receiver array have to steer at the same time to form the wanted two-way pattern that rejects the grating lobes. Conventional methods to obtain the wanted two-way patterns include implementing either RF beamforming or digital beamforming on both the transmitter array and the receiver array. However, these conventional methods significantly increase the overall hardware intensity and cost. MIMO is a novel technique that utilises the benefit of the combined patterns. Generally, MIMO radar systems transmit orthogonal waveforms. On the receiver, the signals from different transmitter antennas are separated based on their orthogonality. Then, the transmitter pattern and the receiver pattern can be formed at the same time on the receiver. In an FMCW radar, the orthogonal waveforms can be realised by using phase modulation or time-division multiplexing (TDM). With phase modulation, all the transmitter channels can transmit simultaneously. On the receiver, the same phase codes are used to demodulate the received signal and separate the signal from each transmitter channel. On the other hand, the TDM method requires each transmitter antenna transmit at its own time slot. The receiver channels separate signals from different transmitter antenna channels based on different time. Since TDM methods do not require modulation or demodulation, it has much simpler hardware architecture than the phase modulation method. However, transmitting signals at different time slots increases the chirp repetition period of an FMCW radar, which decreases the maximum unambiguous Doppler frequency in range-Doppler processing.

In analysing the antenna array properties of a MIMO radar, using the concept of virtual elements is very convenient. As has been mentioned, the two-way pattern is the multiplication of the transmitter pattern and the receiver pattern. It is also known that for an equally spaced array, the radiation pattern is the Fourier transform of the distribution of the array elements. Thus, the two-way pattern can also be obtained by performing Fourier transform for a virtual array, which is the convolution of the transmitter array distribution and the receiver array distribution. Figure 9.17 shows several examples of virtual array. Figure 9.17(a) is the same array configuration shown in Figure 9.15, which has four transmitter antennas and four receiver antennas. The virtual array has eight antenna elements with $\lambda/2$ spacing. Figure 9.17(b) is an "L"-shaped array configuration, which has eight transmitter elements placed horizontally and eight receiver elements placed vertically. With this array configuration, an 8×8

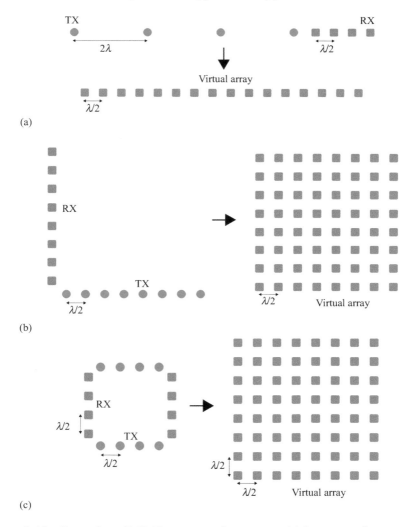

Figure 9.17   Examples of MIMO array configurations. (a) Linear configuration. (b) "L" shape configuration. (c) Square configuration

planar array can be synthesised, which has the capability to perform beam steering on both the horizontal plane and the vertical plane. Figure 9.17(c) is a square array configuration, which has two horizontal lines of transmitter elements and two vertical lines of receiver elements. An 8×8 planar virtual array can be formed with this configuration [6,16,17].

9.4   Radar link budget analysis

In evaluating the detection capability of an FMCW radar, a link budget analysis can be performed. For an FMCW radar shown in Figure 9.18, the power at different locations can be expressed as follows:

a.   Effective radiation power at the transmitter antenna:

$$PG_t \tag{9.31}$$

b.   Signal radiation power density at the target location:

$$\frac{PG_t}{4\pi R^2} \tag{9.32}$$

c.   Signal power reflected by the target:

$$\frac{PG_t\sigma}{4\pi R^2} \tag{9.33}$$

d.   Signal radiation power density at the receiver antenna:

$$\frac{PG_t\sigma}{(4\pi)^2 R^4} \tag{9.34}$$

e.   Signal power received by the receiver antenna:

$$\frac{PG_t\sigma A}{(4\pi)^2 R^4} \tag{9.35}$$

where P is the transmitted power, G_t is the transmitter antenna gain, σ is the radar cross section (RCS) of the target, R is the range, and A is the effective aperture of the receiver antenna.

The noise of the receiver can be calculated as

$$N = kT_0BF_n \tag{9.36}$$

where $k = 1.38 \times 10^{-23}$ J/K is Boltzmann's constant, T_0 is the noise temperature, B is the receiver bandwidth, and F_n is the noise figure of the receiver.

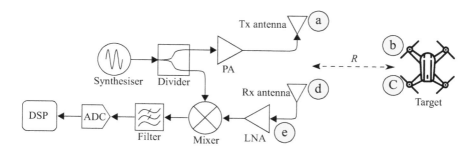

Figure 9.18   Link budget

Thus, the overall *SNR* of the radar is

$$SNR = \frac{PG_t\sigma A}{(4\pi)^2 R^4 kT_0 BF_n L} \tag{9.37}$$

where L is the extra loss or gain introduced in the receiver.

Assume the minimal detectable *SNR* is $SNR_{\min}$, then the maximum detection range $R_{\det}$ for a target with RCS σ can be obtained as

$$R_{\det} = \sqrt[4]{\frac{PG_t\sigma A}{(4\pi)^2 SNR_{\min} kT_0 BF_n L}} \tag{9.38}$$

which is the well-known radar equation.

9.5   FMCW radar signal processing

In this section, the basic signal processing procedures for an FMCW radar is introduced. The measurement data from an FMCW radar prototype are used to demonstrate the procedures. The radar prototype used in the measurement is a C-band short-range FMCW radar, shown in Figure 9.19. Table 9.2 lists the basic parameters of the radar prototype. The centre frequency of this radar prototype is 6 GHz and its bandwidth is 320 MHz. The chirp repetition of the radar prototype is 3.5 ms. The baseband signal of the radar prototype is sampled by the audio card of a computer. The sampling rate

Figure 9.19   FMCW C-band radar prototype [4]

Table 9.2   Parameters of the C-band radar prototype

Centre frequency	6 GHz	Transmitted power	8 dBm
Antenna directivity	11.8 dBi	Beamwidth (E-plane)	40.3°
Beamwidth (H-plane)	44.4°	Bandwidth	320 MHz
Chirp repetition period (T)	3.5 ms	Sampling rate (f_s)	192 ksps
Range resolution	0.67 m	Maximum range	16.4 m

is 192 ksps. With these configurations, the range resolution of the radar is 0.67 m with a maximum range of 16.4 m.

9.5.1 Range processing

The C-band radar prototype samples two-channel data, as shown in Figure 9.20. The first channel is the chirp trigger pulse which aligns the chirp sequences on the transmitter. The other channel is the beat signal sequence. To process the beat signal, the beat signal sequence is sliced into sections, i.e. Beat 1, Beat 2, ..., and Beat N, according to the trigger sequence. The edges of each beat signal sections need to be discarded to avoid edging effects of chirps. An example of raw baseband data at C-band is illustrated in Figure 9.21.

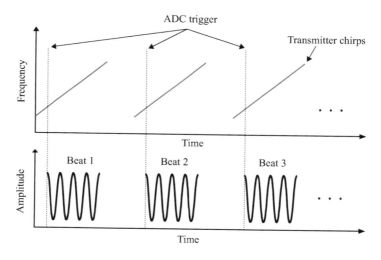

Figure 9.20   Formats of the beat signals

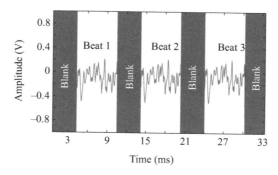

Figure 9.21   Discarding of samples in each FMCW period

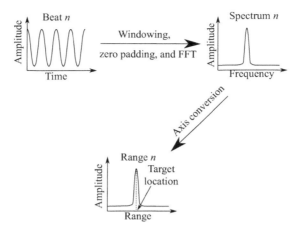

Figure 9.22   Obtaining of range profile

The procedure to obtain a range profile from a section of the beat signal, Beat *n*, is illustrated in Figure 9.22. First, a window is applied to this time-domain beat signal. A Chebyshev window and a Taylor window are the most widely used windows for radar signal processing. A Chebyshev window provides the narrowest possible main lobe for a specified sidelobe level. A Taylor window allows trade-offs between the main lobe width and the sidelobe level. A Taylor window also avoids edge discontinuity which is an issue for a Chebyshev window. After applying the windows and necessary zero padding, fast Fourier transform (FFT) is then used to convert the time-domain signal into the frequency domain. The windows are usually used to achieve a certain sidelobe rejection performance, and zero padding is used to improve FFT frequency resolution. Since the frequencies of the beat signal and the ranges of the targets correspond one-to-one, the range profile can be obtained through a simple axis conversion.

Figure 9.23 shows an experiment with a human target walking back and forth in front of the radar prototype. The horizontal axis corresponds to the range of the human target, and the vertical axis is the measurement time. The bright curve is the range history of the human target.

9.5.2   Range-Doppler processing

In a crowded environment, strong clutter from stationary surroundings makes it very difficult for an FMCW radar to discriminate wanted moving targets. Range-Doppler imaging in an FMCW radar can be used to extract Doppler information from moving targets. These 2D diagrams help to isolate moving targets from stationary clutter.

Figure 9.24 details the major steps to obtain the range-Doppler maps. The first step of range-Doppler imaging is obtaining and arranging the raw data matrix, which consists of arrays of beat signals. The time interval along the horizontal axis in the raw

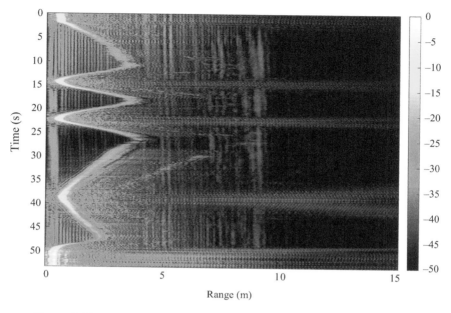

Figure 9.23 Range-slow-time map for a pedestrian. Adapted from [18]

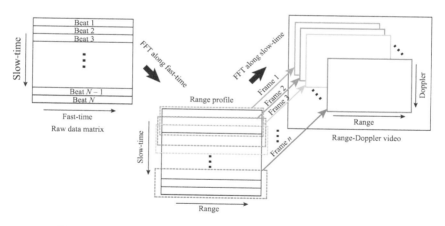

Figure 9.24 Obtaining the video of range-Doppler maps [3]

data matrix is $1/f_s$, where f_s is the sampling rate of the ADC. $1/f_s$ is also called "fast-time." On the vertical axis, the time interval is T, which is the chirp repetition period. T is also named as "slow-time." The range profile can be easily obtained by performing FFT along the fast-time after windowing and zero padding. A series of range-Doppler diagrams can be obtained by windowing the range profile and performing FFT along

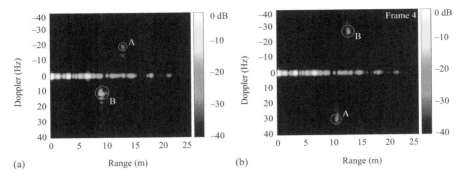

Figure 9.25   Examples of range-Doppler maps for two pedestrians [4]

the slow-time. These range-Doppler diagrams can be combined into a video for further processing.

Figure 9.25 shows an experiment of range-Doppler imaging with two human subjects (Subjects A and B) walking in front of the C-band radar prototype. Figure 9.25(a) shows the measurement when Subject A was walking toward the radar and Subject B was walking away from the radar. On the other hand, Figure 9.25(b) illustrates the time when Subject B was walking toward the radar and Subject A was leaving the radar. It can be clearly seen that the target has a negative Doppler frequency when it is moving toward the radar. Moreover, the target has a positive Doppler frequency when it is moving away from the radar. All the stationary targets in this experiment, such as the clutter from walls and the ground, have zero Doppler. In addition, micro-Doppler from the human subjects can also be observed surrounding the signatures of the moving subjects. Researchers have been investigating the potential applications of range-Doppler imaging in various areas, such as fall detection, gesture recognition, and active shooter detection.

9.5.3   Micro-Doppler

The micro-Doppler effect indicates the Doppler effects introduced by the vibrations and motions of other mechanical structures of a target undergoing micro-motion dynamics, other than the bulk motion of the target [1]. Micro-Doppler has been widely used in various applications, including identifying different types of vehicles, identifying the models and types of drones, and categorising activities of humans. Doppler radar has been used in obtaining the micro-Doppler signatures for gesture recognition, fall detection, etc. However, since Doppler radar, which transmits a single-tone signal, does not have the capability to obtain the range of a target, micro-Doppler features and Doppler signatures from multiple moving targets in different ranges are overlapped, which makes it almost impossible to discriminate the features in these scenarios. On the other hand, the range information can be easily obtained with an FMCW radar after range FFT for beat signals. The Doppler signatures and micro-Doppler features can also be extracted by range-Doppler processing for an

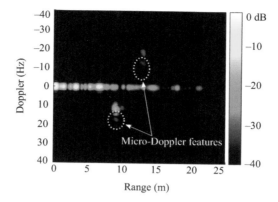

Figure 9.26   Micro-Doppler features of walking human subjects

FMCW radar. With the capabilities to obtain both the range and Doppler information of targets, FMCW radar is an optimal type of radar systems for micro-Doppler applications. Figure 9.26 illustrates the micro-Doppler features obtained from range-Doppler processing of the two walking human subjects in Figure 9.25. The bright dots in Figure 9.26 indicate the moving of the human bodies. A spreading in the Doppler response can be observed around the signatures of the human subjects, indicating the micro-Doppler features of moving gaits.

9.6   Applications of micro-Doppler effects

In this section, several emerging applications of micro-Doppler are introduced.

9.6.1   Gesture recognition

Advanced human–machine interfaces have been using optical cameras and image processing algorithms [19]. It has been extensively studied in the area of computer science. However, optical camera-based solutions and their algorithms are very high demanding for computational resources and are sensitive to ambient light. These two issues hinder a wide use of camera-based gesture recognition, especially for portable devices with limited computational capabilities. Doppler radar sensors show promising performance in detecting motion properties of targets [20]. Simple algorithms, such as time–frequency analysis, can be used in human gesture recognition. The advantages for a Doppler radar in gesture recognition include high integration and low demand for computational resources. However, one of the major issues for Doppler radar is that it is unable to separate motions from multiple targets. On the other hand, FMCW radar provides range information from targets which helps to separate motions from targets in different ranges. FMCW radar can have almost the same hardware architecture as a Doppler radar, and its signal processing is a little more complex than that of a Doppler radar. However, compared with camera-based

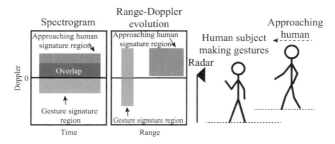

Figure 9.27   *Illustration of a human making gestures while another human subject is approaching the radar, and corresponding spectrogram and range-Doppler image regions [21]*

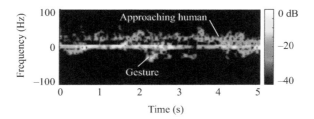

Figure 9.28   *Spectrogram of two moving targets after conventional time-frequency micro-Doppler analysis [21]*

solutions, FMCW radar still requires much less computational resources. Figure 9.27 demonstrates a complex case when two moving targets are in front of an FMCW radar [21]. In this scenario, one human subject is making gestures and the other person is approaching the radar. For a Doppler radar, the signatures of the two moving activities are overlapped, which makes it very difficult to separate. However, for an FMCW radar, it is quite straightforward to separate the two moving targets based on their ranges, and to recognise the gestures based on their micro-Doppler and micro-range signatures. The experiment has been taken with two human subjects illuminated by a 5.8 GHz Doppler radar prototype and a 5.8 GHz FMCW radar prototype [4]. In this experiment, one person was making gestures in front of the radar, while the other was walking toward the radar. The results of the experiment match well with the analysis. The Doppler radar failed to separate the two motions, shown in Figure 9.28, and the FMCW radar can successfully separate the two targets based on their ranges and identify the gesture of one of the subjects by its micro-Doppler and micro-range features, shown in Figure 9.29.

Google's project Soli [22] is one of the best-known works for human gesture recognition with FMCW radar based on micro-Doppler features and machine learning. The Soli sensor was designed and developed at 60 GHz with antenna-in-package (AiP) technology by Infineon Technologies. The Soli sensor has been demonstrated to be a promising technology that can enhance and improve user interaction experience.

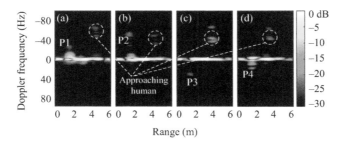

Figure 9.29   *Range-Doppler frames of gesture recognition with two moving targets.*
            (a) Phase 1. (b) Phase 2. (c) Phase 3. (d) Phase 4 [21]

9.6.2   Fall detection

With population aging, the health of the elderly has become a major concern in the society. Accidental fall is one of the most dangerous threats to the health of seniors. To minimise the harm of a fall accident, it is essential to immediately alert family or caretakers to give a necessary assistance. Researchers have been working on different approaches to detect and identify fall accidents from other daily activities. These approaches include accelerometer-based solutions [23], camera-based solutions [24–26], and radar-based systems [27–29]. Accelerometer-based techniques, as well as other wearable devices, require subjects to wear sensors all the time. The batteries of these kinds of sensors need to be recharged or replaced frequently. Camera-based solutions use algorithms to extract motion features of human subjects. However, these algorithms usually require high computational resources and do not have sufficient accuracy in relatively complex environments. Moreover, camera-based solutions require the targets to be fully within the line of sight. Obstacles with even partial blockage can fail the detection for camera-based solutions. Doppler radar has also been used for fall detection by detecting micro-Doppler effects of fall accidents. Unlike wearable solutions, Doppler radar does not need subjects to wear anything, which makes it more user-friendly. In addition, compared with camera-based solutions, Doppler radar is not demanding for computational resources and it is able to penetrate obstacles. However, Doppler radar is not capable to isolate the fall incident with the presence of other moving activities.

Fall detection with FMCW radar has been investigated in [29]. Besides the advantages of low requirements for computational resources and penetration capability, FMCW radar provides absolute range information of targets which helps to isolate multiple moving targets. Range-Doppler imaging methods have been used to process the FMCW radar signal. By analysing the changes of the RCS, range, and Doppler on range-Doppler images during the movement of a subject, fall accidents can be distinguished from normal activities. Figure 9.30 illustrates the case when a human subject falls toward the radar. The falling event can be divided into four featured phases based on the changes of velocity, RCS, and range of the human subject. At P1, the human subject starts to fall, with relatively low speed. Thus the signature in range-Doppler

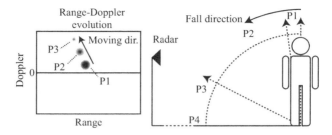

Figure 9.30 Illustration of the actions of a human subject and the corresponding range-Doppler image evolutions when the human subject falls toward the radar [29]

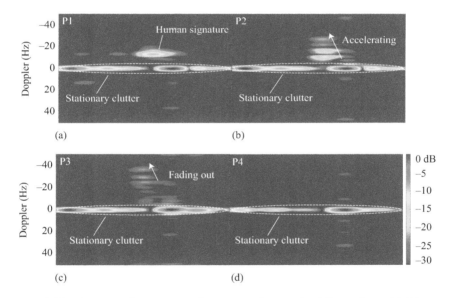

Figure 9.31 Frames of range-Doppler images during the fall incident. (a) P1. (b) P2. (c) P3. (d) P4 [29]

image has low Doppler frequency. At time P2, the signature of the human subject has a sudden Doppler change, which is caused by the acceleration during the fall incident. When the human subject continues to fall, the RCS decreases due to the large tilt angle of the human subject at P3. Finally, the human subject hits the ground, and the signature in the range-Doppler image disappears. Experiment result shown in Figure 9.31 matches well with the analysis at the four time instances. The results of this research may enable FMCW radar-based long-term remote fall detection without requiring seniors to wear any sensing device in home and nursing environments.

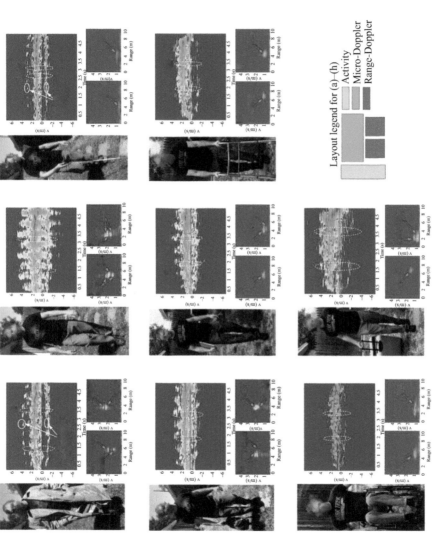

Figure 9.32 Micro-Doppler and range-Doppler for different activities. (a) Walking with a concealed rifle. (b) Walking with natural arm swings. (c) Walking with a cane for a blind person. (d) Walking while holding a gym bag. (e) Walking while carrying a laptop. (f) Walking with a walker. (g) Moving in a wheelchair. (h) Walking with a rolling suitcase [32]

9.6.3 Human activity categorising

Micro-Doppler features have been used to characterise and differentiate humans, animals, and vehicles [2]. Researchers have found that it is possible to use micro-Doppler features to estimate human motion and determine if the human subject is carrying a weapon [30]. The number of active shooting incidents rises year by year according to the latest US Federal Bureau of Investigation (FBI) report. The number of active shooting incidents in the period of 2014–15 is six times higher than that during the 2000–01 period. Current technologies for detecting active shooters include acoustic gunshot identification and infrared camera gunfire flash detection. The main problem of these two solutions is that they can only be triggered after a weapon is fired [31]. Although they can help to reduce the response time for the first responders, they do not prevent those incidents from happening. Thus, researchers are exploring radar-based approaches to detect shooters armed with concealed rifle/shotgun before the shooter draws the weapon. Micro-Doppler and range-Doppler signatures have been used in active shooter detection in [32] with a portable radar system. In this work, the authors analyse the special gaits and gesture characteristics of a person walking with a concealed rifle. An artificial neural network is adopted to classify a person walking with a concealed weapon, as well as seven other activities. Figure 9.32 shows the measured results and analysis of micro-Doppler and micro-range features among different activities. The classification results showed an average accuracy of 99.21% for differentiating a subject walking with a concealed rifle and other seven similar activities.

9.7 Summary

This chapter has provided an overview of the radar fundamentals, system design, prototyping, and signal processing of FMCW radars for micro-Doppler detection. The design principles of each part of an FMCW radar, including the radar transceiver, link budget, antenna design, etc., have been introduced and discussed. Besides the hardware design, range-Doppler and micro-Doppler effects in radar signal processing have also been discussed in detail. Finally, a brief introduction has been given in this chapter to cover the emerging applications for range-Doppler and micro-Doppler effects.

References

[1] Chen VC, Li F, Ho SS, and Wechsler H. Micro-Doppler effect in radar: phenomenon, model, and simulation study. *IEEE Transactions on Aerospace and Electronic Systems*. 2006;42(1):2–21.

[2] Tahmoush D, Silvious J, and Clark J. An UGS radar with micro-Doppler capabilities for wide area persistent surveillance. In: Ranney KI, Doerry AW, editors. Proceedings SPIE. vol. 7669. International Society for Optics and Photonics; 2010. pp. 766904-1–766904-11.

[3]   Peng Z, Muñoz-Ferreras JM, Gómez-García R, *et al.* 24-GHz Biomedical radar on flexible substrate for ISAR imaging. In: IEEE MTT-S International Wireless Symposium (IWS). IEEE; 2016. pp. 1–4.

[4]   Peng Z, Muñoz-Ferreras JM, Tang Y, *et al.* A portable FMCW interferometry radar with programmable low-IF architecture for localization, ISAR imaging, and vital sign tracking. *IEEE Transactions on Microwave Theory and Techniques.* 2017;65(4):1334–1344.

[5]   Feger R, Wagner C, Schuster S, *et al.* A 77-GHz FMCW MIMO radar based on an SiGe single-chip transceiver. *IEEE Transactions on Microwave Theory and Techniques.* 2009;57(5):1020–1035.

[6]   Bleh D, Rosch M, Kuri M, *et al.* W-band time-domain multiplexing FMCW MIMO radar for far-field 3-D imaging. *IEEE Transactions on Microwave Theory and Techniques.* 2017;65(9):3474–3484.

[7]   Stove AG. Linear FMCW radar techniques. *IEE Proceedings F – Radar and Signal Processing.* 1992;139(5):343–350.

[8]   Ayhan S, Scherr S, Bhutani A, *et al.* Impact of frequency ramp nonlinearity, phase noise, and SNR on FMCW radar accuracy. *IEEE Transactions on Microwave Theory and Techniques.* 2016;64(10):3290–3301.

[9]   Peng Z, Muñoz-Ferreras JM, Tang Y, *et al.* Portable coherent frequency-modulated continuous-wave radar for indoor human tracking. In: IEEE Topical Conference on Biomedical Wireless Technologies, Networks, and Sensing Systems (BioWireleSS). IEEE; 2016. pp. 36–38.

[10] Visser HJ. *Array and Phased Array Antenna Basics.* Hoboken, NJ: John Wiley & Sons; 2006.

[11] Peng Z, Ran L, and Li C. A κ-band portable FMCW radar with beamforming array for short-range localization and vital-Doppler targets discrimination. *IEEE Transactions on Microwave Theory and Techniques.* 2017;65(9):3443–3452.

[12] Harris FJ. On the use of windows for harmonic analysis with the discrete Fourier transform. *Proceedings of the IEEE.* 1978;66:51–83.

[13] Lee MS, and Kim YH. Design and performance of a 24-GHz switch-antenna array FMCW radar system for automotive applications. *IEEE Transactions on Vehicular Technology.* 2010;59(5):2290–2297.

[14] Lee W, Kim J, and Yoon YJ. Compact two-layer Rotman lens-fed microstrip antenna array at 24 GHz. *IEEE Transactions on Antennas and Propagation.* 2011;59(2):460–466.

[15] Hong W, Lin CW, Lin YD, *et al.* Design of a novel planar Bulter matrix beamformer with two-axis beam-switching capability. In: Asia-Pacific Microwave Conference Proceedings, APMC. vol. 5. IEEE; 2005. pp. 1–4.

[16] Peng Z, and Li C. A portable K-band 3-D MIMO radar with nonuniformly spaced array for short-range localization. *IEEE Transactions on Microwave Theory and Techniques.* 2018;66(11):5075–5086.

[17] Zhuge X, and Yarovoy AG. A sparse aperture MIMO-SAR-based UWB imaging system for concealed weapon detection. *IEEE Transactions on Geoscience and Remote Sensing.* 2011;49(1):509–518.

[18] Peng Z, and Li C. A portable 24-GHz FMCW radar based on six-port for short-range human tracking. In: IEEE MTT-S International Microwave Workshop Series on RF and Wireless Technologies for Biomedical and Healthcare Applications (IMWS-BIO). IEEE;2015. pp. 81–82.

[19] Pavlovic VI, Sharma R, and Huang TS. Visual interpretation of hand gestures for human–computer interaction: A review. *IEEE Transactions on Pattern Analysis and Machine Intelligence*. 1997;19(7):677–695.

[20] Fan T, Ma C, Gu Z, *et al.* Wireless hand gesture recognition based on continuous-wave Doppler radar sensors. *IEEE Transactions on Microwave Theory and Techniques*. 2016;64(11):4012–4020.

[21] Peng Z, Li C, Muñoz-Ferreras JM, *et al.* An FMCW radar sensor for human gesture recognition in the presence of multiple targets. In: IEEE MTT-S International Microwave Bio Conference (IMBIOC); 2017. pp. 1–3.

[22] Lien J, Gillian N, Karagozler ME, *et al.* Soli: Ubiquitous gesture sensing with millimeter wave radar. *ACM Transactions on Graphics*. 2016;35(4):1–19.

[23] Juan Cheng, Xiang Chen, and Minfen Shen. A framework for daily activity monitoring and fall detection based on surface electromyography and accelerometer signals. *IEEE Journal of Biomedical and Health Informatics*. 2013;17(1):38–45.

[24] Yu M, Naqvi SM, Rhuma A, *et al.* One class boundary method classifiers for application in a video-based fall detection system. *IET Computer Vision*. 2012;6(2):90.

[25] Bian ZP, Hou J, Chau LP, *et al.* Fall detection based on body part tracking using a depth camera. *IEEE Journal of Biomedical and Health Informatics*. 2015;19(2):430–9.

[26] Vaidehi V, Ganapathy K, Mohan K, *et al.* Video-based automatic fall detection in indoor environment. In: 2011 International Conference on Recent Trends in Information Technology (ICRTIT). IEEE; 2011. pp. 1016–1020.

[27] Garripoli C, Mercuri M, Karsmakers P, *et al.* Embedded DSP-based telehealth radar system for remote in-door fall detection. *IEEE Journal of Biomedical and Health Informatics*. 2015;19(1):92–101.

[28] Wu Q, Tao W, Zhang YD, *et al.* Radar-based fall detection based on Doppler time–frequency signatures for assisted living. *IET Radar, Sonar & Navigation*. 2015;9(2):164–172.

[29] Peng Z, Muñoz-Ferreras JM, Gómez-García R, *et al.* FMCW radar fall detection based on ISAR processing utilizing the properties of RCS, range, and Doppler. In: IEEE MTT-S International Microwave Symposium (IMS). IEEE; 2016. pp. 5–7.

[30] Tahmoush D, and Silvious J. Radar polarimetry for security applications. In: The 7th European Radar Conference; 2010. pp. 471–474.

[31] Smith T. Weapon location by acoustic-optic sensor fusion; 2001.

[32] Li Y, Peng Z, Pal R, *et al.* Potential active shooter detection based on radar micro-Doppler and range-Doppler analysis using artificial neural network; 2018.

Chapter 10

Digital-IF CW Doppler radar and its contactless healthcare sensing

Heng Zhao[1], Biao Xue[1], Li Zhang[1], Jiamin Yan[1],
Hong Hong[1], and Xiaohua Zhu[1]

According to the definition of the Doppler effect, walking and running can generate the frequency shift when the human target is moving forth or back with respect to the radar sensor. Compared with walking and running, the periodic body movement (i.e. breathing and heartbeat) and non-periodic body movement cause the micro-Doppler effect. The vital sign-induced micro-Doppler effect, which is also called vital Doppler [1], has been widely used for radar-based non-contact vital sign detection.

Various types of radar sensors have been developed for non-contact vital sign detection including single-carrier continuous-wave (CW) Doppler radar, frequency-modulated CW radar, stepped frequency CW radar and ultra-wideband radar [2–6]. The homodyne CW Doppler radar is widely used due to its simple structure and low cost [2]. However, it encounters several problems such as quadrature channel imbalance and DC offset [2].

Fortunately, the digital receiver with direct intermediate-frequency (IF)-to-digital conversion (IF sampling) is a suitable alternative [7–13]. In this chapter, the principle of the digital-IF Doppler radar is introduced. Then, the radio frequency (RF) layer, IF layer and optimised baseband signal processing are discussed.

Recently, developed from the conventional application of vital sign detection, the digital-IF Doppler radar is widely applied for healthcare sensing. The digital-IF Doppler radar can not only work as a single sensor but also serve as a sensor in the multi-sensor network. After the introduction of radar sensor, two applications, i.e. non-contact beat-to-beat blood pressure (BP) estimation and multi-sensor-based sleep-stage classification, are briefly introduced.

10.1 Principles of digital-IF Doppler radar

Figure 10.1 shows the simplified block diagram of a CW Doppler radar with digital-IF architecture. In general, the digital-IF Doppler radar has three layers: the RF layer, the digital-IF layer and the baseband layer.

[1]School of Electronic and Optical Engineering, Nanjing University of Science and Technology, Nanjing, China

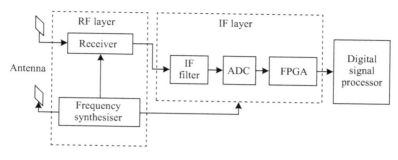

Figure 10.1   Simplified block diagram of the digital-IF CW Doppler radar

The main function of the RF layer is generating, transmitting and receiving RF signals as well as the first-stage down-conversion. As shown in Figure 10.1, the RF layer can be directly divided into the transmitting chain and the receiving chain. In the transmitting chain, the frequency synthesiser generates the transmitting RF signal, which is

$$T(t) = \cos[2\pi ft + \phi(t)] \tag{10.1}$$

where f is the carrier frequency and $\phi(t)$ is the phase noise.

For the ideal model of radar-based non-contact vital sign detection, the region of interest of a human body is the chest wall. Assume the tiny displacement induced by heartbeat, respiration and body motion is $x(t)$, then the total time-varying displacement can be regarded as

$$d(t) = d_0 + x(t) \tag{10.2}$$

where d_0 is the constant distance between radar and chest wall of a human subject. The emitted electromagnetic wave is reflected by the chest wall, and then received by the receiving antenna. As a result, the received RF signal is

$$
\begin{aligned}
R(t) &= T\left[t - \frac{2d(t)}{c}\right] \\
&= \cos\left[2\pi f\left(t - \frac{2d(t)}{c}\right) + \phi\left(t - \frac{2d(t)}{c}\right)\right]
\end{aligned}
\tag{10.3}
$$

where c is the propagation speed of electromagnetic waves. It is known that the displacement of chest wall is phase modulated in the received RF signal. Note that the amplitude variation is ignored in (10.4). Since the displacement on the chest wall is much smaller than the distance between radar and chest wall, the phase noise can be regarded as constant. As a result, the received RF signal can be written as

$$R(t) = \cos\left[2\pi f\left(t - \frac{2d(t)}{c}\right) + \phi\left(t - \frac{2d_0}{c}\right)\right] \tag{10.4}$$

As we know, for a typical homodyne receiver, the received RF signal is down-converted to the baseband signal directly. However, unlike homodyne architecture, in heterodyne architecture there exists an IF between the RF and baseband. The RF signal is first mixed with a local oscillator (LO) to generate the IF signal. In the second stage, the IF signal is mixed with another LO to be down-converted to baseband. As a result, the analogue IF signal is [14]:

$$R_{IF}(t) \approx \cos\left[2\pi f_{IF}t + \theta + \frac{4\pi x(t)}{\lambda} + \Delta\phi\right] \tag{10.5}$$

where f_{IF} is the IF frequency, $\Delta\phi(t) = \phi(t) - \phi(t - 2d_0/c)$ is the residual phase noise and
$\theta = 4\pi d_0/\lambda + \theta_0$ is the constant phase shift dependent on the nominal distance to the target d_0.

Developed from traditional heterodyne architecture, the digital-IF Doppler radar employs direct IF sampling with a heterodyne architecture, which means that the IF signal is directly sampled by a high-performance analogue-to-digital converter (ADC). As a result, the subsequent signal processing is fulfilled in the digital domain. The digitalised IF signal can be written as the periodic sampling version of (10.5), which is

$$R_{IF}(n) \approx \cos\left[2\pi f_{IF}(nT) + \theta + \frac{4\pi x(nT)}{\lambda} + \Delta\phi(nT)\right] \tag{10.6}$$

where $n = 1, 2, 3, \ldots,$ N is the number of sampling points and T is the sampling period. Due to the limitation of sampling rate of ADC, the IF is usually chosen as several tens of megahertz. Since the null/optimum point problem occurs in single-channel receiver, the quadrature receiver has been utilised to mitigate this problem [15]. As mentioned before, the second-stage down-conversion and quadrature decomposition are both realised in the digital domain without using any analogue mixers and phase shifters. Therefore, the digital in-phase and quadrature baseband outputs are

$$B_I(n) = \cos\left[\theta + \frac{4\pi x(n)}{\lambda} + \Delta\phi(n)\right] \tag{10.7}$$

$$B_Q(n) = \sin\left[\theta + \frac{4\pi x(h)}{\lambda} + \Delta\phi(n)\right] \tag{10.8}$$

It can be seen that the desired displacement is involved in the quadrature baseband signals. When only one channel is considered, unwanted harmonics and interferences will complicate the spectrum. The best solution to extract the phase information is combining both channels. Thus, the arctangent demodulation is proposed [16], which is

$$\psi(n) = \arctan\left[\frac{B_Q(n) - b}{B_I(n) - a}\right] + F$$

$$= \theta + \frac{4\pi x(n)}{\lambda} + \Delta\phi(n) \tag{10.9}$$

where F is a multiple of $\pi/2$, which can successfully unwrap the phase signal. Then θ and $\Delta\varphi(n)$ can be removed by a mean removing operation as

$$\psi_0(n) = \psi(n) - \frac{1}{N}\sum_{n=1}^{N}\psi(n) \tag{10.10}$$

10.2   Overview of RF layer

As shown in Figure 10.1, the RF layer in the digital-IF Doppler radar includes a pair of antennas, an analogue receiver and a frequency synthesiser. The patch antenna is widely used for CW Doppler radar. A pair of 2×2 patch arrays are used in [9,10]. The centre frequency is 2.475 GHz. The total size is 161 mm $\times$ 161 mm. The estimated beamwidth is approximately 40°. The maximum gain reaches 12.8 dB at 2.475 GHz.

Designing a frequency synthesiser is an important issue for a digital-IF Doppler radar system. The frequency synthesiser is required to provide different types of signals for both analogue and digital components. Figure 10.2 presents the specific block diagram of the frequency synthesiser used in [9,10]. In order to guarantee the coherence of the whole radar, a temperature-compensated crystal oscillator is selected as the reference signal. Then a four-port power splitter is used to divide the input reference signal into different parts. One channel is fed to the digital-IF layer working as the reference clocks. Meanwhile, the other two channels are directly used in the RF layer. One is the transmitted RF signal, and the other one serves as the LO signal for the mixer. Two phase-locked loop (PLL) circuits are used with the reference inputs.

The PLL is a device which causes one signal to synchronise with another in phase. The PLL consists of a phase detector, a loop filter, a voltage-controlled oscillator (VCO) and a feedback divider. In its basic configuration, a PLL compares the phase of a reference signal to the phase of an adjustable feedback signal. When the comparison is in steady state, and the output frequency and phase are matched to the incoming frequency and phase of the error detector, the PLL is locked. As a result, if the frequency of the input signal is f_{in}, the frequency divider ratio is N and the output frequency is Nf_{in} while synchronising in phase.

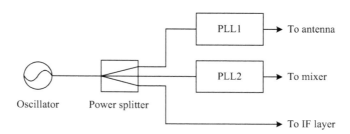

Figure 10.2   Specific block diagram of the frequency synthesiser used in [9,10]

In the receiver, the reflected signal is received by the receiving antenna, then amplified by a low-noise amplifier. The amplified RF signal then passes through a filter to suppress out-of-band noise. Then the filtered signal is mixed with the LO to produce the analogue IF signal, which is further processed in the IF layer.

10.3 Implementation of digital-IF layer

Conventional heterodyne radar receiver is constructed of analogue circuits and components. Typically, an ADC is used to digitalise the baseband signals after down-conversion. Then the vital sign information can be extracted by using digital signal processing. It is known that the analogue components or circuits have the disadvantages of temperature sensitivity and uncertainties in component values, which greatly limits the detection performance of radar system [17].

In the past two decades, the high-performance ADC technique has been developed rapidly. Moreover, owing to the development of software-defined radio (SDR) technique and field-programmable gate array (FPGA), the modern approach to constructing the digital-IF layer is programming the FPGA based on the SDR technique. Compared with the specialised digital circuits, it has the advantages of flexibility, hardware-timed speed, reliability and parallelism.

Figure 10.3 shows a typical scheme of the digital-IF layer. The first stage is an IF filter to eliminate the out-of-band noise. It should be noted that the filtered analogue IF signal is digitalised directly by a high-speed ADC so that the subsequent signal processing is implemented in the digital domain. The digital down-converter (DDC) module consists of digital quadrature demodulation and decimation. In order to eliminate the null/optimal detection point problem [18], the quadrature demodulation is utilised in the DDC module. The digital low-pass filter (LPF) in each channel maintains the baseband component and removes the high-frequency component due to the multiplication. The decimation can significantly reduce the computational load of the subsequent digital signal processor. For instance, the sampling rate for direct IF sampling reaches 100 MHz. However, the vital sign signals occupy a very narrow band (usually less than 5 Hz) in the spectrum. Thus, the 100 Hz baseband sampling rate satisfies the Nyquist sampling theorem, which is adequate for baseband signal

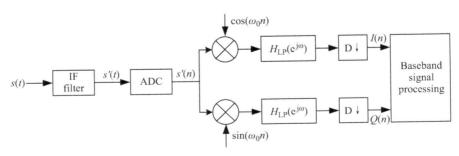

Figure 10.3   Block diagram of the digital-IF layer including ADC and DDC

processing. As a result, a 10^6 decimation rate is efficient for storage and baseband signal processing.

However, there are still two problems that need to be further taken into consideration in the digital-IF layer. First, the IF signal is typically a band-limited signal with its carrier frequency around several tens of megahertz. According to the Nyquist Sampling theorem, should the sampling frequency of ADC be greater or at least equal to twice the highest frequency of the IF signal? Second, the analogue down-converter circuit is replaced by DDC in the digital radar receiver. How to realise DDC in the digital domain?

10.3.1 Direct IF sampling

According to the Nyquist sampling theorem, if the input signal needs to be adequately recovered from the sampled sequence without distortion, the sampling rate should be greater or at least twice the highest frequency of the input signal. If an IF signal has a frequency band of $[f_L, f_H]$, the f_s should be at least $2f_H$. Suppose the IF is 70 MHz, the sampling rate should be at least 140 MHz, which is very high for common ADC devices. Even if the ADC has such a high sampling rate, the computational load of subsequent digital signal processing is also very heavy at such a high data rate.

The bandpass sampling theorem perfectly answers the first problem we raised in the last section [19]. Figure 10.4(a) shows the general case of bandpass sampling. The shaded components represent the original spectrum of IF signal. Assume the IF signal $x(t)$ whose frequency is within $[f_L, f_H]$, its centre frequency f_0 and bandwidth B are

$$f_0 = \frac{f_H + f_L}{2} \tag{10.11}$$

$$B = f_H - f_L \tag{10.12}$$

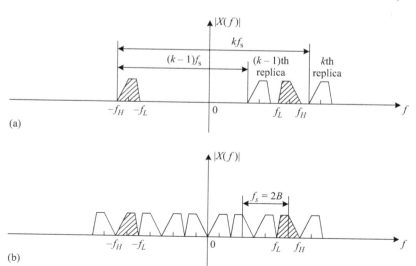

Figure 10.4   Spectrum of bandpass sampling without aliasing

Obviously, if the spectral aliasing does not happen, the positive component between kth replica of the negative component and $(k-1)$th replica of the negative component should be

$$2f_H \leq kf_s \tag{10.13}$$

$$(k-1)f_s \leq 2f_L \tag{10.14}$$

Substituting (10.13) into (10.14), the restriction of sampling frequency becomes

$$\frac{2f_H}{k} \leq f_s \leq \frac{2f_L}{k-1} \tag{10.15}$$

Moreover, k is a positive integer given by

$$1 \leq k \leq \left\lfloor \frac{f_H}{B} \right\rfloor \tag{10.16}$$

where $\lfloor \bullet \rfloor$ denotes the integer operation. The equality conditions of (10.16) hold only if the band is confined to the interval (f_L, f_H), and a frequency component at $f = f_H$ or $f = f_L$ will be aliased. The bandpass sampling theorem states that it is not necessary to sample the IF signal at twice the highest frequency.

Figure 10.4(b) presents a special case in which the theoretical minimum sampling rate $f_s = 2B$ is considered. The replicas of negative and positive components interweave in the spectrum at such sampling frequency. In this case, f_L and f_H should be located at the multiples of B to avoid aliasing.

For example, considering the digital-IF Doppler radar for vital sign detection in [9], the IF f_{IF} and the sampling frequency f_s are 75 and 100 MHz, respectively. Typically, the vital sign signals occupy an extremely narrow frequency band when compared with the IF. Suppose the frequency band concerned is [74–76] MHz, the bandwidth B is 2 MHz. According to (10.16), we have $2 \leq k \leq 38$. When $k = 2$ is considered, the sampling frequency without aliasing is

$$76 \leq f_s \leq 148 \tag{10.17}$$

Therefore, 100 MHz is capable of sampling the 75 MHz IF signal.

10.3.2 Digital quadrature demodulation

In order to avoid the null/optimal detection point problem, the analogue quadrature demodulation including a phase shifter, two mixers and two analogue LPFs is widely used in the homodyne Doppler radar. Similarly, the digital quadrature demodulation is fulfilled in the digital domain. In contrast to the analogue devices, the precision of digital calculation is only limited by the device itself, and the orthogonality of the quadrature signals can be guaranteed.

The first method to realise the digital quadrature demodulation is using quadrature digital mixers. As shown in Figure 10.3, the input digitalised IF signal is multiplied by two quadrature LO signals. The mixers are replaced by digital multipliers. Generally, the digitalised LO signals $\cos(\omega_0 n)$ and $\sin(\omega_0 n)$ can be generated by the numerically controlled oscillator (NCO). The NCO consists of a phase accumulator and a lookup table. The phase accumulator is responsible for calculating the instant phase when a new frequency control word is coming. The lookup table stores pre-calculated values of the sampled sine waveform in the read-only memory (ROM). The incoming phase serves as an instant address to read the ROM. The output waveform is a sine waveform with different successive phase deviation. Finally, the LPF preserves the fundamental frequency component and filters out the second harmonics due to multiplication.

The accuracy of the NCO is fundamentally limited by the phase accumulator register size, the lookup table size and the accuracy of the clock that drives the system. The drawback of this method is that the high sampling frequency leads to a very high order of LPF, which is difficult to design. Moreover, the NCO needs a large ROM to restore the sine waveform.

An alternative method to realise the digital quadrature demodulation is based on the polyphase filter. Assume that the input IF signal is

$$x(t) = a(t) \cos [2\pi f_0 t + \phi(t)] \tag{10.18}$$

Based on the bandpass sampling theorem, the sampling frequency is

$$f_s = \frac{4f_0}{2m + 1} \tag{10.19}$$

where $m = 0, 1, 2, \ldots$. As a result, the sampled IF signal is

$$
\begin{aligned}
x(n) &= a(n) \cos \left[2\pi \frac{f_0}{f_s} n + \phi(n) \right] \\
&= a(n) \cos \phi(n) \cos \left(\frac{2m+1}{2} \pi n \right) - a(n) \sin \phi(n) \sin \left(\frac{2m+1}{2} \pi n \right) \\
&= I(n) \cos \left(\frac{2m+1}{2} \pi n \right) - B(n) \sin \left(\frac{2m+1}{2} \pi n \right)
\end{aligned} \tag{10.20}
$$

where

$$I(n) = a(n) \cos \phi(n) \tag{10.21}$$

$$Q(n) = a(n) \sin \phi(n) \tag{10.22}$$

Here, (10.21) and (10.22) are the in-phase and quadrature-phase baseband signals, respectively. If the odd position and even position of $x(t)$ are decimated, (10.20) becomes

$$
\begin{aligned}
x(2n) &= I(2n) \cos [(2m + 1)\pi n] \\
&= I(2n)(-1)^n
\end{aligned} \tag{10.23}
$$

$$
\begin{aligned}
x(2n + 1) &= -Q(2n + 1) \sin \left[\frac{2m+1}{2} \pi (2n + 1) \right] \\
&= Q(2n)(-1)^n
\end{aligned} \tag{10.24}
$$

$$I'(n) = I(2n) = x(2n)(-1)^n \tag{10.25}$$

$$Q'(n) = Q(2n + 1) = x(2n + 1)(-1)^n \tag{10.26}$$

It is known that the second decimation of quadrature baseband signals $I(n)$ and $Q(n)$ can be easily obtained from (10.25) and (10.26). Therefore, the digital quadrature demodulation can be achieved by using the structure in Figure 10.5. The spectra of $I'(n)$ and $Q'(n)$ are

$$F_{I'}(e^{j\frac{\omega}{2}}) = \frac{1}{2}F_I(e^{j\frac{\omega}{2}}) \tag{10.27}$$

$$F_{Q'}(e^{j\frac{\omega}{2}}) = \frac{1}{2}F_Q(e^{j\frac{\omega}{2}})e^{j\frac{\omega}{2}} \tag{10.28}$$

where $F_{\bullet}$ denotes the Fourier transform. As we can see in (10.27) and (10.28), a delay factor $e^{j\frac{\omega}{2}}$ exists between the two spectra, which corresponds to a half sampling point in the time domain. Thus, two filters are needed to eliminate the time delay between the quadrature channels. It is proved that the delay filters $H_I(e^{j\frac{\omega}{2}})$ and $H_Q(e^{j\frac{\omega}{2}})$ can be

$$\begin{aligned} H_I &= 1 \\ H_Q &= e^{j\frac{\omega}{2}} \end{aligned} \tag{10.29}$$

or

$$\begin{aligned} H_I &= e^{j\frac{\omega}{4}} \\ H_Q &= e^{j\frac{3\omega}{4}} \end{aligned} \tag{10.30}$$

Table 10.1 shows the coefficients of the four filters. We can choose a pair of filters $h_0(n)$ and $h_2(n)$ (or $h_1(n)$ and $h_3(n)$) as the delay correction filters. At last, the LPF is utilised to filter out the high-frequency component.

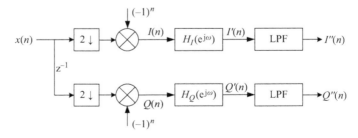

Figure 10.5 Basic structure of digital quadrature demodulator based on polyphase filter

Table 10.1   Coefficients of the delay correction filters

Coefficients	$h_0(n)$	$h_1(n)$	$h_2(n)$	$h_3(n)$
$h(0)$	0	$3.2568132\mathrm{E}^{-5}$	$8.3456202\mathrm{E}^{-5}$	$4.5833163\mathrm{E}^{-5}$
$h(1)$	$-3.0689215\mathrm{E}^{-4}$	$-1.6234104\mathrm{E}^{-3}$	$-4.2909505\mathrm{E}^{-3}$	$-8.1703486\mathrm{E}^{-3}$
$h(2)$	$-1.199357\mathrm{E}^{-2}$	$-1.204163\mathrm{E}^{-2}$	$-4.1642617\mathrm{E}^{-3}$	$1.5687875\mathrm{E}^{-2}$
$h(3)$	$4.8911806\mathrm{E}^{-2}$	$9.0803854\mathrm{E}^{-2}$	0.133363	0.1652476
$h(4)$	0.1767598	0.1652476	0.133363	$9.0803847\mathrm{E}^{-2}$
$h(5)$	$4.8911810\mathrm{E}^{-2}$	$1.5687877\mathrm{E}^{-2}$	$-4.1642617\mathrm{E}^{-3}$	$-1.2041631\mathrm{E}^{-2}$
$h(6)$	$-1.1993569\mathrm{E}^{-2}$	$-8.1703495\mathrm{E}^{-3}$	$-4.2909505\mathrm{E}^{-3}$	$-1.6234104\mathrm{E}^{-3}$
$h(7)$	$-3.0689215\mathrm{E}^{-4}$	$4.5833163\mathrm{E}^{-5}$	$8.3456202\mathrm{E}^{-5}$	$3.2568132\mathrm{E}^{-5}$

10.3.3   Decimation

As we mentioned before, the down-converted signal has an extremely high sampling rate, which leads to a heavy computational load for subsequent digital signal processing. Thus, the decimation is indispensable after down-conversion. Assume that the input signal $x(n)$ is decimated by D, the decimated signal is

$$x_D(m) = x(mD) \tag{10.31}$$

If the sampling frequency is f_s, the spectrum of the baseband signal without aliasing is within $f_s/2$. As a result, the sampling frequency after decimation is f_s/D and the spectrum without aliasing is within $f_s/2D$. The signal whose spectrum is over $f_s/2D$ suffer from aliasing.

In order to avoid the spectral aliasing, the anti-aliasing filter should be used before decimation. However, if the decimation ratio is very high, designing such an LPF is very difficult due to the narrow transitional band and high filter order. Typically, the decimation is realised by a multi-stage structure, which is called multi-stage decimation. The decimation is cascaded by a series of combinations of LPFs and decimators. As a result, the final decimation ratio is

$$D = \prod_{n=1}^{N} D_n, \tag{10.32}$$

where D_n, $n = 1, 2, \ldots, N$, is the decimation ratio of each stage.

The challenge of the multi-stage decimation is how to design efficient LPFs to eliminate aliasing in each stage. Two kinds of anti-aliasing filters are widely used in the digital-IF Doppler radar, being the cascaded integrator-comb (CIC) filter and the half-band filter. The half-band filter is suitable for dealing with decimation and interpolation whose ratio is powers of 2. The CIC filter is usually responsible for the

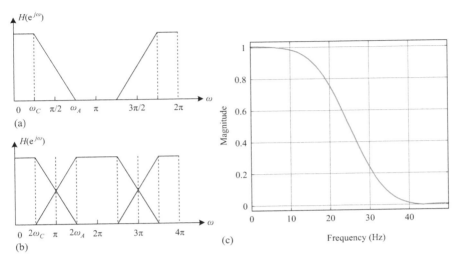

Figure 10.6 (a) *Ideal magnitude response of half-band filter before decimation. (b) Ideal magnitude response of half-band filter after decimation. (c) Simulated magnitude response of half-band fitter*

stage whose decimation ratio is not a power of 2. Therefore, the decimation ratio D can be decomposed as

$$D = \left(\prod_{n=1}^{N} D_n \right) \times 2^M \qquad (10.33)$$

where D_n are the decimation ratios that are not the powers of 2, and M is an integer. It means that the multi-stage decimation is composed of N stages of CIC filters and M stages of half-band filters.

The half-band filter is a finite impulse response filter which has the equal bandwidth and ripple in passband and stopband. As a result, the half-band filter has the properties of

$$H(e^{j\omega}) = 1 - H(e^{j(\pi - \omega)}) \qquad (10.34)$$

$$H(e^{j\frac{\pi}{2}}) = 0.5 \qquad (10.35)$$

$$h(n) = \begin{cases} 1 & n = 0 \\ 0 & n = \pm 2, \pm 4 \end{cases} \qquad (10.36)$$

The half-band filter is computationally efficient due to the zero response at the even points except the zero point. Figure 10.6(a) and (b) shows magnitude response of a half-band filter before and after twice decimation, respectively. It is apparent that signal whose frequency within $[2\omega_C, \pi]$ is still aliased after twice decimation.

However, no aliasing appears in the frequency band $[0, 2\omega_C]$. Thus, the signal inside $[0, 2\omega_C]$ can be recovered without distortion.

The half-band filter can be designed by using the window function method. Figure 10.6(c) shows the simulated magnitude response of 11th-order half-band filter using Hamming Window. The simulation is implemented in the MATLAB FDAtool®. The sampling frequency is 100 Hz. The coefficients are $[0.0051, 0, -0.0422, 0, 0.2903, 0.5, 0.2903, 0, -0.0422, 0.0051]$. It can be seen that the attenuation at cut-off frequency is 6 dB.

Another efficient anti-aliasing filter is the CIC filter. The impulse response of the CIC filter is

$$h(n) = \begin{cases} 1 & 0 \le n \le D - 1 \\ 0 & Others \end{cases} \tag{10.37}$$

where D is the decimation ratio. The z-Transform of (10.37) is

$$H(z) = \sum_{n=0}^{D-1} h(n)z^{-n}$$

$$= \frac{1 - z^{-D}}{1 - z^{-1}} \tag{10.38}$$

$$= H_1(z)H_2(z)$$

where

$$H_1(z) = \frac{1}{1 - z^{-1}} \tag{10.39}$$

$$H_2(z) = 1 - z^{-D} \tag{10.40}$$

Obviously, $H_1(z)$ is a typical integrator. The frequency response of $H_2(z)$ is

$$H_2(e^{j\omega}) = 1 - e^{-j\omega D}$$

$$= 2je^{-\frac{j\omega D}{2}} \sin\left(\frac{\omega D}{2}\right) \tag{10.41}$$

It can be found that $H_2(z)$ has a comb-like magnitude response. Its zeros locate at $\omega = 2n\pi/D$, where $n = 1, 2, \ldots$. As a result, the final frequency response of CIC filter is

$$H(e^{j\omega}) = H_1(e^{j\omega}) \cdot H_2(e^{j\omega})$$

$$= \frac{1}{1 - e^{-j\omega}} \cdot \left[2je^{-\frac{j\omega D}{2}} \sin\left(\frac{\omega D}{2}\right) \right] \tag{10.42}$$

$$= D \mathrm{sinc}\left(\frac{\omega D}{2}\right) \mathrm{sinc}^{-1}\left(\frac{\omega}{2}\right)$$

When $\omega = 0$, $|H(e^{j\omega})| = D$. The bandwidth of main lobe is $2\pi/D$. It can be proved that when $D \gg 1$, the amplitude difference between the main lobe and the first side

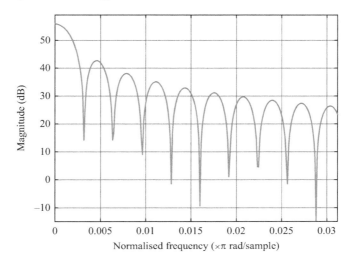

Figure 10.7 Simulated magnitude response of a CIC filter

lobe is 13.46 dB. In order to achieve high attenuation, CIC filters can be cascaded. As a result, the frequency response becomes

$$H_N(e^{j\omega}) = \left[D\mathrm{sinc}\left(\frac{\omega D}{2}\right) \mathrm{sinc}^{-1}\left(\frac{\omega}{2}\right) \right]^N \tag{10.43}$$

where N is the number of stages. The final attenuation is $13.46N$ dB. For a bandwidth factor b, the corresponding signal bandwidth is $b(2\pi/D)$. It is proved that the attenuation in the stopband compared with the main lobe can be approximated as $-20\lg b$ when $b \ll 1$ and $D \gg 1$. Thus, the attenuation of N-stage cascaded CIC filter becomes $-20N\lg b$.

A brief example is given to show one of the CIC filters used in the digital-IF radar in [9,10]. The decimation ratio is 625. The simulated magnitude response is shown in Figure 10.7. It can be seen that the attenuation between the main lobe and first side lobe is about 13 dB.

10.4 DC offset calibration technique

As we mentioned before, the baseband layer contains two steps: the DC offset calibration and the arctangent demodulation. For a digital-IF receiver, the amplitude imbalance can be negligible. However, the DC offset still exists in the quadrature channels. As a result, the baseband quadrature signals can be reformulated as

$$B_I(n) = \cos\left[\theta + \frac{4\pi x(n)}{\lambda} + \Delta\phi(n)\right] + DC_I \tag{10.44}$$

$$B_Q(n) = \sin\left[\theta + \frac{4\pi x(n)}{\lambda} + \Delta\phi(n)\right] + DC_Q \tag{10.45}$$

where DC_I and DC_Q represent the DC offsets in the in-phase (I) and quadrature-phase (Q) channels, respectively. The DC offset reduces the linearity and demodulation accuracy.

Using the quadrature baseband signals as the axes, the ideal signals without DC offsets can formulate a trajectory which is a part of the circle in the constellation diagram. The ideal circle centre is located at the origin, whereas the DC offset leads to the position change of the circle centre. Therefore, the DC offset estimation can be modelled as a circle-fitting problem.

Considering the measured signals from quadrature channels I_n and Q_n, where $1 \le n \le N$ is sampling points, the residual can be constructed as

$$d_n = (I_n - a)^2 + (Q_n - b)^2 - r^2 \tag{10.46}$$

where a and b denote the estimated DC offsets in quadrature channels and r is the estimated amplitude. The circle fitting problem is to minimise the residual sequence $\mathbf{d} = [d_1, d_2, \ldots, d_N]$ as

$$\min\mathbf{d} = \min\|Ax - y\| \tag{10.47}$$

where

$$A = \begin{bmatrix} 2I_1 & 2Q_1 & 1 \\ \cdot & \cdot & \cdot \\ \cdot & \cdot & \cdot \\ \cdot & \cdot & \cdot \\ 2I_n & 2Q_n & 1 \end{bmatrix}, x = \begin{bmatrix} a \\ b \\ r^2 - a^2 - b^2 \end{bmatrix}, y = \begin{bmatrix} I_1^2 + Q_1^2 \\ \cdot \\ \cdot \\ I_n^2 + Q_n^2 \end{bmatrix} \tag{10.48}$$

Several methods have been proposed to solve the minimisation problem [16,20]. However, the conventional method, l_2 minimisation, is quite sensitive to outliers with large residuals because it always blindly considers all measurements, including outliers. When outliers in measurement appear (such as very significant noise or errors), the results will be off from the optimal solution.

Recently, the compressed sensing (CS)-based methods are widely used in DC offset estimation [21,22]. In practice, most measurements are accurate and the number of outliers is relatively small, and can be considered as sparse. In that case, the CS-based method is used to explore the circle with maximum matching points, which is

$$\min\mathbf{d}_{l_0} = \min\|Ax - y\|_{l_0} \tag{10.49}$$

Equation (10.49) is a l_0 minimisation-based error correction problem. However, the l_0 minimisation is an NP hard problem, which is generally replaced by the l_1-norm minimisation. In [22], the $l_p(0 \le p \le 1)$ minimisation-based method is proposed to estimate the DC offsets a and b from the two channels. It is proved to outperform l_1 minimisation in many situations since it is closer to l_0 minimisation.

In order to solve the l_p minimisation-based error correction, the iterative reweighed l_1 (IRL1) algorithm is utilised. The IRL1 algorithm is used to iteratively solve the weighted l_1 minimisation, which is

$$\mathbf{d}^{(l)} = \min_{d \in \mathbb{R}^n} \|\mathbf{W}^{(l)}(Ax - y)\|_{\ell_1} \tag{10.50}$$

where l is the iteration count and $\mathbf{W}^{(l)} = \mathrm{diag}(w_1^{(l)}, w_2^{(l)}, \dots, w_n^{(l)})$ is the suggested weights.

At the beginning, we start with the initial weight $w_i^{(0)} = 1$ for $1 \le i \le n$. Then the weights are updated as

$$w_i^{(l+1)} = \frac{1}{(|d_i^{(l)}| + \varepsilon)^{1-p}} \tag{10.51}$$

where ε is a positive real number close to zero, for ensuring the non-zero denominator when $d_i^{(l)}$ is zero.

The iteration ends when it encounters convergence or l attains the maximum iteration count. Finally, using (10.49), we can obtain the estimated DC offsets accurately. Figure 10.8 shows an example of DC offset calibration using l_p minimisation-based method. It can be seen that the signal strength is highly increased after the DC offset calibration.

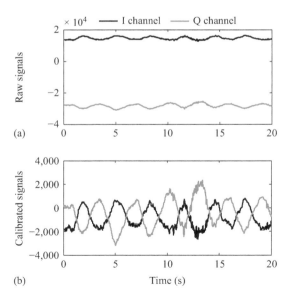

Figure 10.8   *(a) Baseband quadrature signals before DC offset calibration. (b) Baseband quadrature signals after DC offset calibration*

10.5 Applications to healthcare sensing

The digital-IF CW Doppler radar can measure the vital sign signals accurately. Extended from the conventional vital sign detection, it can be further used for healthcare sense. Considering the healthcare sensing applications, the digital-IF CW Doppler radar can not only be used individually, but also cooperate with other sensors. In this section, two examples are introduced. One is contactless beat-to-beat BP estimation using a single radar sensor [23], the other is multi-sensor-based sleep-stage classification.

10.5.1 Contactless beat-to-beat BP estimation using Doppler radar

In our daily lives, hypertension becomes a major cardiovascular risk factor. In particular, increasing BP and BP variability (BPV) are related to a higher incidence of target organ damage [24]. As we know, the most widely used device for BP monitoring is sphygmomanometry which uses a cuff to collapse and then releases the artery. Due to its operational principle, the sphygmomanometry is unable to measure BP continuously. Thus, the beat-to-beat BP and its variability cannot be extracted. Moreover, invasive nature limits its use to critically ill patients.

Recently, pulse wave analysis is a hot research topic in biomedical engineering. For the pulse wave signal, one of the most useful applications is the beat-to-beat BP measurement based on the pulse transit time (PTT) [25]. According to the definition, PTT is the time duration when a pulse wave travels between two arterial sites. It usually uses two attached sensors on the human body to measure the pulse waves on the different arterial sites, then the PTT is calculated between the two pulse waves.

As mentioned before, the CW Doppler radar has the ability to capture tiny movement on the human body surface with sub-millimetre accuracy. The pulse wave signal is detectable by the Doppler radar theoretically, and the BP can be further calculated. In [26], a CW Doppler radar with its antenna placed at the sternum was utilised to capture the aortic arterial pulsation. Combined with the carotid pulse wave from the bioimpedance, the systolic BP (SBP) estimation was realised using the PTT. However, the wearable sensor is uncomfortable for long-term monitoring. In [27], the pulse wave signal was measured from the carotid, the venous and the ventricle by a CW radar. However, the BP measurement was not further investigated.

Therefore, a contactless BP measurement system is proposed to realise both beat-to-beat SBP and diastolic BP (DBP) measurements. In the proposed system, the digital-IF Doppler radar is used to capture the aortic pulse wave signal. Then the PTT is extracted from the radar-detected signal. Finally, using the relationship between PTT and BPs, both static and beat-to-beat BPs are obtained.

The digital-IF Doppler radar is used to capture the aortic pulse wave on the abdomen induced by the vasomotion of the central aortic artery. As shown in Figure 10.9, the human subject was in a supine position on a bed. The Doppler radar was fixed on a cantilever with its beam facing the subject's abdomen. According to the detection theory of the digital-IF Doppler, the pulse wave signal is involved in the phases of quadrature baseband outputs.

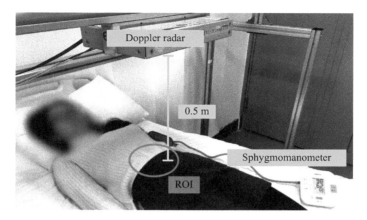

Figure 10.9   Experimental set-up of the contactless BP measurement using Doppler radar. ROI refers to the region that is covered by the radar signal.

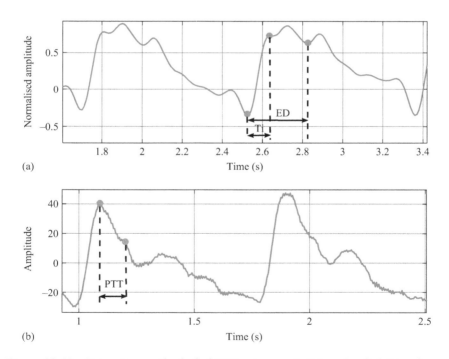

Figure 10.10   Extraction method of cf-PTT using aortic (a) and radial (b) pulse waves

The radar-detected aortic pulse wave signal is plotted in Figure 10.10(a). Many fluctuations can be found in this time-domain waveform. According to [28], it is demonstrated that the carotid-femoral PTT (cf-PTT) can be extracted in one single

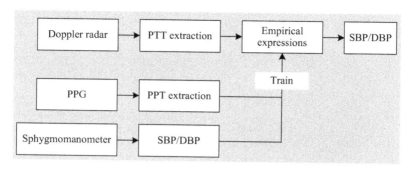

Figure 10.11   Signal processing diagram of contactless beat-to-beat BP measurement

cycle. The cf-PTT is defined as the half duration between the first systolic peak and the dicrotic notch. As shown in Figure 10.10(a), the cf-PTT can be calculated as

$$PTT = \frac{ED - T_i}{2} \tag{10.52}$$

It is proved that the cf-PTT can also be derived from the radial pulse wave [15]. As shown in Figure 10.10(b), the reference cf-PTT for calibration is defined as the time interval between the first two systolic peaks in the radial pulse signal [29]. Therefore, if the relationship between cf-PTT and BP (both SBP and DBP) can be determined in advance through the radial pulse wave signal and sphygmomanometer, BP can be obtained from the radar-detected aortic pulse wave signal. As a result, the contactless beat-to-beat blood pressure can be calculated.

The corresponding processing diagram is shown in Figure 10.11. According to [26], SBP is linearly proportional to the cf-PTT, which can be written as

$$SBP = a \times PTT_{cf} + b \tag{10.53}$$

where a and b are the linear parameters of SBP.

Unlike SBP, it is difficult to find a linear relationship between DBP and cf-PTT. According to [29], Moens–Kortewegs formula expresses pulse wave velocity (PWV) in terms of vessel dimensions, blood density and arterial wall elasticity. PWV is inversely proportional to the PTT, which can be substituted into the Moens–Kortewegs formula. It is found that the elastic modulus of the arterial wall increases exponentially with mean arterial pressure (MBP). MBP is often approximated as

$$MBP = \frac{1}{3}SBP + \frac{2}{3}DBP \tag{10.54}$$

Table 10.2 Result of Doppler radar-based BP measurements using the empirical expressions

Subjects	SBP (mmHg)			DBP (mmHg)		
	Radar	Reference	Error	Radar	Reference	Error
S_1	110	109	1	72	69	3
S_2	98	100	2	54	58	2
S_3	126	128	2	78	80	3
S_4	107	110	3	64	69	5
S_5	119	122	3	76	74	0
S_6	121	119	2	70	71	1
S_7	117	118	1	77	80	3
S_8	112	110	2	72	69	3
S_9	106	106	0	66	65	1
S_{10}	96	99	3	54	58	4

Therefore, based on the Moens–Kortewegs formula and Hughes' finding, the relationship between DBP and PTT can be written as [29]:

$$DBP = \frac{SBP_0}{3} + \frac{2DBP_0}{3} + A \ln \frac{PTT_0}{PTT_{cf}} - \frac{(SBP_0 - DBP_0)}{3} \frac{PTT_0^2}{PTT_{cf}^2} \tag{10.55}$$

where A is the subject-dependent parameter. SBP_0, DBP_0 and PTT_0 are the parameters from a calibration procedure. In practice, these parameters are both subject-dependent.

Two experiments were carried out to evaluate the performance of the proposed scheme. Ten healthy human subjects (six females and four males) were involved in the experiments. First, the PTT-BP calibration for each subject is carried out individually. A photoplethysmograph (PPG) device (HKG-07C) was used to capture the radial pulse signal. Meanwhile, a sphygmomanometer (OMRON HEM-7052) was used to measure SBP and DBP as reference. At last, using the reference PTT and BP, the subject-independent empirical expressions (10.53 and 10.55) are calculated. The cf-PTT is defined as the average value of cf-PTTs extracted from ten successive cycles of radial pulse during the BP measurement. The data were analysed using MATLAB®.

Since SBP is linear to PTT, the linear regression method is used to estimate the parameters in (10.53). In order to determine the subject-dependent parameter A, a set of $[SBP_0, DBP_0, PTT_0]$ is selected as the referenced parameters. Then, using the measured cf-PTT and DBP, the parameter A is uniquely determined.

Table 10.2 presents the results of both SBP and DBP measurements from the Doppler radar using the empirical expressions, along with the references. It is evident that the errors for both SBP and DBP measurements are all less than 5 mmHg,

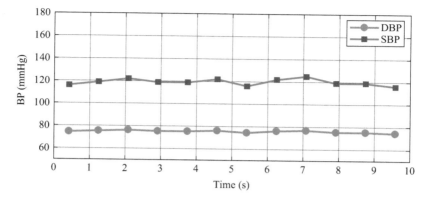

Figure 10.12   Beat-to-beat BP measurement using Doppler radar

which shows that the empirical expression-based method is suitable for SBP and DBP measurements.

In the last experiment, the beat-to-beat BP of S_3 is extracted. A sequence of beat-to-beat PTTs is extracted first from the 10-s radar-detected aortic pulse wave. Then, the beat-to-beat BP is obtained using (10.53) and (10.55). Figure 10.12 shows the result of beat-to-beat SBP and DBP. The fluctuations can be found in both SBP and DBP measurements. In the SBP result, the value fluctuates around 120 mmHg, while the reference is 120 mmHg. In the DBP result, the value fluctuates around 76 mmHg, while the reference is 77 mmHg. Since the sphygmomanometer has to spend 10 s to measure once, it cannot provide the beat-to-beat BP measurement. Moreover, the variability in SBP is larger than one in DBP. The results show the feasibility of Doppler radar in contactless beat-to-beat BP monitoring.

10.5.2   Multi-sensor-based sleep-stage classification

With the accelerated pace of life, the sleep health issues of modern people have received much attention under the influence of intense work pressures and irregular life schedules [30]. Quality and effective sleep are not only related to the health of the individual but also directly affects the spirit and efficiency of work. Sleep medicine has become an essential part of modern medicine. The monitoring of sleep and exploring to improve sleep quality and efficiency are essential topics in sleep medicine.

The polysomnography (PSG) is a gold standard for assessing sleep quality and diagnosing sleep disorders in sleep medicine. PSG can measure the patient's sleep time, quality, efficiency and sleep stages by monitoring the physiological parameters during the whole night. However, there are some limitations of this device, such as limitation of the patients' activity by the electrodes, affecting the quality of sleep. To overcome these issues, various non-contact sensors, such as microphone and Doppler radar, are utilised to achieve sleep monitoring [31–35]. Due to the limitation of single sensors, we suggest that the fusion of non-contact microphone sensor and radar sensor

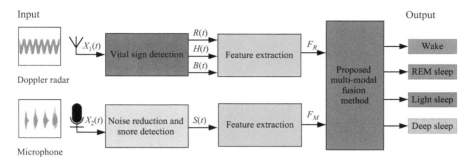

Figure 10.13   Overall architecture of the sleep-stage estimation system

data-based advanced signal processing algorithms would likely enhance the accuracy of sleep-stage classification.

In this example, a novel machine learning-based scheme is presented for the classification of sleep stages using the radar-microphone prototype. The digital-IF radar serves as the primary sensor to record the heartbeat, respiration and body movement signals. Synchronously, a microphone sensor detects acoustic signals, including snoring sounds and breathing sounds. The sensor fusion algorithm determines sleep stages using vital sign signals related to sleep, such as breathing, heartbeat, body movement and snoring sound. After extracting the selected features from the multi-modal sensors, the sleep stages were classified by signal processing using statistical medical knowledge and a uniquely trained machine learning algorithm. A medical expert at Nanjing University of Science and Technology Hospital confirmed reference devices and standard score for PSG data collected from the subjects.

The phase of baseband signals involves physiological signals related to sleep. The respiration and heartbeat are the so-called vital sign signals, variable within the different sleep stages. The scheme flow chart is shown in Figure 10.13. The demodulated signal $S(t)$ is filtered by two bandpass filters (0.1–0.6 and 0.8–2 Hz) to obtain respiratory signal $R(t)$ and heartbeat signal $H(t)$, respectively. Meanwhile, we utilise the energy spectrum method to determine body movements [32]. The sampling rate of the demodulated signal $S(t)$ is 100 Hz. We divide the $S(t)$ into continuous non-overlapping epochs S_k of 60 s. In the kth epoch, the Riemann integral $A_k(i)$ of S_k with respect to the 10 s period is defined as

$$A_k(i) = \sum_{t=t_k-i*1000}^{t_k-(i-1)*1000} |S_k|\, dt \tag{10.56}$$

where t_k denotes the time of S_k. $A_k(i)(i = 1, 2, \ldots, 6)$ is calculated every 10 s. A body movement is identified only if there is a change of more than an order of magnitude of $A_k(i)$. The detected movements are validated against the posture information provided by the PSG system. Our algorithm correctly detects over 95% of the body movement events.

Table 10.3   List of features extracted from both radar sensor and microphone [11,12,36]

Types	Features from Doppler radar	Features from microphone
Body movement	Body movement index	–
Respiratory	Respiration per minute (RPM) Variance of RPM Amplitude difference accumulation of respiratory signal REM parameter Sample entropy	Breathing Cycle RPM Variance of RPM Breathing variance parameters Root mean square error Approximate entropy Energy difference accumulation
Heartbeat	Heartbeat per minute (HPM) Variance of HPM Amplitude difference accumulation of heartbeat signal Deep parameter	–
Sleep pattern	Timestamp	Timestamp

Since the digital-IF Doppler radar can acquire various physiological parameters, new feature sets have been extracted from the respiration, heartbeat and body movement signals. The body movement mostly occurs in the Wake and Light Sleep. It is also an indicator for discriminating different sleep stages. Table 10.3 summarises 11 features used in [11,12] including 1 body movement feature, 5 respiration features, 4 heartbeat features and 1 sleep pattern feature. Without loss of generality, the epoch for determining sleep stages is 60 s with non-overlapping window.

The preprocessing of the microphone-based method consists of three basic parts: noise reduction and snore detection, and feature extraction, as shown in Figure 10.13. The noise reduction algorithm is used to get clean breathing sound and snore sound. The Wiener-filter algorithm emphasises non-stationary events by tracking prior signal-to-noise ratio (SNR) [37]. The vertical box control algorithm (V-box) is used with an adaptive energy threshold to detect audio events automatically. Table 10.3 summarises seven features including six respiration features and one non-linear feature to describe the audio candidate period, intensity, variation and non-linearity [36].

The purpose of multi-sensor information fusion is to produce more reliable and accurate information representation. Compared with single sensor, multi-sensor signals often contain redundant and complementary information, which is useful for diagnosis. Multi-sensor information fusion has been widely used in the automatic target recognition, robotics and image processing. According to the fusion operation level, it can be divided into data-level fusion, feature-level fusion and decision-level fusion. Data-level fusion is performed on the original observation information, and multi-source signals are fused. Feature-level fusion is carried out on the feature space. By reducing the dimension of input data, multiple feature vectors are merged into one or more main vectors. As a result, the data set is compressed, while the necessary

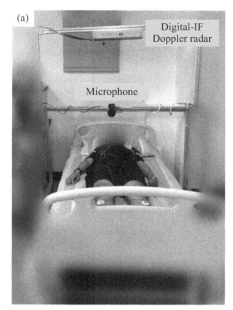
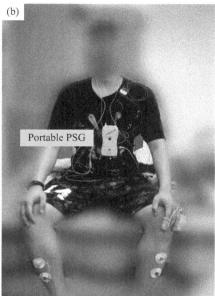

Figure 10.14   Experimental set-up of the non-contact sleep-stage estimation using Doppler radar and microphone. (a) Sit position. (b) Sleep position

information retains fewer data. Feature-level fusion can achieve considerable information compression and facilitate real-time data processing [38]. Therefore, it is considered to be more effective than the other two methods.

The multiple features are extracted from the snore sound signals $S(t)$, vital sign signals (including the respiratory signals $R(t)$, the heartbeat signals $H(t)$ and the body movement signals $B(t)$). Then, two feature sets are built based on data collected from the microphone and Doppler radar, respectively. Since the size of two feature sets is different, the serial feature fusion scheme is utilised in this study, which works by merely concatenating two sets of feature vectors into a single feature vector.

The sleep staging results after fusion of multiple sensors were analysed and compared, including the results of each part of the sub-model and the total model. In the experiment, only radar and audio sensors are used. One is signal processing and feature extraction through radar echo information; the other is signal processing and feature extraction through the acquisition of breath sound and snore sound. Although these two sensors are entirely different in type, they also share some of the same target and characteristic parameters. For the decision-level fusion method, two sub-classifiers based on subspace K-nearest neighbours (KNN) are used for preliminary sleep-stage estimation [39]. The final classifier results are determined from the previous prediction results based on the Naïve Bayes algorithm.

The data for this example come from a study in [11,12]. An illustration of the experimental set-up is shown in Figure 10.14. The digital-IF Doppler radar was located

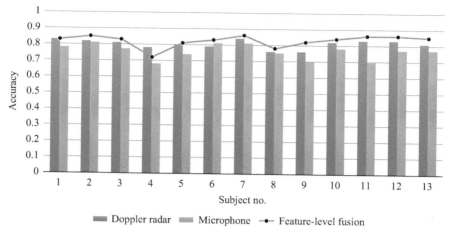

Figure 10.15   Sleep estimation results using single sensors and feature-level fusion in terms of accuracy

on the top of healthy subjects with its beam facing the human subject from 11:30 p.m. to 5:30 a.m. The distance between radar and subject was 0.8 m. The portable PSG SOMNOCheck 2 RK was used as a reference. Thirteen human subjects (eleven men and two women) aged from 21 to 26 years were involved in the experiments. The experiment was approved by the institutional review board of the Nanjing Integrated Traditional Chinese and Western Medicine Hospital. During the experiments, the room brightness and temperature were controlled to be suitable for sleeping. In this study, a total of 12,124 epochs were analysed and the tenfold validation was used to evaluate the performance.

In statistics and machine learning, ensemble methods use multiple learning algorithms to obtain better predictive performance that could be obtained from any of the constituent learning algorithms alone [40]. Unlike a statistical ensemble in statistical mechanics, which is usually infinite, a machine learning ensemble only refers to a real finite set of alternative models, but typically allows for much more flexible structure to exist among those alternatives. In our experiment, we have chosen the subspace KNN [39] as the ensemble algorithm.

The sleep-stage classification results of the single sensors are shown in Figure 10.15. The preliminary results of both the radar-based and the microphone-based classifier are good with the accuracy above 60%. The performance of the classifier from the radar data is superior compared with the microphone for all subjects. The multi-modal sensor results using feature-level fusion are illustrated in Figure 10.15. The sleep-stage accuracy of the feature-level fusion system was up to 82.0% on average. The decision-level fusion shows poor performance with the average accuracy of 80.0%.

10.6 Summary

This chapter contains two parts to introduce the digital-IF CW Doppler radar and its healthcare sensing applications. In the first part, the RF layer, IF layer and baseband signal processing of digital-IF CW Doppler radar are introduced. In particular, the design principle of digital-IF layer is well discussed. In the second part, two examples including contactless beat-to-beat blood pressure estimation and multi-sensor-based sleep-stage classification are introduced. These examples show that the digital-IF CW Doppler radar can not only work as a single sensor but also serve as a sensor node in the multi-sensor network for healthcare sensing applications.

References

[1]   Peng Z, Muñoz-Ferreras JM, Tang Y, *et al.* A portable FMCW interferometry radar with programmable low-IF architecture for localization, ISAR imaging, and vital sign tracking. *IEEE Transactions on Microwave Theory and Techniques.* 2017;65(4):1334–1344.

[2]   Massagram W, Lubecke VM, Høst-Madsen A, and Boric-Lubecke O. Assessment of heart rate variability and respiratory sinus arrhythmia via Doppler radar. *IEEE Transactions on Microwave Theory and Techniques.* 2009;57(10):2542–2549.

[3]   Wang G, Muñoz-Ferreras J, Gu C, Li C, and Gómez-Garcia R. Application of linear-frequency-modulated continuous-wave (LFMCW) radars for tracking of vital signs. *IEEE Transactions on Microwave Theory and Techniques.* 2014;62(6):1387–1399.

[4]   Wang H, Dang V, Ren L, *et al.* An elegant solution: An alternative ultra-wideband transceiver based on stepped-frequency continuous-wave operation and compressive sensing. *IEEE Microwave Magazine.* 2016;17(7):53–63.

[5]   Schleicher B, Nasr I, Trasser A, and Schumacher H. IR-UWB radar demonstrator for ultra-fine movement detection and vital-sign monitoring. *IEEE Transactions on Microwave Theory and Techniques.* 2013;61(5):2076–2085.

[6]   Bi S, Liu X, and Matthews D. An experimental study of 2-D cardiac motion pattern based on contact radar measurement. In: 2015 IEEE 16th Annual Wireless and Microwave Technology Conference (WAMICON); 2015. pp. 1–4.

[7]   Sun L, Li Y, Hong H, Xi F, Cai W, and Zhu X. Super-resolution spectral estimation in short-time non-contact vital sign measurement. *Review of Scientific Instruments.* 2015;86(4):105–133.

[8]   Sun L, Hong H, Li Y, *et al.* Noncontact vital sign detection based on stepwise atomic norm minimization. *IEEE Signal Processing Letters.* 2015;22(12):2479–2483.

[9]   Zhao H, Hong H, Miao D, *et al.* A noncontact breathing disorder recognition system using 2.4-GHz digital-IF Doppler radar. *IEEE Journal of Biomedical and Health Informatics.* 2019;23(1):208–217.

[10] Zhao H, Hong H, Sun L, Li Y, Li C, and Zhu X. Noncontact physiological dynamics detection using low-power digital-IF Doppler radar. *IEEE Transactions on Instrumentation and Measurement.* 2017;66(7): 1780–1788.

[11] Hong H, Zhang L, Gu C, Li Y, Zhou G, and Zhu X. Noncontact sleep stage estimation using a CW Doppler radar. *IEEE Journal on Emerging and Selected Topics in Circuits and Systems.* 2018;8(2):260–270.

[12] Hong H, Zhang L, Zhao H, *et al.* Microwave sensing and sleep: Noncontact sleep-monitoring technology with microwave biomedical radar. *IEEE Microwave Magazine.* 2019;20(8):18–29.

[13] Le Kernec J, Fioranelli F, Ding C, *et al.* Radar signal processing for sensing in assisted living: The challenges associated with real-time implementation of emerging algorithms. *IEEE Signal Processing Magazine.* 2019;36(4):29–41.

[14] Gu C, Li C, Lin J, Long J, Huangfu J, and Ran L. Instrument-based noncontact Doppler radar vital sign detection system using heterodyne digital quadrature demodulation architecture. *IEEE Transactions on Instrumentation and Measurement.* 2010;59(6):1580–1588.

[15] Zhang Y, Zheng Y, Ma Z, and Sun Y. Radial pulse transit time is an index of arterial stiffness. *Hypertension Research.* 2011;34(2):884–887.

[16] Park B, Boric-Lubecke O, and Lubecke V. Arctangent demodulation with DC offset compensation in quadrature Doppler radar receiver systems. *IEEE Transactions on Microwave Theory and Techniques.* 2007;55(5):1073–1079.

[17] Wu Y, and Li J. The design of digital radar receivers. *IEEE Aerospace and Electronic Systems Magazine.* 1998;13(1):35–41.

[18] Li C, Xiao Y, and Lin J. Experiment and spectral analysis of a low-power*Ka*-band heartbeat detector measuring from four sides of a human body. *IEEE Transactions on Microwave Theory and Techniques.* 2006;54(12):4464–4471.

[19] Vaughan R, Scott N, and White D. The theory of bandpass sampling. *IEEE Transactions on Signal Processing.* 1991;39(9):1973–1984.

[20] Zakrzewski M, Raittinen H, and Vanhala J. Comparison of center estimation algorithms for heart and respiration monitoring with microwave Doppler radar. *IEEE Sensors Journal.* 2012;12(3):627–634.

[21] Xu W, Gu C, Li C, and Sarrafzadeh M. Robust Doppler radar demodulation via compressed sensing. *Electronics Letters.* 2012;48(22):1428–1430.

[22] Zhao H, Hong H, Sun L, Xi F, Li C, and Zhu X. Accurate DC offset calibration of Doppler radar via non-convex optimisation. *Electronics Letters.* 2015;51(16):1282–1284.

[23] Zhao H, Gu X, Hong H, Li Y, Zhu X, and Li C. Non-contact beat-to-beat blood pressure measurement using continuous wave Doppler radar. In: 2018 IEEE/MTT-S International Microwave Symposium – IMS; 2018. pp. 1413–1415.

[24] Dawson S, Manktelow B, Robinson T, Panerai R, and Potter J. Which parameters of beat-to-beat blood pressure and variability best predict early outcome after acute ischemic stroke? *Stroke.* 2000;31(2):463–468.

[25] Pitson D, and Stradling J. Value of beat-to-beat blood pressure changes, detected by pulse transit time, in the management of the obstructive sleep apnoea/hypopnoea syndrome. *The European Respiratory Journal.* 1998;12(3):685–692.

[26] Buxi D, Redouté J, and Yuce MR. Blood pressure estimation using pulse transit time from bioimpedance and continuous wave radar. *IEEE Transactions on Biomedical Engineering.* 2017;64(4):917–927.

[27] Will C, Shi K, Schellenberger S, *et al.* Local pulse wave detection using continuous wave radar systems. *IEEE Journal of Electromagnetics, RF and Microwaves in Medicine and Biology.* 2017;1(2):81–89.

[28] Qasem A, and Avolio A. Determination of aortic pulse wave velocity from waveform decomposition of the central aortic pressure pulse. *Hypertension.* 2008;51(2):188–195.

[29] Poon C, and Zhang Y. Cuff-less and noninvasive measurements of arterial blood pressure by pulse transit time. In: 2005 IEEE Engineering in Medicine and Biology 27th Annual Conference; 2005. pp. 5877–5880.

[30] Kushida C, Littner M, Morgenthaler T, *et al.* Practice parameters for the indications for polysomnography and related procedures: An update for 2005. *Sleep.* 2005;28(4):499–523.

[31] Kurihara Y, and Watanabe K. Sleep-stage decision algorithm by using heartbeat and body-movement signals. *IEEE Transactions on Systems, Man, and Cybernetics-Part A: Systems and Humans.* 2012;42(6):1450–1459.

[32] Kagawa M, Sasaki N, Suzumura K, and Matsui T. Sleep stage classification by body movement index and respiratory interval indices using multiple radar sensors. In: 37th Annual International Conference of the IEEE Engineering in Medicine and Biology Society (EMBC), 2015. pp. 7606–7609.

[33] Lin F, Zhuang Y, Song C, *et al.* SleepSense: A noncontact and cost-effective sleep monitoring system. *IEEE Transactions on Biomedical Circuits and Systems.* 2017;11(1):189–202.

[34] Dafna E, Tarasiuk A, and Zigel Y. Sleep-wake evaluation from whole-night non-contact audio recordings of breathing sounds. *PLOS One.* 2015;10(2):e0117382.

[35] Dafna E, Halevi M, Or DB, Tarasiuk A, and Zigel Y. Estimation of macro sleep stages from whole night audio analysis. In: 38th Annual International Conference of IEEE Engineering in Medicine and Biology Society (EMBC), 2016. pp. 2847–2850.

[36] Xue B, Deng B, Hong H, Wang Z, Zhu X, and Feng DD. Non-contact sleep stage detection using canonical correlation analysis of respiratory sound. *IEEE Journal of Biomedical and Health Informatics.* 2020;24(2):614–625

[37] Scalart P, and Filho JV. Speech enhancement based on a priori signal to noise estimation. In: 1996 IEEE International Conference on Acoustics, Speech, and Signal Processing Conference Proceedings (vol. 2), Atlanta, GA, USA. IEEE; 1996. pp. 629–632.

[38]   Yang J, Yang Jy, Zhang D, and Lu Jf. Feature fusion: Parallel strategy vs. serial strategy. *Pattern Recognition*. 2003;36(6):1369–1381.
[39]   Ho TK. Nearest neighbors in random subspaces. In: Joint IAPR International Workshops on Statistical Techniques in Pattern Recognition (SPR) and Structural and Syntactic Pattern Recognition (SSPR). Berlin, Heidelberg: Springer; 1998. pp. 640–648.
[40]   Zhou ZH. *Ensemble Methods: Foundations and Algorithms*. Boca Raton, FL: Chapman and Hall/CRC; 2012.

Chapter 11

L1-norm principal component and discriminant analyses of micro-Doppler signatures for indoor human activity recognition

Fauzia Ahmad¹ and Panos P. Markopoulos²

This chapter considers L1-norm-based principal component analysis (PCA) and linear discriminant analysis (LDA) of micro-Doppler signatures for classification of human gross motor activities in an indoor environment. Subspaces derived by L1-norm PCA (L1-PCA) of the associated micro-Doppler signatures can be employed for indoor activity monitoring. Alternatively, L1-norm LDA (L1-LDA) can identify low-rank subspaces whereon micro-Doppler signatures from distinct activities are most differentiable. Both L1-PCA and L1-LDA are more robust than their standard counterparts, namely PCA and LDA, exhibiting resistance against outliers that may be present among the training data, e.g., due to mislabelling. Whether L1-PCA is performed or L1-LDA is applied, we show in this chapter that the L1 variants exhibit classification performance similar to their standard counterparts when the training data are nominal/clean, while they provide enhanced performance when the training micro-Doppler signatures are outlier-corrupted.

11.1 Introduction

An increasing number of elderly all over the world are opting for aging in place. As such, there is a growing interest in assisted living technologies that provide activity monitoring, e.g., in the form of fall detection for the elderly, ensuring them more years of life in good health in their own homes and residences. Also, there is renewed interest in the healthcare industry to remotely monitor the well-being of patients outside of a hospital setting on a routine basis. Provision of remote access to patients' vital signs and physical activities to healthcare providers promises to reduce hospital readmissions due to relapse after a treatment and the likelihood

¹Department of Electrical and Computer Engineering, Temple University, Philadelphia, PA, USA
²Department of Electrical and Microelectronic Engineering, Rochester Institute of Technology, Rochester, NY, USA

Table 11.1   Comparison of human activity recognition methods

Method	Automated feature extraction	Training data required	Training complexity	Memory required
Physical feature-based classifier	No	Low	Low[1]	Low
Nonphysical feature-based classifier	Yes	Moderate	Moderate	Low
Deep learning-based classifier	Yes	High	High	High

[1]These methods, however, require ad hoc tuning of parameters and thresholds.

of occurrence of costly major medical episodes. Among various modalities being considered for remote monitoring of patients and the elderly, radar has gained impetus [3–15], due to its insensitivity to lighting variations and robustness against visual obstructions—two commonplace characteristics of indoor monitoring environments. Moreover, the privacy-aware nature of radar sensing offers the means for activity monitoring without visually observing the human subjects. Furthermore, radar is a non-wearable technology which can be useful for older patients or patients with cognitive impairments.

Radar-based human activity recognition methods for in-home monitoring focus predominantly on Doppler or continuous-wave (CW) radar technology. Several different features extracted from micro-Doppler signatures associated with indoor human gross motor activities have been considered for classification. These motion-discriminating features can be broadly classified into two main categories, namely, physical and nonphysical features. The former are associated directly with physical attributes of human motion, e.g., extreme Doppler frequency captures the maximum velocity associated with a human activity [15,16]. Such features are typically extracted manually from recorded micro-Doppler signatures and the process involves ad hoc tuning of parameters and thresholds. On the other hand, nonphysical features can be extracted in an automated fashion by optimisation of some numerical criterion [6,9], e.g., by means of PCA and LDA [9,17]. Existing works have revealed that PCA and LDA of micro-Doppler signatures yield superior indoor activity classification performance as compared to commonly used physical ones [1,9,17,18]. Deep learning techniques have also been employed for human activity recognition, but they require large amounts of training data and memory and impose a much higher computational load [11,12]. The pros and cons of these broadly categorised activity recognition methods are summarised in Table 11.1.

Despite the documented success of PCA and LDA in machine learning and pattern recognition applications, including feature extraction from micro-Doppler signatures, both methods are known to be sensitive to the presence of irregular points or outliers among the training data, resulting in significantly diminished performance [19,20].

The sensitivity of PCA and LDA against the outlying training points derives from the squared emphasis placed on the contribution of each and every training point in the optimisation metric. For in-home radar monitoring, the diverse and infrequent nature of some activities of interest, e.g., falls, may require long-term data collections in nursing homes or private residences to capture sufficient training data. As such, outliers may arise therein due to a variety of causes, including radar-sensor malfunctions, intermittent/unexpected non-focal activity interference, and mislabellings in the training data sets.

Recent research in signal processing and machine learning has shown that L1-norm variants of PCA and LDA, called L1-norm PCA (L1-PCA) and L1-LDA, attain strong resistance against outliers among the training data. The robustness of L1-PCA and L1-LDA is attributed to the linear emphasis placed by the L1-norm optimisation metric on each data point [19–27]. Motivated by the outlier-resistant nature of L1-PCA and L1-LDA, we present in this chapter two indoor human activity recognition methods. The first method performs activity classification by means of L1-PCA-derived subspaces of associated micro-Doppler signatures, while the second method is based on L1-LDA of micro-Doppler signatures [1,2,18]. Using real radar data of falling, sitting, walking, and bending motions and assuming presence of outliers due to mislabelling, we demonstrate that each L1-norm-based method offers superior activity classification performance over its standard counterpart when the training micro-Doppler signature data sets are corrupted, while it exhibits comparable performance to its standard counterpart in case of nominal/clean training data.

The chapter is organised as follows. The radar signal model is described in Section 11.2. L1-PCA-based indoor human activity classification approach is presented in Section 11.3 along with illustrative examples based on real data. Section 11.4 describes the L1-LDA-based motion recognition method with supporting real-data examples. The trade-offs between robustness and training complexity for PCA, LDA, and their L1-norm variants are discussed in Section 11.5. Concluding remarks are provided in Section 11.6.

11.2 Radar signal model

For a CW radar, the baseband return due to a moving point target has the general form [2]:

$$s(t) = a(t) \exp\{-jp(t)\} \tag{11.1}$$

where $a(t)$ is the range-dependent amplitude and $p(t)$ is the phase. The derivative of $p(t)$ provides the corresponding Doppler frequency. An extended target, such as a human, can be viewed as an ensemble of point targets and, accordingly, its return can be expressed as a summation of the individual returns of the point targets that compose it. The use of a continuous waveform eliminates the ability to separate these scatterers in range. As such, the target's micro-Doppler signature is the summation of the micro-Doppler responses of each scattering centre.

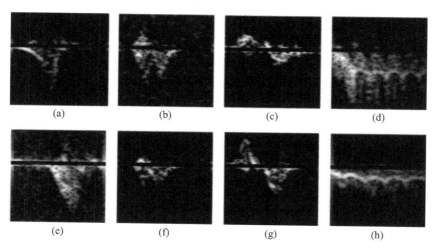

Figure 11.1   Example micro-Doppler signatures of indoor human activities:
Falling, (a) and (e); Sitting, (b) and (f); Bending, (c) and (g); and
Walking, (d) and (h). (Parts (a)–(c) and (e)–(g) adopted from [1],
© 2017 IEEE)

Similarly, the Doppler signature associated with a moving human subject derives from the superposition of component Doppler frequencies. At the same time, human activities typically produce time-varying Doppler frequencies. Time–frequency (TF) processing can be used to reveal the associated instantaneous frequency signatures [28–30]. These TF signal representations and the corresponding micro-Doppler signatures characterise the dynamics of each human gross motor activity. The most commonly considered TF representation is the short-time Fourier transform (STFT). The STFT, $S(t,f)$ of signal $s(t)$ is defined as

$$S(t,f) = \int s(t-\tau)w(\tau)e^{-j2\pi f\tau}d\tau \tag{11.2}$$

where $w(\tau)$ denotes a window function that determines the trade-off between time and frequency resolutions. In this chapter, we employ the *spectrogram*, which is the energetic version of the STFT in (11.2).

Figure 11.1 depicts two instances each of micro-Doppler signatures for four different types of activities, namely, sitting, falling, bending to pick up an object, and walking. In each signature, the horizontal dimension denotes time whereas the vertical axis represents frequency. The DC component of each signature has been removed, as it does not contain any useful information for activity classification. These micro-Doppler signatures are based on real data collected in a laboratory environment using a 6 GHz CW radar. All of the motions were performed along the line of sight of the radar. The two instances correspond to two different persons and are provided to show the variability in the same type of activity across different human subjects.

11.3 L1-PCA-based classification

In this section, we develop the problem formulation of L1-PCA of micro-Doppler signatures and describe an efficient solution to the resulting optimisation problem. Examples based on real-data experiments are also provided that illustrate the robustness of L1-PCA against outliers in the training data.

11.3.1 L1-norm PCA

Consider K indoor activity classes and assume $N_{\text{tr},k}$ Doppler signatures are available from class $k \in \{1, 2, \ldots, K\}$ for training. The nth training signature for class k is in matrix form, $\mathbf{S}_{k,n} \in \mathbb{R}^{N_t \times N_f}$, which is obtained by sampling the energetic version of (11.2) in N_t time bins and N_f frequency bins of interest. Then, the nth vector signature for class k is obtained as $\mathbf{s}_{k,n} = \text{vec}(\mathbf{S}_{k,n}) \in \mathbb{R}^{D \times 1}$, where $D = N_t N_f$ and $\text{vec}(\cdot)$ returns the column-wise vectorisation of its matrix argument. Alternatively, row-wise vectorisation of each Doppler signature could be considered. The $N_{\text{tr},k}$ signatures from class k are organised as a training matrix:

$$\mathbf{S}_k = [\mathbf{s}_{k,1}, \mathbf{s}_{k,2}, \ldots, \mathbf{s}_{k,N_{\text{tr},k}}] \in \mathbb{R}^{D \times N_{\text{tr},k}} \tag{11.3}$$

For class k, we define the zero-centred training matrix as

$$\mathbf{S}_k^{(zc)} = \mathbf{S}_k - \mathbf{m}_k \mathbf{1}_{N_{\text{tr},k}}^{\top} \tag{11.4}$$

where $\mathbf{1}_{N_{\text{tr},k}}$ denotes the all-ones vector of length $N_{\text{tr},k}$,

$$\mathbf{m}_k = \frac{1}{N_{\text{tr},k}} \mathbf{S}_k \mathbf{1}_{N_{\text{tr},k}} \tag{11.5}$$

is the training sample mean of the kth class, and the superscript "$\top$" denotes matrix transpose.

At this point, it is worth understanding how standard PCA is formulated. Broadly speaking, given $\mathbf{S}_k^{(zc)}$, PCA seeks a small number of orthogonal directions or principal components (PCs) that define a subspace wherein data presence of the class k is maximised. Mathematically, PCA approximates the training data matrix $\mathbf{S}_k^{(zc)}$ by the low-rank matrix $\mathbf{Q}\mathbf{X}_k^{\top} \in \mathbb{R}^{D \times N_{\text{tr},k}}$, where $\mathbf{Q}^{\top}\mathbf{Q} = \mathbf{I}_d$ and $d \leq \text{rank}(\mathbf{S}_k^{(zc)}) \leq \min(N_{\text{tr},k}, D)$, so that the L2-norm of the approximation error is minimised. That is, PCA is formulated as [31]:

$$(\mathbf{Q}_{L2,k}, \mathbf{X}_k) = \underset{\mathbf{Q} \in \mathbb{R}^{D \times d};\ \mathbf{Q}^{\top}\mathbf{Q} = \mathbf{I}_d;\ \mathbf{X}_k \in \mathbb{R}^{N_{\text{tr},k} \times d}}{\text{argmin}} \left\| \mathbf{S}_k^{(zc)} - \mathbf{Q}\mathbf{X}_k^{\top} \right\|_2^2 \tag{11.6}$$

where the L2-norm ($\|\cdot\|_2^2$) returns the sum of the squared entries of its matrix argument. Observing that, for any given $\mathbf{Q}$, $\mathbf{X}_k = (\mathbf{S}_k^{(zc)})^{\top}\mathbf{Q}$ minimises the error in (11.6), $\mathbf{Q}_{L2,k}$ can be determined as solution to the equivalent projection maximisation problem:

$$\mathbf{Q}_{L2,k} = \underset{\mathbf{Q} \in \mathbb{R}^{D \times d};\ \mathbf{Q}^{\top}\mathbf{Q} = \mathbf{I}_d}{\text{argmax}} \left\| \mathbf{Q}^{\top}\mathbf{S}_k^{(zc)} \right\|_2^2 \tag{11.7}$$

Accordingly, $\mathbf{X}_{L2,k} = (\mathbf{S}_k^{(zc)})^\top \mathbf{Q}_{L2,k}$. PC matrix $\mathbf{Q}_{L2,k}$ consists of the d-dominant left singular-vectors of $\mathbf{S}_k^{(zc)}$, obtained with standard singular value decomposition (SVD) [32].

Following the formulation in (11.7), the d L1-norm PCs (L1-PCs) of class k can be computed as

$$\mathbf{Q}_k = \underset{\mathbf{Q} \in \mathbb{R}^{D \times d}; \ \mathbf{Q}^\top \mathbf{Q} = \mathbf{I}_d}{\arg\max} \left\| \mathbf{Q}^\top \mathbf{S}_k^{(zc)} \right\|_1 \tag{11.8}$$

where $\| \cdot \|_1$ returns the L1-norm of its matrix argument (summation of the absolute values of all its entries). It was shown in [19] that, if

$$\mathbf{B}_{\text{opt}} = \underset{\mathbf{B} \in \{\pm 1\}^{N_{\text{tr},k} \times d}}{\arg\max} \left\| \mathbf{S}_k^{(zc)} \mathbf{B} \right\|_* \tag{11.9}$$

where nuclear norm $\| \cdot \|_*$ returns the sum of the singular values of its matrix argument, then L1-PCA in (11.8) is solved by

$$\mathbf{Q}_k = \Phi(\mathbf{S}_k^{(zc)} \mathbf{B}_{\text{opt}}) \tag{11.10}$$

where, for any tall matrix $\mathbf{A} \in \mathbb{R}^{m \times n}$ with SVD $\mathbf{A} \overset{\text{SVD}}{=} \hat{\mathbf{U}} \hat{\mathbf{D}}_{n \times n} \hat{\mathbf{V}}^\top$, $\Phi(\mathbf{A}) = \hat{\mathbf{U}} \hat{\mathbf{V}}^\top$. Therefore L1-PCA in (11.8) can be cast as an equivalent combinatorial optimisation problem over antipodal binary variables ("bits") in $\{\pm 1\}$. Several algorithms have been proposed for solving the L1-PCA optimisation problem [19,21,23]. In this chapter, we employ efficient L1-PCA through bit flipping (L1-BF) [21,23]. We note that in both PCA and its L1-norm variant, the number of PCs is tunable and depends on the data. To simplify, in this chapter we choose $d = K - 1$.

11.3.2 L1-PCA through bit flipping

L1-BF is an efficient, near-optimal L1-PCA calculator. It is based on optimal single-bit-flipping iterations and has a computational cost comparable to that of standard PCA [23]. L1-BF initiates a bit matrix $\mathbf{B}(0) \in \{\pm 1\}^{N_{\text{tr},k} \times d}$ and executes a sequence of optimal single-bit-flipping iterations, across which the metric in (11.9) monotonically increases. Specifically, at each iteration, L1-BF examines all $N_{\text{tr},k} d$ bits and recognises the single one that, when flipped, offers the highest increase to the metric of (11.9). A pseudocode for the calculation of the d L1-PCs of $\mathbf{S}_k^{(zc)}$ by means of L1-BF is provided in Figure 11.2.

11.3.3 Classifier

Having computed the L1-PCs for the K activity classes, $\{\mathbf{Q}_k\}_{k=1}^K$, each new vectorised micro-Doppler signature $\mathbf{s} \in \mathbb{R}^{D \times 1}$ is assigned to one of the K activity classes as follows. For the kth class, $\mathbf{s}$ is first zero-centred and then its distance from the kth subspace is measured as

$$d_k(\mathbf{s}) = \left\| (\mathbf{I}_D - \mathbf{Q}_k \mathbf{Q}_k^\top)(\mathbf{s} - \mathbf{m}_k) \right\|_2 \tag{11.11}$$

L1-BF for the calculation of d L1-PCs of $\mathbf{S}_k^{(zc)}$

Input: $\mathbf{S}_k^{(zc)} \in \mathbb{R}^{D \times N_{tr,k}}$ of rank r, $d \le r$

1:    $(\mathbf{U}, \mathbf{D}_{r \times r}, \mathbf{V}) \leftarrow \mathrm{svd}(\mathbf{S}_k^{(zc)})$
2:    $\mathbf{Y} \leftarrow \mathbf{D}\mathbf{V}^\top$, $\mathbf{v} \leftarrow [\mathbf{V}]_{:,1}$, $\mathbf{B} = \mathrm{sgn}(\mathbf{v1}_d^\top)$
3:    $\mathbf{B}_{bf} \leftarrow \mathrm{bf}(\mathbf{Y}, \mathbf{B}, d)$
4:    $(\hat{\mathbf{U}}_{D \times d}, \hat{\mathbf{D}}_{d \times d}, \hat{\mathbf{V}}_{d \times d}) \leftarrow \mathrm{svd}(\mathbf{S}_k^{(zc)} \mathbf{B}_{bf})$

Output: $\mathbf{Q}_k \leftarrow \hat{\mathbf{U}}\hat{\mathbf{V}}^\top$

Function bf$(\mathbf{Y}_{r \times N_{tr,k}}, \mathbf{B}_{N_{tr,k} \times d}, d \le r)$

1:    $\omega \leftarrow d\|\mathbf{Y}[\mathbf{B}]_{:,1}\|_2$
2:    $\mathscr{L} \leftarrow \{1, 2, \ldots, d\}$
3:    while true (or terminate at $N_{tr,k}d$ iterations)
4:      for $x \in \mathscr{L}$, $l \leftarrow \lceil \frac{x}{N_{tr,k}} \rceil$, $m \leftarrow x - N_{tr,k}(l-1)$
5:        $\tilde{\mathbf{B}} \leftarrow \mathbf{B}$, $\tilde{B}_{m,l} \leftarrow -B_{m,l}$
6:        $(\mathbf{U}, \mathbf{S}_{d \times d}, \mathbf{V}^\top) \leftarrow \mathrm{svd}(\mathbf{Y}\tilde{\mathbf{B}})$
10:      $a_{m,l} \leftarrow \sum_{j=1}^{d} S_{j,j}$
11:      $(n, k) \leftarrow \mathrm{argmax}_{m,l: (l-1)N_{tr,k}+m \in \mathscr{L}} \, a_{m,l}$
12:      if $\omega < a_{n,k}$,
13:        $B_{n,k} \leftarrow -B_{n,k}$, $\omega \leftarrow a_{n,k}$,
14:        $\mathscr{L} \leftarrow \mathscr{L} \setminus \{(k-1)N_{tr,k}+n\}$
15:      elseif $\omega \ge a_{n,k}$ and $|\mathscr{L}| < N_{tr,k}d$, $\mathscr{L} \leftarrow \{1, 2, \ldots, N_{tr,k}d\}$
16:      else, break
17:    Return $\mathbf{B}$

Figure 11.2   L1-BF algorithm for the calculation of d L1-PCs of a rank-r zero-centred class-k data set $\mathbf{S}_k^{(zc)}$ of $N_{tr,k}$ samples of dimension D; svd(·) returns the SVD of its matrix argument. (Adopted from [1], © 2017 IEEE)

Having measured $d_k(\mathbf{s})$ for every $k \in \{1, 2, \ldots, K\}$, the classifier decides that the micro-Doppler signature $\mathbf{s}$ has captured activity from the lth class if

$$l = \operatorname*{argmin}_{k \in \{1,2,\ldots,K\}} d_k(\mathbf{s}) \tag{11.12}$$

The classifier in (11.12) is the L1-norm-based alternative to the popular nearest subspace classifier (NSC) [33], which for any k would calculate $\mathbf{Q}_{L2,k}$ by means of standard PCA using (11.7) and employ $\mathbf{Q}_{L2,k}$ in place of $\mathbf{Q}_k$ for computing the distances in (11.11).

11.3.4   Illustrative example

We consider recognition of falling, bending, and sitting activity classes. We use micro-Doppler signatures based on real data collected with a 6 GHz CW radar in a laboratory environment. Two instances from each class are illustrated in Figure 11.1. One female and four males, with ages ranging from 24 to 60 years and heights ranging from 1.62 to 1.85 m, performed each activity six times. The falling, sitting, and bending activities were performed at a distance of 2.75 m along the radar line of sight. The feed point

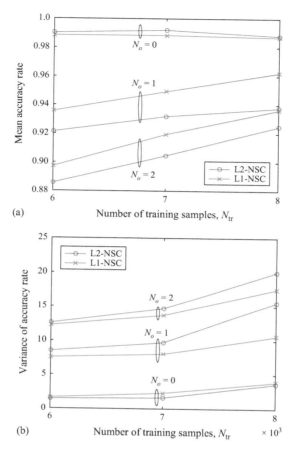

Figure 11.3   (a) Average and (b) variance of accuracy rate for standard and L1-norm NSC. (Adopted from [1], © 2017 IEEE)

of the antenna was positioned 1 m above the floor and a sampling rate of 1 kHz was used for data collection.

For illustration, out of a collection of $N = 10$ signatures from each class, $N_{tr,1} = N_{tr,2} = N_{tr,3} = N_{tr} = 6, 7$, and 8 signatures per class are used for subspace training and $N_{te} = N - N_{tr}$ are used for testing. The subspace dimensionality is set to $d = 2$. We assume that $N_o = 0, 1$, or 2 training signatures from bending class are mistakenly labelled as signatures from sitting class, and vice versa; that is, the training data sets of bending and sitting classes are corrupted by N_o outliers due to mislabelling. For each combination of N_{tr} and N_o, we conduct 1,000 classification experiments over distinct training/evaluation data sets using standard NSC and the L1-norm NSC. We calculate the mean accuracy rate (number of correct classifications versus number of testing signatures classified) for both classifiers and plot it in

Figure 11.3(a) versus N_{tr} for the three different values of N_o (three curves per classifier). In Figure 11.3(b), we plot the corresponding variance of the measured accuracy rates. We observe that, when there is no mislabelling ($N_o = 0$) and the training points are nominal for each class, both methods perform equally well exhibiting high mean accuracy rate. However, for $N_o = 1$ and 2, the L1-norm NSC method exhibits higher mean accuracy rate than standard NSC, with lower variance, for all values of N_{tr}.

11.4 L1-LDA-based activity classification

In this section, we provide a mathematical description of standard LDA and L1-LDA-based classifiers. Conceptually, we use discriminant analysis to find the linear subspace wherein micro-Doppler signatures of different activity classes are most differentiable. Then, when an unlabelled micro-Doppler signature is collected, we project it on the computed subspace and classify it based on its average distance from the projected labelled signatures from each class.

11.4.1 Problem formulation

Assume K activity classes with $\mathbf{S}_k$ given by (11.3) denoting the training signature matrix of the kth class. Consider an orthonormal basis $\mathbf{U} \in \mathbb{R}^{D \times d}$, $\mathbf{U}^\top \mathbf{U} = \mathbf{I}_d$, spanning a d-dimensional subspace for $d < D$. The value of d, in general, depends on the data and is tunable. In this work, we choose $d = K - 1$. Projecting the nth vectorised training Doppler signature, $\mathbf{s}_{k,n}$, from class k on $\mathbf{U}$, we obtain $\mathbf{U}^\top \mathbf{s}_{k,n} \in \mathbb{R}^{d \times 1}$. The mean of the projected samples from class k is, accordingly, $\mathbf{U}^\top \mathbf{m}_k$, with $\mathbf{m}_k$ given by (11.5). Likewise, the mean of all projected samples (from all classes) is $\mathbf{U}^\top \mathbf{m}$, where

$$\mathbf{m} = \frac{1}{N} \sum_{k=1}^{K} N_{tr,k} \mathbf{m}_k \tag{11.13}$$

Standard LDA strives to find a basis $\mathbf{U}$ that offers high-class dispersion (i.e., scattering of projected-sample means) $\sum_{k=1}^{K} N_{tr,k} \|\mathbf{U}^\top (\mathbf{m}_k - \mathbf{m})\|_2^2$ and low aggregate intra-class scattering $\sum_{k=1}^{K} \sum_{n=1}^{N_{tr,k}} \|\mathbf{U}^\top (\mathbf{s}_{k,n} - \mathbf{m}_k)\|_2^2$. Introducing the matrix of weighted zero-centred means

$$\mathbf{A}_b = [(\mathbf{m}_1 - \mathbf{m})N_{tr,1}, \dots, (\mathbf{m}_K - \mathbf{m})N_{tr,K}]$$

and the matrix of per-class zero-centred data

$$\mathbf{A}_w = [\mathbf{S}_1^{(zc)}, \dots, \mathbf{S}_K^{(zc)}]$$

LDA can be written as

$$\mathbf{U}_{LDA} = \underset{\mathbf{U} \in \mathbb{R}^{D \times d}; \ \mathbf{U}^\top \mathbf{U} = \mathbf{I}_d}{\operatorname{argmax}} F_{L2}(\mathbf{U}) \tag{11.14}$$

where $F_{L2}(\mathbf{U}) = \|\mathbf{U}^\top \mathbf{A}_b\|_2^2 \|\mathbf{U}^\top \mathbf{A}_w\|_2^{-2}$. Defining inter-class dispersion matrix $\mathbf{S}_{inter} = \mathbf{A}_b \mathbf{A}_b^\top$ and intra-class scatter matrix $\mathbf{S}_{intra} = \mathbf{A}_w \mathbf{A}_w^\top$, the solution to LDA in (11.14)

L1-LDA Basis Calculation

Input: $\mathbf{A}_b$ and $\mathbf{A}_w$, initial $\mathbf{U}$, $\beta \in [0,1]$

1:   **for** $n = 1, 2, \ldots$ (until convergence)
2:       $\mathbf{U}_{old} \leftarrow \mathbf{U}$
3:       $\mathbf{G} \leftarrow \mathbf{A}_b \mathrm{sgn}(\mathbf{A}_b^{\top}\mathbf{U}_{old}) - F_{L1}(\mathbf{U}_{old})\mathbf{A}_w \mathrm{sgn}(\mathbf{A}_w^{\top}\mathbf{U}_{old})$
4:       $m \leftarrow 1$
5:       $\mathbf{U} \leftarrow P(\mathbf{U}_{old} + \beta^m \mathbf{G})$
6:       If $F_{L1}(\mathbf{U}) < F_{L1}(\mathbf{U}_{old})$, set $m \leftarrow m+1$ and repeat step (5)

Output: $\mathbf{U}_{L1} \leftarrow \mathbf{U}$

Figure 11.4   *L1-LDA algorithm [26] for approximately solving (11.15). (Adopted from [2], © 2019 IEEE)*

is given by the eigenvalue decomposition of $(\mathbf{S}_{\mathrm{intra}} + a * \mathbf{I}_D)^{-1}\mathbf{S}_{\mathrm{inter}}$, where a is 0 if $\sum_{k=1}^{K} N_{\mathrm{tr},k} > D$, or $0 < a \ll 1$ otherwise, to guarantee invertibility.

L1-LDA can be formulated straightforwardly from (11.14) by substituting the L2-norm with the L1-norm; that is, the L1-LDA basis is derived as

$$\mathbf{U}_{L1} = \operatorname*{argmax}_{\mathbf{U} \in \mathbb{R}^{D \times d}; \; \mathbf{U}^{\top}\mathbf{U} = \mathbf{I}_d} F_{L1}(\mathbf{U}) \tag{11.15}$$

where $F_{L1}(\mathbf{U}) = \|\mathbf{U}^{\top}\mathbf{A}_b\|_1 \|\mathbf{U}^{\top}\mathbf{A}_w\|_1^{-1}$. We present next an algorithm for solving L1-LDA in (11.15), as introduced in [26].

11.4.2   L1-LDA algorithm

First, $\mathbf{U}$ is arbitrarily initialised to a feasible $\mathbf{U}^{(0)}$; then, it is updated iteratively. For $i = 1, 2, \ldots$, $\mathbf{U}^{(i)}$ is updated as

$$\mathbf{U}^{(i)} = P(\mathbf{U}^{(i-1)} + \beta^m \mathbf{G}^{(i)}) \tag{11.16}$$

where β is a fixed parameter in $[0, 1]$ and, for any matrix $\mathbf{A}$, $P(\mathbf{A}) = \mathbf{A}(\mathbf{A}^{\top}\mathbf{A})^{-\frac{1}{2}}$. $\mathbf{G}^{(i)}$ is defined as

$$\mathbf{G}^{(i)} = \mathbf{A}_b \mathrm{sgn}(\mathbf{A}_b^{\top}\mathbf{U}^{(i-1)}) - F_{L1}(\mathbf{U}^{(i-1)})\mathbf{A}_w \mathrm{sgn}(\mathbf{A}_w^{\top}\mathbf{U}^{(i-1)}) \tag{11.17}$$

so that $\mathrm{Tr}(\mathbf{G}^{(i)\top}\mathbf{U}^{(i-1)}) = 0$. Here, $\mathrm{sgn}(\cdot)$ denotes the sign function, $\mathrm{Tr}(\cdot)$ returns the trace of its matrix argument, and the parameter m in (11.16) is the minimum positive integer that satisfies $F_{L1}(\mathbf{U}^{(i)}) \geq F_{L1}(\mathbf{U}^{(i-1)})$. Iterations terminate when the subspace of the basis sequence converges, which is practically attained when there is no value for integer m such that $F_{L1}(\mathbf{U}^{(i)}) \geq F_{L1}(\mathbf{U}^{(i-1)})$. Alternatively, an upper bound can be set for the total number of iterations. A pseudocode of the presented algorithm is offered in Figure 11.4.

11.4.3   Classifier

When a discriminative basis $\mathbf{U}_{L1}$ is trained, the classifier projects any testing point $\mathbf{s}$ on $\mathbf{U}_{L1}$ and then classifies it by means of any standard classification algorithm. For

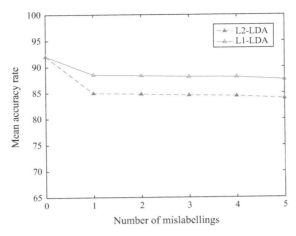

Figure 11.5   Mean accuracy rate versus number of mislabellings per class, for standard LDA and L1-LDA. (Adopted from [2], © 2019 IEEE)

example, employing the nearest-centroid classification (NCC), the classifier decides that **s** belongs to activity class l if l minimises:

$$d_{\text{L1-LDA}}(\mathbf{s}; l) = \left\| \mathbf{U}_{\text{L1}}^{\top} (\mathbf{s} - \mathbf{m}_l) \right\|_2 \tag{11.18}$$

Note that the standard version of the nearest-centroid classifier would employ $\mathbf{U}_{\text{LDA}}$ of (11.14) in place of $\mathbf{U}_{\text{L1}}$ for computing the distances in (11.18).

11.4.4   Illustrative example

We consider recognition of falling, bending, sitting, and walking activities. We use all 30 micro-Doppler signatures from each class for the experiments described in Section 11.3.4. Five different human subjects participated in the experiments, who repeated each activity 6 times. While falling, sitting, and bending motions were performed at a distance of 2.75 m from the radar, the walking motion was carried out between 1.5 and 4.5 m along the radar line of sight. Out of the $N = 30$ signatures collected from each class, 20 signatures are used for training and $N_{\text{te}} = 10$ are used for testing. The training data set was augmented by means of the SMOTE method [34], resulting in overall $N_{\text{tr}} = 40$ measurements per class. The subspace dimensionality is set to $d = 3$. We assume that $N_o = 0, 1, \ldots, 5$ training signatures from falling class are mistakenly labelled as signatures from sitting class and vice versa. For each value of N_o, we conduct 200 LDA and L1-LDA classification experiments over distinct training/evaluation data sets. Then, we calculate the mean accuracy rate for the two classifiers and plot it in Figure 11.5, versus the number of mislabellings per class, N_o. We observe that when there is no mislabelling ($N_o = 0$) and the training points are nominal for each class, LDA and L1-LDA perform equally well, exhibiting high mean accuracy rate of about 92%. However, for all nonzero numbers of mislabellings, L1-LDA outperforms the standard LDA.

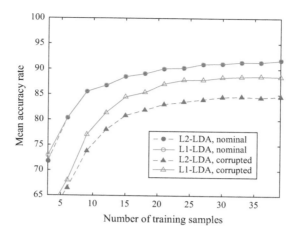

Figure 11.6   *Mean accuracy rate versus number of training data per class, for*
standard LDA and L1-LDA subspaces of micro-Doppler signatures.
The number of mislabellings takes values $N_o = 0, 1$. (Adopted
from [2], © 2019 IEEE)

Next, in Figure 11.6, we examine the mean accuracy rate as a function of the number of training samples per class, $N_{tr} = N_{tr,1} = N_{tr,2} = 3, 6, \ldots, 39$. The number of mislabellings takes values $N_o = 0$ (nominal training sets) and 1 (corrupted training sets). For each combination of N_{tr} and N_o, we conduct 200 classification experiments over distinct training/evaluation data sets. Again, we clearly observe that for $N_o = 0$, both L1-LDA and LDA provide almost identical performance. However, for $N_o = 1$, L1-LDA markedly outperforms the standard LDA.

11.5   Discussion

Table 11.2 provides the robustness versus training complexity trade-offs for PCA, LDA, and their L1-norm variants. For complexity calculations, we assume $D \geq N_{tr,k} = N_{tr} = \text{rank}(\mathbf{S}_k) > d, \forall k$ and L1-PCA is by means of the L1-BF algorithm presented in Section 11.3.2. L1-PCA is more robust than standard PCA, but it has one additional complexity term that is cubic in N_{tr} (which, here, is the smaller dimension) as compared to PCA. L1-LDA is more robust than LDA. However, the runtime of LDA is not affected as much as that of L1-LDA by increase in d. We stress that the complexity varies between standard methods and their L1-norm variants only for the training part. When testing a new sample, the computational load is identical as we employ the exact same procedure for both standard and L1-norm methods (NSC with K d-dimensional subspaces for both PCA and L1-PCA; NCC for K centroids on a d-dimensional subspace for both LDA and L1-LDA). The last column in Table 11.2 implies that LDA methods actively strive to separate the classes during training, whereas the PCA methods just represent each class compactly by means of its PCs.

Table 11.2   Comparison of PCA, LDA, and their L1-norm variants

Method	Outlier suppression	Training complexity	Active class discrimination
PCA (NSC)	No	$O(N_{tr}^2 KD)$	No
LDA (NCC)	No	$O(D^3)$	Yes
L1-PCA (NSC)	Yes	$O(N_{tr}^2 KD + N_{tr}^2 Kd^2(d^2 + N_{tr}))$	No
L1-LDA (NCC)	Yes	$O(N_{tr} KDdQ)$	Yes

[1]Q denotes the number of L1-LDA iterations which, in practice, is set as a function of $\max(N_{tr}, D)$.

For high-dimensional micro-Doppler signatures, both LDA and L1-LDA can have high computational complexity. A potential means of avoiding this issue is through the use of PCA for data dimensionality reduction prior to the application of discriminant analysis [35]. In order to exploit the outlier-resistant nature of the L1-norm-based PC and discriminant analyses, two possibilities seem plausible for discriminant analysis-based activity classification in case of high-dimensional data. The first option involves L1-PCA followed by standard LDA-based classifier, whereas the second possibility employs L1-PCA prior to L1-LDA. However, the advantages of one option over the other are not obvious. Also, more importantly, there are no theoretical guarantees that L1-PCA will not cause any loss of important discriminatory information, owing to the fact that the optimisation criteria of PC and discriminant analyses are considerably different. These remain open problems for L1-norm-based human activity classification.

11.6   Conclusion

This chapter has addressed indoor human activity classification using automated feature extractions through L1-PCA and L1-LDA of associated micro-Doppler signatures. First, L1-PCA of micro-Doppler signatures was described to enable extraction of the subspace for each class wherein data presence of that class is maximised. The L1-PCA-based classifier was applied to human activity recognition problem, and its superiority over standard PCA of micro-Doppler signature was demonstrated for the case when the outliers are present in the training data due to mislabellings. Next, L1-LDA classification of human activities was presented and its performance was validated using real-data experiments. It was demonstrated that unlike its standard counterpart, L1-LDA offers robustness against outliers.

References

[1]   Markopoulos PP, and Ahmad F. Indoor human motion classification by L1-norm subspaces of micro-Doppler signatures. In: Proceedings of the IEEE Radar Conference; 2017 May 8–12; Seattle, WA. IEEE; 2017. pp. 1807–1810.

[2] Markopoulos PP, Zlotnikov S, and Ahmad F. Adaptive radar-based human activity recognition with L1-norm linear discriminant analysis. *IEEE J Electromagn, RF, Microw Med Biol.* 2019;3(2):120–126.

[3] Amin MG. *Radar for Indoor Monitoring: Detection, Classification, and Assessment.* Boca Raton, FL: CRC Press; 2017.

[4] Ahmad F, Narayanan RM, and Schreurs D. Application of radar to remote patient monitoring and eldercare. *IET Radar, Sonar, & Navig.* 2015;9(2): 115–115.

[5] Ahmad F, Cetin AE, Ho KCD, and Nelson J. Signal processing for assisted living: developments and open problems [from the Guest Editors] *IEEE Signal Process Mag.* 2016;33(2):25–26.

[6] Liu L, Popescu M, Skubic M, Rantz M, Yardibi T, and Cuddihy P. Automatic fall detection based on Doppler radar motion signature. In: 2011 5th International Conference on Pervasive Computing Technologies for Healthcare (PervasiveHealth) and Workshops 2011 May 23–26; Dublin, Ireland. IEEE; 2011. pp. 222–225.

[7] Gadde A, Amin MG, Zhang YD, and Ahmad F. Fall detection and classifications based on time-scale radar signal characteristics. In: Proc. SPIE 9077, Radar Sensor Technology XVIII, 907712 (29 May 2014); https://doi.org/10.1117/12.2050998

[8] Su BY, Ho KC, Rantz M, and Skubic M. Doppler radar fall activity detection using the wavelet transform. *IEEE Trans Biomedical Eng.* 2015;62(3): 865–875.

[9] Jokanovic B, Amin MG, Ahmad F, and Boashash B. Radar fall detection using principal component analysis. In: Proc. SPIE 9829, Radar Sensor Technology XX, 982919 (12 May 2016); https://doi.org/10.1117/12.2225106.

[10] Erol B, Amin MG, Boashash B, Ahmad F, and Zhang YD. Wideband radar based fall motion detection for a generic elderly. In: 2016 50th Asilomar Conference on Signals, Systems and Computers; 2016 Nov 6–9; Pacific Grove, CA. IEEE; 2016. pp. 1768–1772.

[11] Jokanovic B, Amin M, and Ahmad F. Radar fall motion detection using deep learning. In: Proceedings of the IEEE Radar Conference; 2016 May 2–6; Philadelphia, PA. IEEE; 2016. pp. 1–6.

[12] Seyfioglu MS, Ozbayoglu AM, and Gurbuz SZ. Deep convolutional autoencoder for radar-based classification of similar aided and unaided human activities. *IEEE Trans Aerosp Electronic Syst.* 2018;54(4):1709–1723.

[13] Amin MG, Ahmad F, Zhang YD, and Boashash B. Human gait recognition with cane assistive device using quadratic time–frequency distributions. *IET Radar, Sonar, & Navig.* 2015;9(9):1224–1230.

[14] Tomii S, and Ohtsuki T. Falling detection using multiple Doppler sensors. In: Proceedings of the IEEE International Conference on e-Health Networking, Applications and Services; 2012 Oct 10–13; Beijing, China. IEEE; 2012. pp. 196–201.

[15] Amin MG, Zhang YD, Ahmad F, and Ho KCD. Radar signal processing for elderly fall detection: The future for in-home monitoring. *IEEE Signal Process Mag.* 2016;33(2):71–80.

[16] Wu Q, Zhang YD, Tao W, *et al.* Radar-based fall detection based on Doppler time-frequency signatures for assisted living. *IET Radar, Sonar, & Navig.* 2015;9(2):173–183.

[17] Mobasseri B, and Amin MG. A time-frequency classifier for human gait recognition. In: Kumar BVKV, Prabhakar S, Ross AA, editors. Proceedings of the SPIE Biometric Tech. for Human Identification VI; Volume 7306; 2009 Apr 13–17; Orlando, FL. SPIE; 2009. p. 730628.

[18] Markopoulos PP, and Ahmad F. Robust radar-based human motion recognition with L1-norm linear discriminant analysis. In: Proceedings of the IEEE International Microwave Biomedical Conference; 2018 Jun 14–15; Philadelphia, PA. IEEE; 2018. pp. 1–3.

[19] Markopoulos PP, Karystinos GN, and Pados DA. Optimal algorithms for L1-subspace signal processing. *IEEE Trans Signal Process.* 2014;62(10): 5046–5058.

[20] Zhong F, and Zhang J. Linear discriminant analysis based on L1-norm maximization. *IEEE Trans Image Process.* 2013;22(8):3018–3027.

[21] Kundu S, Markopoulos PP, and Pados DA. Fast computation of the L1-principal component of real-valued data. In: Proceedings of the 39th IEEE International Conference on Acoustics, Speech, and Signal Processing; 2014 May 4–9; Florence, Italy. IEEE; 2014. pp. 8028–8032.

[22] Kundu S, Markopoulos PP, and Pados DA. L1-fusion:Robust linear-time image recovery from few severely corrupted copies. In: Proceedings of the IEEE International Conference on Image Processing; 2015 Sep 27–30; Quebec City, Canada; pp. 1225–1229.

[23] Markopoulos PP, Kundu S, Chamadia S, *et al.* Efficient L1-norm principal-component analysis via bit flipping. *IEEE Trans Signal Process.* 2017;65(16): 4252–4264.

[24] Tsagkarakis N, Markopoulos PP, Sklivanitis G, *et al.* L1-norm principal-component analysis of complex data. *IEEE Trans Signal Process.* 2018;66(12): 3256–3267.

[25] Markopoulos PP, Dhanaraj M, and Savakis A. Adaptive L1-norm principal-component analysis with online outlier rejection. *IEEE J Sel Topics Signal Process.* 2018;12(6):1131–1143.

[26] Zhong F, and Zhang J. A non-greedy algorithm for L1-norm LDA. *IEEE Trans Image Process.* 2017;26(2):684–695.

[27] Wang H, Lu X, Hu Z, *et al.* Fisher discriminant analysis with L1-norm. IEEE Trans Cybern. 2014;44(6):828–842.

[28] Otero M. Application of a continuous wave radar for human gait recognition. In: Proceedings of the SPIE Signal Processing, Sensor Fusion, and Target Recognition XIV Conference; Volume 5809; 2005 May 25; Orlando, FL, USA.

[29] Kim Y, and Ling H. Human activity classification based on micro-Doppler signatures using a support vector machine. *IEEE Trans Geosci Remote Sens.* 2009;47(5):1328–1337.

[30] Orovic I, Stankovic S, and Amin MG. A new approach for classification of human gait based on time-frequency feature representations. *Signal Process.* 2011;91(6):1448–1456.

[31] Hotelling H. Analysis of a complex of statistical variables into principal components. *J Ed Psych.* 1933;24:417–441.

[32] Eckart C, and Young G. The approximation of one matrix by another of lower rank. *Psychometrika.* 1936;1:211–218.

[33] Lee K, Ho J, and Kriegman D. Acquiring linear subspaces for face recognition under variable lighting. *IEEE Trans Patt Anal Machine Intell.* 2005;27(5): 684–698.

[34] Chawla NV, Bowyer KW, Hall LO, *et al.* SMOTE: Synthetic minority over-sampling technique. *J Artif Intell Research.* 2002;16:321–357.

[35] Yang J, and Yang JY. Why can LDA be performed in PCA transformed space? *Pattern Recognition.* 2003;36(2):563–566.

Chapter 12

Micro-Doppler signature extraction and analysis for automotive application

René Petervari[1], Fabio Giovanneschi[1], and María A. González-Huici[1]

12.1 Introduction and state of the art

Radar is becoming a key technology for advanced driving assistance systems and autonomous vehicles, thanks to its robustness to weather and lighting conditions, and its capacity to detect motion. This justifies the effort in developing high-resolution and low-cost automotive radar systems along with efficient algorithms for target detection, localisation and classification.

In the research area of traffic safety, there has been a growing interest in automotive radar-based solutions towards protection of vulnerable road users like pedestrians and cyclists. According to statistics, a high percentage of road accidents involve pedestrians and cyclists, most of them occurring in urban areas. In 2016, a total of 5,320 pedestrians died in road accidents in the EU (excluding Lithuania and Slovakia), which amounts to 21% of all road fatalities in Europe [1]. The same year in the US, 5,987 pedestrians and 840 cyclists were killed in crashes with motor vehicles, which represented the highest number of pedestrians and cyclists killed in a single year since 1990 and 1991, respectively [2]. In this context, potential use of micro-Doppler features has attracted considerable attention in the recent years, and numerous publications have appeared in the scientific literature, especially in the field of pedestrian detection and recognition.

Various models of a human's gait are available in the micro-Doppler literature [3–6] including references where some other activities besides walking are analysed [7,8]. For the application case of road and urban scenarios, there are various contributions which examine and compare the micro-Doppler features of pedestrians and cyclists by using experimental data taken with 76–77 GHz automotive radar sensors as well as by computer simulations [9–11]. The main findings from these analyses have been summarised in Table 12.1.

The first observation that needs to be pointed out is that the magnitude and spectrum are illumination geometry dependent. Due to the radial measurement principle

---

[1]Department for Cognitive Radar, Fraunhofer FHR, Wachtberg, Germany

Table 12.1   Characteristics of micro-Doppler contributions from pedestrians and cyclists

	Pedestrians	Cyclists
Micro-Doppler	Micro-Doppler signature constructed by the contributions from each individual body component, i.e. torso, arm, foot, etc.	Micro-Doppler comprising components from, e.g. wheels, pedals/legs as well as the bike frame and the human body on the bike
Magnitude and spectrum width	Feet have the largest magnitude variation spectrum width as they have the largest relative speed. Legs and arms have also a larger magnitude and spectrum width than the torso	The width of the wheels' spectrum is much wider than the spectrum of the pedal/legs and the bike frame. The Micro-Doppler magnitude of the wheels is significantly larger than the magnitude of the bike frame
Strength/ energy of the pattern	The torso component is the strongest, while that of feet is much weaker. Limbs show less reflecting power with more fluctuation in scattering characteristics compared to the torso	The reflection from pedals/legs is much stronger than that from the wheels, where the lower part of the wheel micro-Doppler is much weaker than the upper part
Pattern cycle	The pattern cycle depends on the gait cycle. Duration of the cycle empirically around $D_{\text{cycle}} = 1.346/\sqrt{(v_{\text{tgt}})}$, where v_{tgt} is the relative target speed in m/s (normalised by the height of the thigh) [7], hence the faster the pedestrian, the shorter the cycle	The pattern cycle is related to the rotation cycles of the wheels and pedals/legs. Wheel micro-Doppler might be invisible in a far range due to the attenuation of the echo signal in far range as well as radar illumination geometry. Pedal/leg micro-Doppler is much stronger than that of wheels implying that pedal/leg micro-Doppler is also a promising feature for cyclist classification

of microwave radar, the measured velocity of a laterally moving object near the radar's centre line shows very small radial velocities. Hence, the measurable velocity of a detected object depends on its angle, its longitudinal and lateral position, its velocity and its direction of motion. Theoretically, these components disappear for an angle of zero degree to the radar boresight. However, if the radar sensor has enough resolution, the small radial components of the pedestrian's motion can still be recognised in the range-Doppler spectrum. It has been demonstrated that a laterally moving pedestrian still shows a small velocity spread even at a position of zero degree to the sensor's centre line caused by the motion of the pedestrian's limbs [11]. Analogous to the characteristics of a laterally walking pedestrian, a laterally moving cyclist shows a small but recognisable velocity spread at a position of zero degree to the sensor's boresight. This allows distinguishing between laterally moving cyclists and stationary targets in front of the sensor. In any case, a certain signal-to-noise ratio (SNR) is required to ensure robust detection and classification.

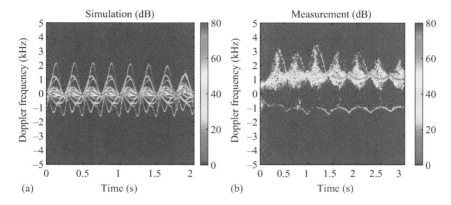

Figure 12.1 *An example of a simulated (a) and measured spectrogram (b) from a walking pedestrian*

Figure 12.1 shows an example of a simulated and measured spectrogram of a walking pedestrian.

The micro-Doppler patterns usually cannot directly be used for target classification. Several approaches to micro-Doppler feature extraction have been documented where the aim is to generate feature sets which are useful for a subsequent classification stage to identify the type of motion present in the original micro-Doppler observation. In many cases, the first stage of feature extraction has been to perform time–frequency (TF) analysis, typically via short-time Fourier transform (STFT), of the Doppler signal (spectrogram generation), in order to highlight time-varying information. Signatures suitable for classification include extent, range-rate and various statistics of the micro-Doppler spread such as the mean of the Doppler signal, standard deviation (STD) of the Doppler signal, total bandwidth of the Doppler signal, normalised STD of the Doppler signal, average Doppler frequency and periodicities in the Doppler signal, among others [8]. The extent in the range-Doppler map for classification of pedestrians is used in [12]. The authors in [13] employed the physical extent and Doppler spread from a single range-Doppler and direction-of-arrival measurement. In [14], Doppler spread measurements originating from scans at different frequencies are considered. A currently popular approach for micro-Doppler-based radar target classification is based on features generated by singular value decomposition (SVD) or principal component analysis (PCA) that can be directly applied to processed target spectrograms [15,16] or to some of the above-mentioned features [17]. SVD is a powerful approach that can be used to reduce data dimensionality, and it is crucial for techniques such as PCA and independent component analysis.

Most of these strategies, however, focus on single targets in well-known environments. Therefore, they use continuous-wave (CW) radar without range resolution. In highly dynamic automotive scenarios, however, it is crucial to resolve targets at different ranges and frequency-modulated CW (FMCW) radars are typically used. In [18], the authors describe a signal processing technique suitable for FMCW radar sensors using chirp sequence modulation, allowing observation of slow-moving objects with high resolution capability in range and velocity through autoregressive linear prediction. An additional but essential part of an FMCW-based classification system in a multi-target environment must be a tracking algorithm.

A two-stage target classification system is presented in [19], where initially, the Doppler spectrum and range profile are extracted from the radar echo signal and applied in the recognition system and in a second step, additional features are calculated from the tracker which are fed back to the target recognition system to improve the system performance and to increase the probability of correct classification. In [20], measurements were conducted with a commercial multi-channel LFMCW radar. Furthermore, algorithms were used that produce not only radar images but also a description of the scenario on a higher level. After conventional spectrum estimation and thresholding, a clustering stage that combines individual detections and generates representations of each target individually is presented followed by a Kalman filter-based multi-target tracking block. The tracker allows each target to be followed and its properties to be collected over time. With this method of jointly estimating tracks and characteristics of each individual target in a scenario, inputs for classifiers can be generated.

Another important criterion for the application to road and urban environments is real-time capability. In the next sections, a system for a real radar sensor is presented that incorporates the popular spectrogram and SVD approach in [5] and the idea of the FMCW tracking and classification system of [20] in real time. Unlike in [20], a probability hypothesis density (PHD) tracking method and association filter is considered. The main novelty of the approach presented in this chapter lies in the design focus on real-time applicability and environmental robustness, which is crucial for automotive radar application. The description of the methods is presented in the first part of the chapter followed by the experimental validation with data from a 94 GHz FMCW radar system. The results obtained from the application of a support vector machine (SVM) classifier to the extracted features [21] show that a running pedestrian can be successfully discriminated from a cyclist, a wheelchair or a pedestrian with rollator with an accuracy over 90% while it is also possible to classify between a single pedestrian moving or running longitudinally or transversally with respect to the radar with a similar accuracy.

12.2   Micro-Doppler analysis in automotive radar

In the subsequent paragraphs, the required procedures for micro-Doppler analysis are presented. These procedures encompass methods for target detection, tracking and spectrogram extraction and strategies for feature extraction and target classification. There are multiple suitable techniques for performing these tasks. However, each technique has its individual advantages and disadvantages. Automotive applications place high demands on efficiency and complexity. The subsequent sections focus on techniques which can obey these demands while others are covered only in less detail.

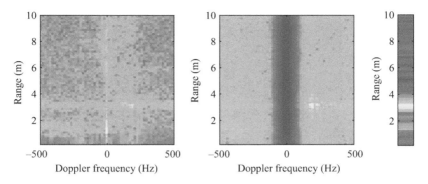

Figure 12.2 *The basic preprocessing steps: 2D-FFT leading to a range-Doppler-map (left); zero-Doppler notch filtering (centre); summing across the Doppler spectrum in every range bin (right)*

12.2.1 Target detection techniques

Micro-Doppler-based classification algorithms profit from the unique motion schemes of different target classes like pedestrians, cyclists, cars, dogs or other animals. In many cases, these schemes become evident when the target is moving, and often moving targets seem to be most relevant. A mature processing technique relies on this latter fact and suppresses stationary targets: moving target indication (MTI).

A simple and effective way for implementation of MTI is the application of a notch filter in the Fourier domain. The data are split up into sequences of N range profiles and transformed across the slow time into the Fourier domain, i.e. the Doppler domain, in every range cell, leading to N entries each of which yields a range-Doppler map (Figure 12.2, left). Ideally, the Doppler is linked to the velocities of the reflecting objects in the considered range cell only. However, there usually is a significant part of the contributing clutter objects in a range cell that are stationary and not moving. Depending on the scenery, the components corresponding to small radial velocities decay with increasing Doppler velocity at different rates. To respect this degree of freedom, a parametric notch filter of exponential form is introduced, such as that used by [22] though for a different purpose:

$$h(f) = -\gamma \exp\left[-\frac{f^2}{2\sigma^2}\right] \tag{12.1}$$

The width of the notch filter is controlled by σ, whereas the depth is defined by γ. The modulus of the filtered spectrogram $S_{filt}(f, t)$ is now calculated from the original spectrogram $S(f, t)$ as follows:

$$\log_{10}\left[|S_{filt}(f, t)|\right] = \log_{10}[|S(f, t)|] \cdot h(f)$$

where the first factor represents the filtered amplitude and the second one the untouched angles in the complex Gaussian plane. The modulus of the range-Doppler-map with zero-Doppler filter is shown in Figure 12.2 (centre).

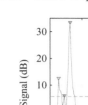
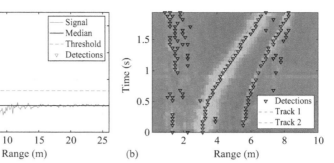

Figure 12.3   (a) Detections extracted from the filtered range-Doppler map
collapsed onto the range axis. (b) The range-Doppler maps collapsed
onto the range axis during 2 s with detections an confirmed tracks

After having applied the MTI filtering and thus having suppressed the stationary part in the radar data, a detection strategy is needed. A straightforward method is to integrate, i.e. sum, the energy across the Doppler domain containing N elements in every range cell (Figure 12.2, right). The resulting scalar is a measure for the accumulated energy of significantly moving objects in the considered range cell.

For instance, by exceeding a predefined threshold, this accumulated energy can indicate the existence of a target in the corresponding range cell (Figure 12.3(a)). This threshold can be derived from the background statistics which include noise and clutter.

Apart from this specific procedure which will be used later in the section about experimental work, there is another popular detection technique: the constant false alarm rate (CFAR) detector. Unlike the present detector, it is applied directly to every cell in the range-Doppler map, the so called *cell under test*, and estimates the noise level from a set of *reference cells* that are separated by a number of *guard cells* from the cell under test. This detector is covered in more detail in standard literature like [23, p. 296].

12.2.2   Tracking techniques

Unlike CW radar, which has no range resolution, the extraction of micro-Doppler features in FMCW radar requires to track a target during its motion through the range cells.

There are multiple tracking filters available to perform this task. The most common one is the Kalman filter. Starting from an initial set of detections in the range/velocity domain, the filter predicts the state for the time of the next measurement. Here, it uses a predefined motion model which determines the form of the filter matrices. This prediction is used to associate new detections with the predicted states. Simplifications may be used; for instance, a detection can only be associated with a single track, thus limiting the number of associations and filter updates drastically. When the association is done, the predicted states are updated with the new information from the current detections and the cycle starts again. The traditional Kalman filter is computationally efficient and is therefore widely applied.

In some situations, the application of a simple association technique can result in difficulties, especially for extended targets, where multiple detections can occur for a single target. In these cases, less simplifications during the data association step are desirable. Possible solutions to this problem are the Joint Probabilistic Data Association (JPDA), as used in [20], the multi-hypothesis tracker (MHT) or the PHD filter with its implementation using Gaussian mixtures (GMs) by [24]. Tracking filters are covered in more detail, for example, in [25]. In all cases, the central prediction and update Kalman scheme including the motion model stays mostly unchanged. The performance of each of these more complex association algorithms depends on the specific implementation and pruning techniques used. In the experimental work included in this chapter, a Gaussian mixture-PHD (GM-PHD) filter was implemented. Figure 12.3(b) shows the corresponding tracks.

12.2.3 Track-based spectrogram extraction

There are different approaches to extract Micro-Doppler features. The most prominent ones are based on the creation of spectrograms. A spectrogram results from a windowed Fourier transform along a time series. For target classification, it makes sense to consider the spectrogram along the target's track. An exact method to extract a time series along a track from a two-dimensional (2D) array, especially in between the grids, is the discrete-time Fourier transform (DTFT):

$$DTFT\,(s_i)\,[f] = \sum_{i=0}^{N-1} s_i \exp\left(2\pi \frac{i}{N}f\right) \tag{12.2}$$

Other than the fast Fourier transform (FFT), which calculates the spectrum on a static equidistant grid, the DTFT does not rely on any grid and provides the Fourier coefficient $DTFT\,[s_i]\,(f)$ at any given Doppler frequency f.

The gridded range profiles on which the tracks are based are calculated by FFT and are a Fourier transform themselves, here of the raw data. This fact enables the possibility of evaluating this Fourier transform at any specific range with a DTFT, because the range is the argument of the Fourier transform of the raw data d_i and it plays the role of the frequency in (12.2). For each range profile with the track $r(t_k)$ at time step k, this yields

$$HRR[r(t_k)] = \sum_{i=0}^{N-1} d_i \exp\left(2\pi \frac{i}{N}r(t_k)\right) \tag{12.3}$$

and the spectrogram S can be calculated as a set of windowed FFTs, i.e. STFTs, along $HRR(r(t_k))$:

$$S[mean(t_i,\ldots,t_j),f_D] = FFT\,\{HRR[r(t_i)],\ldots,HRR[r(t_j)]\}, \quad with \quad 1 < i < j < M \tag{12.4}$$

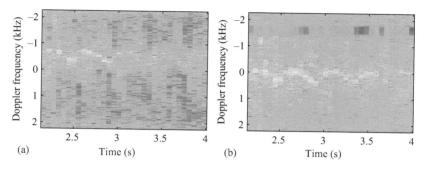

Figure 12.4   (a) Unprocessed spectrogram. (b) Aligned spectrogram with zero-Doppler suppression

where M is the track length in units of time steps and f_D is the Doppler frequency. Depending on the choice of the window length and the overlap with the surrounding windows, spectrograms with different temporal and Doppler resolution can be created. However, the ambiguity range in the Doppler domain only depends on the pulse repetition rate, which is usually the most severe limitation.

Although, this method is desirable in terms of exactness, it requires to save the whole range-Doppler maps over the evaluated time. This is not feasible for automotive applications since unnecessary data in non-required ranges are saved. Therefore, a less exact but more efficient method was proposed in [26]. Here, the window is given by the time and Doppler resolution derived from the range-Doppler map which was already created for the detection process (see Section 12.2.1). The Doppler spectra are extracted from this map only for single range cells in closest proximity to confirmed track states. A given number of successive Doppler spectra of the same target is then assembled to a spectrogram.

12.2.4   Spectrogram processing

In a similar way to the range-Doppler map in Section 12.2.1, the spectrogram has some zero-Doppler contribution due to the homogeneous static background clutter (see Figure 12.4(a)). This is not target specific and always visible in a measured spectrogram. As explained in Section 12.2.1, it can be removed with a notch filter.

Micro-Doppler signatures are distorted by the main Doppler contribution resulting from the centre of mass motion with respect to the radar. A constant velocity results in a offset to the zero-Doppler axis whereas accelerations lead to time-dependent shifts. In case of small accelerations during the considered time, the velocity change can be treated linearly, and for every time step in the spectrogram, both distortions can be mitigated by a cyclic shift in the Doppler domain. The shift in each time step is calculated from the linear approximation of the main Doppler evolution.

In [26], it is solved by first rotating the spectrogram by a set of angles and calculating the sum of the root mean square deviation (RMSD) in every Doppler cell across the whole Doppler domain. The angle which leads to the minimum RMSD sum indicates that the spectrogram is well aligned. The remaining offset can be corrected in a second step by simply determining the root mean square (RMS) in every Doppler cell and interpreting the maximum as the main Doppler. Figure 12.4(b) shows an aligned spectrogram with removed zero-Doppler contribution.

12.2.5 *Feature extraction and classification*

Different classes of dynamic targets have different micro-Doppler signatures, some of them may be clearly visible in the TF analysis provided by the spectrogram and some of them may be more subtle depending on the quality of the data, the condition of the acquisition, the SNR, the orientation of the target and, of course, the type of target itself.

Methods for discriminating micro-Doppler contributions can rely on models of dynamical properties of the targets or be data-driven or a combination of both.

Model-based methods estimate the probability of a radar data set to fit a predefined model. In case of humans, the motion can be characterised by walking, running or crawling kinematics [3]. On the other hand, forward flight and hovering kinematics can represent different drone/helicopter motion dynamics. Model-based methods can provide useful information regarding the target current status, by identifying the most probable target dynamics [27]. Moreover, these approaches implicitly carry out the estimation of target parameters, such as target dimensions and speed. However, they have drawbacks such as long decision-making time, the definition of precise generalised target models with parameters and the 'curse of dimensionality'.

Methods such as template matching, machine learning and deep learning are representative of data-driven classification approaches. Template matching makes decisions using a signature database, which contains the signatures of all the targets of interest. A decision can be found when the input data matches one of the database entries with a certain confidence level. Therefore, the decision-making process can be potentially slow if the database size is considerable. However, it has the advantage that little to no information is lost in the process. An example of template matching applied to micro-Doppler signature classification can be found in [28]. On the other hand, machine learning and deep learning algorithms for target classification learn properties and characteristics from the data and, based on them, predict the class of a set of unlabelled data. The learning mechanism is generally time-consuming and done offline, as it is required to extract features from the data, regardless of whether the learning is supervised or unsupervised. However, due to the fact of high variability across the spectrogram belonging to a single class, these techniques are a currently very popular solution to capture the target-specific variances. Deep learning techniques such as convolutional neural network [29,30] tend to require a larger amount of training data whereas machine learning techniques such as SVM requires generally less training data, especially if the dimension of the feature space can be reduced by traditional signal processing techniques [31].

Feature extraction methods aim to better highlight micro-Doppler contributions by particular transformation of the TF data, empirical evaluation of certain

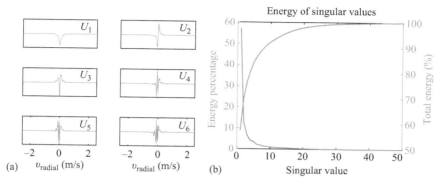

Figure 12.5   The six left singular vectors extracted from a walking pedestrian spectrogram (a) and the energy percentage of the relative singular values (b)

spectrogram parameters and the analysis of principal components. These features can be used as inputs of the state-of-the art data-driven classification approaches to discriminate between different classes of targets. Extracted features may result robust to data variations; this would not only improve the classification accuracy but will also eliminate the need of a very large database of spectrogram signatures. Once the classifier has been trained, the prediction is very fast and done online. In [21], SVM has been applied for classification of human beings using micro-Doppler signatures. In [32], target classification with a pulse-Doppler radar using GM models (GMMs) is presented. State-of-the art feature extraction methods such as the cadence velocity diagram frequency profile (CVDFP) [15] can generate features that benefit from the localisation property of the STFT. This method uses the cadence velocity diagram (CVD), described in [6], which indicates how often different frequencies occur over time within the signal. CVDFP generates a feature vector by summing over cadence frequency for each Doppler frequency bin of the CVD. Other feature extraction techniques rely on image processing such as the Gabor filter in [33], while others use a model-based approach [27]. The cepstrogram, defined as the logarithm of the inverse Fourier transform over the frequencies of the spectrogram, can also produce useful information for micro-Doppler-based classification [34]. Cepstrograms can be used to identify the speed of specific so-called 'events' in the TF domain such as the speed of the rotating blades of UAV and small UAV (drones) [35]. A classification approach based on cepstrum coefficients for feature extraction and a GMM in conjunction with a hidden Markov model (GMM-HMM) was also proposed in [36]. Empirical features of the spectrograms such as bandwidth, mean period, Doppler offset and radar cross section have been used for classifying between armed and unarmed men using a Bayesian classifier as described in [37].

The feature extraction method that we consider in this work is based on the method described in [31] and it is based on SVD. This technique has the advantage of reducing the dimensionality of the feature vectors since the majority of variations in the data can be explained by few low-ordered components while higher-order components account for little variation. The reduced feature set can therefore be seen as a more efficient representation of the original data in which the underlying structure is retained despite the loss of some information. SVD is a matrix factorisation method that decomposes a matrix into three different matrices: singular value matrix S, left singular vectors U and right singular vectors V. The SVD of the aligned spectrogram $A \in \mathbb{R}^{m \times n}$ is calculated as

$$A = U \cdot S \cdot V \tag{12.5}$$

where U and V are the matrices containing the left and right singular vectors, respectively, and S is a matrix with the singular values of A on its diagonal. The left singular vectors U provide information on the motion along the temporal dimension n of the spectrogram. Depending on the magnitude of singular values, it is possible to reduce the total amount of information to be stored (and later processed) by discarding those vectors with negligible contribution to the image. Following the approach of [31], the singular vectors corresponding to the five largest singular values are enough to capture the main motion features. In the later experimental part, the six largest ones will be used. Figure 12.5 shows the eigenvectors corresponding to the first six singular values extracted from a pedestrian spectrogram of a recent measurement campaign, as well as the energy percentage related to every one of them. At a first glance, it is not straightforward to find a direct relationship between this set of orthogonal vectors and the individual micro-Doppler contributions seen in the spectrogram, especially when the number of singular values is large. However, looking at the percentage of energy of each singular value in the whole spectrogram, it is possible to select a reduced number of eigenvectors and singular values and still reconstruct the image with a small loss of information. In this case, 90% of the spectrogram energy is comprised within the first 6 singular values and eigenvectors. Features with high 'energy compaction' [22] can be obtained by calculating the Fourier transform of the set of orthogonal vectors and stacking them together to obtain the so-called *feature vector f* as

$$f = \bigcup_{i=1}^{6} \text{FFT}(U_i) \tag{12.6}$$

where U_i is the left singular vector corresponding to the ith largest singular value. This enables a way to see where most of the signal information in the eigenvectors is concentrated. Figure 12.6 illustrates the magnitude of the first six transformed vectors shown in Figure 12.5. One interesting trend can be seen: the lower the singular value (index), the more the energy concentrates at lower frequencies (FFT components) in the transformed domain. This evidence could be used for defining a cut-off frequency to establish the bandwidth containing most of the relevant information. In this way, the dimension of the vector of features can be reduced, thus reducing the total amount of data processed for classification purposes. In the proposed approach, SVMs are used to classify the feature vectors which are generated from the memory-stored

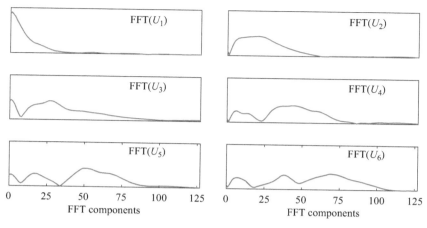

Figure 12.6   FFTs of the six left singular vectors extracted from a walking pedestrian spectrogram

spectrograms for the considered cycle. SVMs are an useful tool for data classification [38–40]. Like neural networks and many other classification approaches, SVM relies on the presence of a training set of known observations and a test set of unknown ones.

Using this approach, it is possible to discriminate between different classes of targets commonly found in urban automotive scenario responses, also according to their speed and direction of movement.

SVMs work in two steps: first, they learn a discriminating functional from the labelled training set, then this functional is used to assign every observation to a particular class. The functional works on a *feature space* of higher dimension with respect to the original space of the observations; here the data can be separated more easily. The second step is to use the obtained functional to classify unknown data. For satisfactory classification results, the training set has to be representative for the test set that we want to classify.

Let $\mathbf{X} \in \mathbb{R}^{m \times l}$ with $x_i \in \mathbb{R}^m$ and $i = 1, \ldots, l$ be a predefined collection of l labelled observations, namely a *training set*, an SVM searches for a functional which will assign a class to any given observation x_i. For this application, the training set is a labelled collection of feature vectors associated to confirmed tracks of different classes of targets. Exemplary spectrograms and the corresponding feature vectors are shown in Figures 12.9 and 12.10, whereas the given observations would be the feature vectors stored in memory. An SVM moves the data into the aforementioned feature space $\mathscr{F}$, by means of a function Φ. In this high-dimensional space, a simple *hyperplane* will suffice for the separation between different classes. The hyperplane can be described by the simple function $f(x) = wx + b$, with b and w being its constitutive parameters.

$$\Phi : \mathbb{R}^n \rightarrow \mathscr{F}, x \rightarrow \Phi(x) \tag{12.7}$$

It is possible to map observations from a general set into a feature space $\mathcal{F}$ without having to compute the mapping explicitly if only dot products are used between vectors in the feature space. These high-dimensional dot products can be computed within the original space, by means of a kernel function; K is a kernel if there exists a mapping Φ such that

$$K(x_i, x_j) = \Phi(x_i)^T \Phi(x_j) \tag{12.8}$$

where x_i and x_j are two observations in the original set. One popular choice for kernel functions is the radial basis function (RBF).

$$K(x_i, x_j) = e^{(-\gamma \|x_i - x_j\|^2)} \tag{12.9}$$

where γ is a free parameter which allows to tune the performance of the classifier.

$f(x) = wx + b$ is the simple hyperplane which separates two different classes in the feature space (with the associated parameters w and b) and $y \in \mathbb{R}^l$ is the indicator vector such that $y_i \in -1, 1$. The SVM solves the following optimisation problem:

$$\underset{w,b,\chi}{\text{minimise}} \quad \frac{1}{2} w^T w + C \sum_{i=1}^{l} \xi_i$$

$$\text{subject to} \quad y_i(w^T \Phi(x_i) + b) \geq 1 - \xi_i$$

$$\text{with} \quad \xi_i \geq 0 \tag{12.10}$$

where $C \geq 0$ is the penalty parameter of the error term. The solution to this problem can be tuned according to the selection of the parameters C and, assuming we are using an RBF kernel, the parameter γ. The correct values for C and γ are usually found by a cross-validation process. Cross-validation consists of splitting the training set (for which we know exactly the labels of each class) into two (or more) subsets. One subset will be used to train the SVM, i.e. to obtain the functional f for each combination pair of C and γ among the empirically selected range of values. The other subset(s) will be used as validation to test the classification accuracy with the obtained functional. By refining the range of values for C and γ, one can find the combination which gives the best classification accuracy for one particular training set. To further improve the selection of the parameters, one can use a k-fold cross-validation where the process is averaged over k different combinations of training and validation vectors. For the experimental part, it was found empirically that a fivefold cross-validation was sufficient; that is, the improvement in accuracy was negligible for larger k.

12.3 Experimental validation

12.3.1 Experimental radar system

For experimental validation, the 94 GHz FMCW radar system 'Dual Use Sensor for Mid-Range Applications' (DUSIM) [41] was used (see Figure 12.7(a)). Its transmit

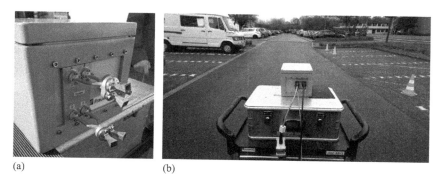

(a)   (b)

Figure 12.7   (a) The DUSIM front end. (b) The field of view

frequency is in the vicinity of 77-GHz band, which is widely used for automotive applications and thus can provide comparable settings. Since the Doppler frequency increases with the transmit frequency, higher frequency systems in the 77 GHz or the 94 GHz are particularly well suited for observing micro-Doppler features. With a transmit power of 100 mW, it lies in the low-power region of radar systems resembling systems used in the automotive context, although it still has significantly more transmit power than an automotive radar which would be around 10 mW (compared to [42]).

The DUSIM system applies an up-down-chirp modulation at a repetition rate of 4.5 kHz and a bandwidth of 1,062.6 MHz; this leads to a range resolution of approximately 15 cm. As it works in a simultaneous transmit and receive mode, it has four receive antennas and a separate transmit antenna. It covers a range up to 26.4 m which is limited by filters. For every pulse, the system stores the real-valued base band signal from which a range profile of the observed scenery can be calculated using a Fourier transform. Originally, the system was developed in the scope of the research project 'DUSIM' to support active protection measures against projectiles and ammunitions jeopardising the vehicle the sensor is placed on.

12.3.2   Experimental set-up

In street traffic, micro-Doppler signatures play an important role for the classification of vulnerable traffic participants. In difficult multi-target scenarios, this classification could increase the situational awareness and thereby foresee possible events and prevent accidents.

For the work described in this chapter, a set of different traffic participants was imitated, namely a pedestrian, a wheelchair user (Figure 12.9(a) and (b)), a rollator user and a cyclist (Figure 12.10(a) and (b)). To obtain a signature data set with reasonable extent for the use of machine learning techniques, each of the four participants moved on controlled paths transversally and longitudinally with respect to the radar line of sight. Each situation was repeated several times. All measurements took place in a car park with occasional cars and pedestrians moving through the scene (Figure 12.7(b)).

12.3.3 Processing

Often there are multiple possibilities to solve a problem and it is not feasible to show them all. Therefore, the experimental study is carried out with a specific processing method from [26] that was developed with respect to real-time performance requirements related to automotive applications.

The basic signal processing and detection starts with the MTI method that was described in detail in Section 12.2.1. In the present clutter environment, it turned out to be more robust than a standard CFAR detector, because the clutter and target often were similar and the estimation of the noise statistics in the reference cells became difficult. Although, in this scheme the Doppler information is neglected and remains unused. The calculations needed in the tracker are simplified significantly by the reduction of dimensionality which pays off in terms of real-time performance.

In the next processing step, the detections from the MTI method are passed to a GM-PHD filter. As already mentioned in Section 12.2.2, the performance of a certain tracking filter mostly depends on their specific implementation and the usage of pruning methods. All of the named filters MHT, PHD and Kalman with JPDA use central elements of the basic Kalman filter and it is just a question of how detailed the association takes place. In the end, the choice fell on the GM-PHD filter since it integrates the association in the tracking process in a consistent way and usually requires less computational effort than an MHT. The PHD filter showed to be fast enough on a single core to provide tracking states before the range-Doppler map of the next cycle is acquired.

With the given tracking states according to Section 12.2.3, the spectrograms can be assembled successively from the already determined (and still available) range-Doppler maps saving computational resources. For the spectrogram processing, the less accurate but more efficient method is mentioned in the last paragraph of Section 12.2.4. Overall, this algorithm is optimised with respect to performance and at the expense of some loss of accuracy. However, this accuracy is not required as the main task of the system is the increase in situational awareness in complex multi-target scenarios; the focus is the real-time classification.

In summary, the algorithm can be separated into two parallel working sections: the continuously working range-Doppler map generation and tracking mode (continuous mode) and the clocked classification mode that classifies all signatures of the current 'clock' in the memory and clears it for acquisition of new signature data during the next 'clock' (clocked mode). Each mode works in cycles and starts again when a cycle and acquisition is completed. A 'clock' consists of several continuous cycles.

Figure 12.8 shows the tracks and extracted spectrograms for two cases where nearby targets merge to a single one and split up, respectively. It shows that the

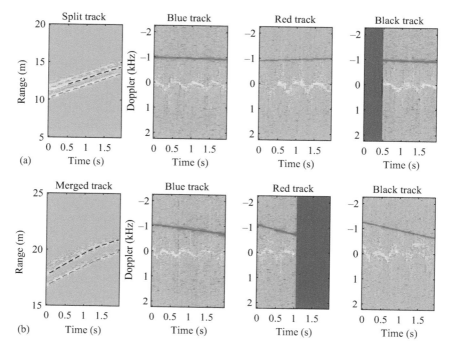

Figure 12.8   Tracking and corresponding spectrograms for challenging tracking scenarios: splitting tracks (a) and merging tracks (b). The dark blue areas indicate times when the track does not exist

tracking algorithm is capable of treating these difficult situations in a reliable way and allows a stable signature extraction.

Figures 12.9 and 12.10 show examples of spectrograms and corresponding feature vectors extracted by the before described processing. Visual inspection gives a first hint that the presented targets can be distinguished based on their micro-Doppler signatures.

12.3.3.1   Classification results

The aligned spectrograms obtained by the process explained in Section 12.2.3 and shown in Section 12.3.3 were used to extract the feature vectors which will become the input to the SVM classifier, as explained in Section 12.3.3. Examples of possible spectrograms and feature vectors coming from the four different classes of targets are shown in Figures 12.9 and 12.10. As it can be seen from the figures, the discrimination of spectrograms is visually difficult for slow-moving targets such as the wheelchair and the rollator (Figure 12.9(b) and (d)). The associated feature vectors (Figure 12.9(e) and (f)), however, show differences that could help the SVM classifier in the discrimination. However, the small differences in pace and orientation of

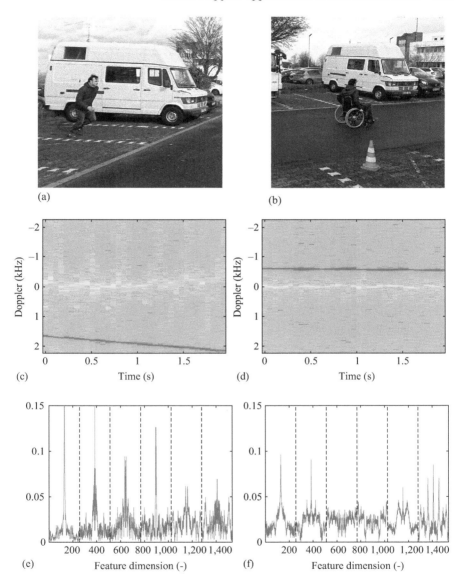

Figure 12.9 Pedestrian and wheelchair user: target image (a)/(b), corresponding spectrogram (c)/(d) and the FFTs of the singular vectors of the spectrogram for the first six singular values assembled into a feature vector (e)/(f)

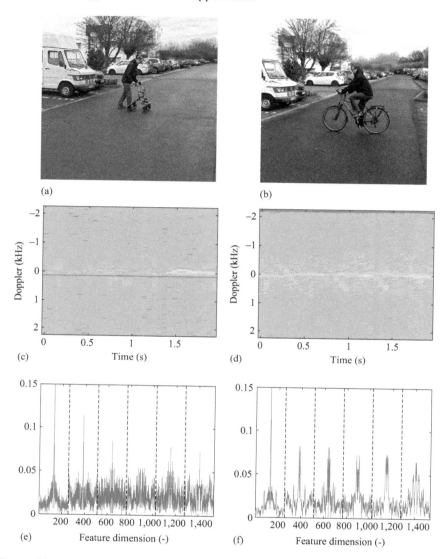

Figure 12.10 Rollator user and cyclist: target image (a)/(b), corresponding spectrogram (c)/(d) and the FFTs of the singular vectors of the spectrogram for the first six singular values assembled into a feature vector (e)/(f)

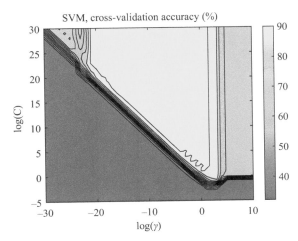

Figure 12.11   *Contour plot in logarithmic scale, showing the accuracy of the*
cross-validation process for different combinations of the two RBF
kernel parameters

the targets, plus the scenario variables that are commonly found in parking lots (i.e. people and cars passing by), can make the classification task extremely challenging.

Before extracting the feature vectors, the entire data set measurements were divided into training set and test set. Each measurement in the training and test set contains many tracks associated to different clocks and to the four different channels of the DUSIM radar. The division into test and training was measurement based to avoid correlation between the test and training data. Classifying tracks in the test set which are coming from the same measurements present in the training set would have probably given better classification results but it will not represent a realistic classification process. Ideally, one would have a general database of feature vectors (stored) extracted from different target responses coming from controlled measurements (such as these ones) or simulations.

The training set was used by the SVM classifier to obtain the right functional, as explained in Section 12.3.3. An RBF kernel was selected and a fivefold cross-validation process was used to obtain the parameters C and Γ, and the number of folds was selected empirically. Cross-validation accuracy varied from 80% to 90% depending on the selection of the measurements included in the training set. The contour plot in Figure 12.11 (according to Case study 2, see later in this chapter) evidences the necessity of correctly selecting the kernel parameters. It is important to keep in mind that this cross-validation process could be a time-consuming task depending on the range of the parameters that have to be explored [39]. However, this has to be done only once; when the functional is obtained, the classification of new data is in real time. The optimal values of C and γ are then used to classify the test set.

Table 12.2   Confusion matrix, Case study 1, transversal movement

	Ped. running (%)	**Cyclist (%)**	**Wheelchair (%)**	**Rollator (%)**
Ped. running	**85.71**	0	0	14.29
Cyclist	5.26	**73.68**	15.79	5.26
Wheelchair	0	11.54	**61.54**	29.92
Rollator	0	8.7	13.04	**78.26**

Table 12.3   Confusion matrix, Case study 1, advancing

	Ped. running (%)	**Cyclist (%)**	**Wheelchair (%)**	**Rollator (%)**
Ped. running	**85**	0	15	0
Cyclist	13.33	**40**	46.66	0
Wheelchair	0	1.31	**84.21**	14.47
Rollator	0	0	29.03	**70.97**

Table 12.4   Confusion matrix, Case study 1, receding

	Ped. running (%)	**Cyclist (%)**	**Wheelchair (%)**	**Rollator (%)**
Ped. running	**87.5**	0	12.5	0
Cyclist	11.76	**29.41**	58.82	0
Wheelchair	1.45	1.45	**84.06**	13.04
Rollator	0	0	13.33	**86.66**

Case study 1: Discriminate pedestrians from other urban targets

A four-class SVM classifier was employed to classify between the following classes: running pedestrian, cyclist, wheelchair and rollator. The classification is repeated for every direction of movement of the targets, either they are advancing, receding or moving transversally with respect to the radar. The differentiation with respect to the direction is not accounted in this test, assuming that the tracker will give us already an information of the orientation of the targets. We show the classification matrices related to the three tests for each one of the movement direction. For this case study, we extracted ten singular vectors (see Section 12.2.5) and used them for the SVM classification. We used the corresponding ten singular values as weights to leverage each singular vector by its energy [31]. We did not perform any filtering on the singular vectors.

Briefly, Tables 12.2–12.4 show preliminary classification results for the proposed approach in the form of confusion matrices. The diagonal elements of these confusion matrices (in bold format) indicate the probabilities of correct classification for each given class. These results show that once the track is identified, SVM discriminates

Table 12.5   Confusion matrix, Case study 2

	Lat./walk (%)	Lat./run (%)	Adv./walk (%)	Adv./run (%)	Rec./walk (%)	Rec./run (%)
Lat./walk	**96.36**	1.82	1.82	0	0	0
Lat./run	8.33	**91.66**	0	0	0	0
Adv./walk	0	0	**94.74**	0	0	5.26
Adv./run	0	0	21.43	**78.57**	0	0
Rec./walk	0	0	4.16	0	**95.83**	0
Rec./run	0	0	0	0	0	**100**

between the four presented urban targets with mixed results, depending on the considered class and the direction of movement. The running pedestrian is always classified with an accuracy around 85% while rollator and wheelchair accuracy varies depending on the orientation of the targets, achieving better results for receding movement. The drop in accuracy for the classification of the cyclist is mainly due to the limited number of measurements that we had for this class. This also depends on the fact that the bicycle track is shorter in time with respect to the other classes, therefore generating less clock intervals on where the spectrograms are extracted.

Case study 2 : Discriminate different movement directions of walking pedestrians

These results are related to the work published in [26]. In this case, we are trying to use the same classification approach to discriminate between contributions from a single pedestrian moving in different directions with respect to the radar, as before, advancing, receding and transversally. Table 12.5 summarises the results of the proposed six-class classifier. For this case study, we extracted six singular vectors (see Section 12.2.5) and used them for the SVM classification. Here, no filtering on singular vectors was performed.

The confusion matrix shows that it is possible to successfully discriminate between different moving directions and two different paces of a single pedestrian. Considering the amount of unwanted contributions present in this measurement campaign (generated from wind, rain, antenna cross-talk, etc.), this is a remarkable result [26] also demonstrating the effectiveness of the GM-PHD filtering and spectrogram alignment procedures. The number of features vector for the runner class (17 for training and 14 for test) was very low and this is the major cause of the drop in accuracy that we experience for that class.

12.4   Practicality discussion

Facing a highly competitive market, automotive radar applications have to respect the limited processing power and memory that are available. Hence, memory and processing efficiency plays an important role in the design of automotive radar applications and is respected in the presented design. A detailed implementation description can be found in [26], however, a short discussion on the practicality of the approach will be done here.

In general, the approach consists of the following parts: detection, tracking and classification. In modern automotive radars, the detection is done typically in two parts. First, a range-Doppler-map is created. Second, a CFAR detector creates detections from this map. In the presented approach, the detection mechanism is changed from CFAR to micro-Doppler-based approach (Section 12.2.1). Since this is done by a very simple, threshold-based algorithm, no significant additional processor or memory load is expected.

The next step comprises the tracking. It uses the actual detections coming from the micro-Doppler-based detection algorithm. Other than a simple Kalman filter with a minimum number of possible target states, the PHD filter considers more hypotheses for the target's motion and thereby requires more processing power. However, with some restrictions on the number of hypotheses very efficient implementations of the PHD filter are possible as shown by the real-time algorithm presented in [26].

The classification part is the most critical one because spectrograms have to be created and therefore Doppler spectra at the targets positions have to be extracted and saved. The processing and memory load results from the desired accuracy. Vice versa, the load can be minimised by reducing the range, the Doppler-resolution and the variation of the duration of spectrograms. It was also proposed that the real-time algorithm presented in [26] does not require multiple cores for calculation, although this would enhance the system capabilities. Due to the reduction of spectrograms to feature vectors, the classification process itself does only require a negligible amount of processing and memory load.

A possible application for the proposed mode would, for instance, be the detection and classification of vulnerable road users in vicinity to an automotive platform. This could increase the situational awareness before turns or similar purposes.

12.5   Conclusion

This chapter addressed the challenges and opportunities that come along with the introduction of micro-Doppler signatures in automotive radar applications. Here, a major interest lies in the development of methods to protect vulnerable road users, i.e. pedestrians, cyclists, etc. As has been shown in various studies, micro-Doppler signatures enable a way to identify these targets, thereby making active protection measures possible.

Apart from the well-studied single-target scenario, this chapter focused on the far more realistic multi-target scenario and presented an experimentally validated method for simultaneous micro-Doppler signature extraction and classification using a 94 GHz FMCW radar. Central to any multi-target classification system is a tracking algorithm as it links signatures with dynamic targets. Furthermore, it has been shown that by solely using micro-Doppler information, the motion of a pedestrian (Section

12.3.3.1, Case 2) can be determined and separated from other vulnerable road users (Section 12.3.3.1, Case 1). Future work might include the fusion of tracking and micro-Doppler signature information to increase the overall classification performance.

In the scope of this chapter, it was not feasible to cover every approach in equal depth. The presented approach is adequate but by no means the only one. Therefore, throughout the chapter other solutions are mentioned and referenced.

References

[1]   Directorate General for Transport. *Traffic Safety Basic Facts on Pedestrians*. Tech. rep. European Commission; 2018.

[2]   National Center for Statistics and Analysis. Tech. rep. U.S. Department of Transportation; 2018.

[3]   Boulic R, Thalmann NM, and Thalmann D. A global human walking model with real-time kinematic personification. *The Visual Computer*. 1990;6(6): 344–358.

[4]   Ritter H, and Rohling H. Pedestrian detection based on automotive radar. In: *IET Int. Conference Radar Systems*; 2007.

[5]   Schubert E, Kunert M, Frischen A, and Menzel W. A multi-reflection-point target model for classification of pedestrians by automotive radar. In: *11th European Radar Conference*, Rome. IEEE; 2014. pp. 181–184.

[6]   Björklund S, Petersson H, Nezirovic A, Guldogan MB, and Gustafsson F. Millimeter-wave radar micro-Doppler signatures of human motion. In: *12th International Radar Symposium (IRS)*, Leipzig. IEEE; 2011. pp. 167–174.

[7]   Fairchild DP, and Narayanan RM. Classification of human motions using empirical mode decomposition of human micro-Doppler signatures. *IET Radar, Sonar & Navigation*. 2014;8(5):425–434.

[8]   Narayanan RM, and Zenaldin M. Radar micro-Doppler signatures of various human activities. *IET Radar, Sonar & Navigation*. 2015;9(9):1205–1215.

[9]   Yan H, Doerr W, Clasen H, *et al.* Micro-Doppler-based classifying features for automotive radar VRU target classification. In: *25th International Technical Conference on the Enhanced Safety of Vehicles (ESV)*; 2017.

[10] Belgiovane D, and Chen C. Micro-Doppler characteristics of pedestrians and bicycles for automotive radar sensors at 77 GHz. In: *European Conference on Antennas and Propagation*; 2017.

[11] Schubert E, Meinl F, Kunert M, and Menzel W. High resolution automotive radar measurements of vulnerable road users – Pedestrians & cyclists. In: *IEEE MTT-S International Conference on Microwaves for Intelligent Mobility (ICMIM)*, Heidelberg. IEEE; 2015. pp. 1–4.

[12] Heuel S, and Rohling H. Pedestrian recognition based on 24 GHz radar. In: *11th International Radar Symposium*; 2010.

[13] Bartsch A, Rasshofer RH, and Fitzek F. Pedestrian recognition using automotive radar sensors. In: *Advances in Radio Science*; 2012.

[14]   Molchanov P, Vinel A, Astola J, and Egiazarian, K. Radar frequency band invariant pedestrian classification. In: *14th International Radar Symposium (IRS)*, Dresden. IEEE; 2013. pp. 740–745.

[15]   Clemente C, Miller A, and Soraghan JJ. Robust principal component analysis for micro-Doppler based automatic target recognition. In: *3rd IMA Conference on Mathematics in Defence*; 2013.

[16]   de Wit JJM, Molchanov P, and Harmanny RIA. Radar micro-Doppler feature extraction using the singular value decomposition. In: *International Radar Conference*; 2014.

[17]   Miller AW, Clemente C, Robinson A, Greig D, Kinghorn AM, and Soraghan JJ. Micro-Doppler based target classification using multi-feature integration. In: *IET Intelligent Signal Processing Conference 2013 (ISP 2013)*, London. IET; 2013. pp. 1–6.

[18]   Andres M, Ishak K, Menzel W, and Bloecher H. Extraction of micro-Doppler signatures using automotive radar sensors. *Frequenz*. 2012;66(11–12): 371–377.

[19]   Heuel S, and Rohling H. Two-stage pedestrian classification in automotive radar systems. In: *12th International Radar Symposium (IRS)*; 2011.

[20]   Wagner T, Feger R, and Stelzer A. Radar signal processing for jointly estimating tracks and micro-Doppler signatures. *IEEE Access*. 2017;5:1220–1238.

[21]   Kim Y, and Ling H. Human activity classification based on micro-Doppler signatures using a support vector machine. *IEEE Transactions on Geoscience and Remote Sensing*. 2009;47(5):1328–1337.

[22]   Molchanov P. Radar target classification by micro-Doppler contributions. Tampereen teknillinen yliopisto Julkaisu-Tampere University of Technology Publication; 1255. 2014.

[23]   Skolnik MI. *Introduction to Radar Systems* (Third Edition). New York, McGraw Hill Book Co, 1980. 2001.

[24]   Vo BN, and Ma WK. The Gaussian mixture probability hypothesis density filter. *IEEE Transactions on Signal Processing*. 2006;54(11):4091.

[25]   Blackman S, and Popoli R. *Design and Analysis of Modern Tracking Systems* (Artech House Radar Library). Artech House. 1999.

[26]   Petervari R, Giovanneschi F, Johannes W, and González-Huici M. A realtime micro-Doppler detection, tracking and classification system for the 94 GHz FMCW radar system DUSIM. In: *2019 16th European Radar Conference (EuRAD)*, Paris. 2019. pp. 193–196.

[27]   Groot SR, Yarovoy AG, Harmanny RIA, and Driessen JN. Model-based classification of human motion: Particle filtering applied to the micro-Doppler spectrum. In: *9th European Radar Conference (EuRAD)*, Amsterdam. 2012. pp. 198–201.

[28]   Smith GE, Woodbridge K, and Baker CJ. Radar micro-Doppler signature classification using dynamic time warping. *IEEE Transactions on Aerospace and Electronic Systems*. 2010;46(3):1078–1096.

[29]   Kim Y, and Moon T. Human detection and activity classification based on micro-Doppler signatures using deep convolutional neural networks. *IEEE Geoscience and Remote Sensing Letters*. 2016;13(1):8–12.

[30] Cao P, Xia W, Ye M, Zhang J, and Zhou J. Radar-ID: Human identification based on radar micro-Doppler signatures using deep convolutional neural networks. *IET Radar, Sonar & Navigation*. 2018;12(7):729–734.

[31] Molchanov P, Egiazarian K, Astola J, Harmanny RIA, and de Wit JJM. Classification of small UAVs and birds by micro-Doppler signatures. In: *European Radar Conference (EuRAD)*, Nuremberg. 2013. pp. 172–175.

[32] Bilik I, Tabrikian J, and Cohen A. GMM-based target classification for ground surveillance Doppler radar. *IEEE Transactions on Aerospace and Electronic Systems*. 2006;42(1):267–278.

[33] Tivive FHC, Phung SL, and Bouzerdoum A. Classification of micro-Doppler signatures of human motions using log-Gabor filters. *IET Radar, Sonar & Navigation*. 2015;9(9):1188–1195.

[34] Harmanny RIA, De Wit JJM, and Cabic GP. Radar micro-Doppler feature extraction using the spectrogram and the cepstrogram. In: *11th European Radar Conference (EuRAD)*. 2014. pp. 165–168.

[35] Harmanny RIA, de Wit JJM, and Premel-Cabic G. Radar micro-Doppler mini-UAV classification using spectrograms and cepstrograms. *International Journal of Microwave and Wireless Technologies*. 2015;7(3-4):469–477.

[36] van Eeden WD, de Villiers JP, Berndt RJ, Nel WAJ, and Blasch E. Micro-Doppler radar classification of humans and animals in an operational environment. *Expert Systems with Applications*. 2018;102:1–11.

[37] Fioranelli F, Ritchie M, and Griffiths H. Classification of unarmed/armed personnel using the NetRAD multistatic radar for micro-Doppler and singular value decomposition features. *IEEE Geoscience and Remote Sensing Letters*. 2015;12(9):1933–1937.

[38] Cristianini N, and Shawe-Taylor J. *An Introduction to Support Vector Machines and Other Kernel-based Learning Methods*. Cambridge: Cambridge University Press; 2000.

[39] Hsu CW, Chang CC, and Lin CJ. A practical guide to support vector classification. 2003. Paper available at http://www.csie.ntu.edu.tw/~cjlin/papers/guide/guide.pdf. 2003.

[40] Chang CC, and Lin CJ. LIBSVM: A library for support vector machines. *ACM Transactions on Intelligent Systems and Technology*. 2011. Available from: https://doi.org/10.1145/1961189.1961199.

[41] Caris M. DUSIM – A mmW sensor to trigger active indirect protection measures. In: *Internationales Symposium Indirekter Schutz (ISIP)*; 2018.

[42] Giammello V, Ragonese E, and Palmisano G. Transmitter chipset for 24/77-GHz automotive radar sensors. In: *IEEE Radio Frequency Integrated Circuits Symposium*; 2010. pp. 75–78.

Conclusion

The rationale of this edited book has been to provide the readers with an overview of the most recent research progress and applications in the field of radar micro-Doppler signatures. It is fascinating to see how this concept, originally born in the defence context for very specific applications, has now been extended to a variety of civilian scenarios from ambient assisted living to the automotive sector and perhaps to original applications, such as the micro-Doppler signatures of farm animals for lameness diagnostics or wild animals for conservation and anti-poaching.

Sometimes, we have been asked what novel or exciting research can take place in radar, a technology that was invented and first demonstrated many decades ago. We would be happy to provide a partial answer with this book, showing the significant number of applications and research challenges related to radar micro-Doppler signatures, their analysis and exploitation. And this is to some extent an inevitably partial answer, as there are many more aspects of current outstanding radar research that for conciseness and focus' reasons are not covered in this book, from detection to tracking algorithms, from cognitive radar to approaches for spectrum sharing between radar and communications, just to mention some.

The contributions in each chapter have explored current and recent research, providing plenty of additional published references to the readers willing to know more details. The follow-up question is then what comes next in terms of future directions, what 'vectors of research' one can expect in the next few years. Answering questions like this is always challenging, as there can always be new disruptive technologies that revolutionise completely the landscape and make any prediction suddenly obsolete. However, we would like to conclude this book highlighting a couple of research directions/challenges that we see prominent in the domain of radar micro-Doppler signatures in the next few years:

- Developing classification techniques that can interpret and process radar micro-Doppler data (or in general radar data) not as a series of individual, separated matrices of data (images), but as a continuous and uninterrupted series of data (temporal sequence). A timely example for micro-Doppler signatures could be moving away from individual spectrograms covering a given time interval and classified by algorithms inspired from image processing (e.g. convolutional neural networks), and focussing on the flow of micro-Doppler data over time using techniques inspired from speech and audio processing (e.g. recurrent networks).
- Complementing micro-Doppler information with other 'domains' of radar data, in particular the range of information that can be made available with increasingly

fine spatial resolution, thanks to the development on mm-wave radar systems with several GHz of bandwidth. There are already examples of published work that consider the radar data as a 'range-time-Doppler cube', or a 4D structure that adds the direction/angle of arrival thanks to multiple-channel radar. As higher operating frequencies and wider bandwidths in the low-THz spectral region become available, echoes from multiple targets and different body or object parts can be resolved directly in the range domain. How this augmented range information can be optimally used for classification is an interesting research trend.

- As deep learning techniques have revolutionised many research fields from image and audio processing to communication, they have an incredible potential to be applied to radar data as well. Unlike images or audio samples though, collecting and labelling an equivalent amount of radar data is unfeasible, which leads to the issue of how to generate enough training data or more in general how to train effectively and efficiently deep learning algorithms for radar data. Classic data augmentation approaches may not work well (e.g. a face rotated by 90° is still a face, a spectrogram rotated of 90° has a completely different physical meaning of its original). Researchers have proposed different approaches, namely using models and simulations to generate synthetic micro-Doppler data, using abundantly available video and motion-capture data to be converted into radar micro-Doppler signatures, and using generative adversarial networks to augment existing, small datasets of experimental radar data. None of these is, at the moment, 'the solution' valid for all situations and classification problems, and there is considerable ongoing research in this domain.

Index

activity recognition, micro-Doppler 44
  classification methods 47
  experiments and results 47
    activity recognition results 50
    analysis of micro-Doppler
      signatures 48–50
    implementation 47
  micro-Doppler feature extraction
    44–6
    data alignment stage 45–6
    data pre-processing stage 46
  signal model 44
adversarial autoencoders (AAEs) 168,
  170–5
artificial neural networks (ANNs) 191,
  202–4
assisted living
  ambient assisted living (AAL)
    context 35–7, 60
  multimodal activity classification for
    198
    experimental setup 198–202
    heterogeneous sensor fusion
      202–8
  sensing for
    radar signal processing 182–5
    wearable sensors 185–6
auto-encoders (AEs) 18–19,
  113–14
automatic feature extraction 188
automatic target recognition (ATR)
  algorithms 249
automotive radar, micro-Doppler
  analysis in 367, 370
  experimental validation
    experimental radar system 379–80

experimental set-up 380–1
processing 381–7
feature extraction and classification
  375–9
practicality discussion 387–8
spectrogram processing 374–5
state of the art 367–70
target detection techniques 371–2
track-based spectrogram extraction
  373–4
tracking techniques 372–3
auxiliary conditional GANs (ACGANs)
  118, 120–1, 125
  architecture 122
  diversity of 124–5
  kinematic fidelity of 121–4
  PCA-based kinematic sifting
    algorithm 125–6
averaged cadence velocity diagram
  (ACVD) 217, 242, 247, 252
  -based feature vector approach
    236–7
average running time 249–53

ballistic missile defence (BMD)
  systems 217–18
ballistic targets (BTs), micro-Doppler
  analysis of 217
  classification framework 233
    classifier 240–1
    feature vector extraction 236–9
  laboratory experiment 230–3
  performance analysis 241
    2D Gabor filter approach 245
    averaged cadence velocity diagram
      (ACVD) approach 242

average running time 249–53
  Krawtchouk (Kr) moments
    approach 243–4
  performance in the presence of the
    booster 245–9
  pseudo-Zernike (pZ) moments
    approach 242–3
  radar return model at radio frequency
    219
    cone plus fins 224–8
    cylinder 228–9
    sphere 229–30
Bayesian classifiers 191, 376
Bayesian sparse signal recovery 77–8
Bayes' theorem 77
beat-to-beat SBP 338, 342
body RCS 269–70
Boulic–Thalmann model 139–41, 149
Butler matrix 303

cadence-velocity diagram (CVD) 10,
  262, 376
cadence velocity diagram frequency
  profile (CVDFP) 376
CAD models 148
carotid-femoral PTT (cf-PTT) 339–40
cascaded integrator-comb (CIC) filter
  332, 334–5
central time–frequency trajectory,
  clustering for 86–8
cepstrograms 175, 286, 376
channel state information (CSI) 36,
  52–3
  CSI-based human activity recognition
    and monitoring system (CARM)
    41, 57
Chebyshev window 300–2, 311
coherent processing interval (CPIs)
  108, 144, 262
competitive fusion networks 192
complementary fusion networks 192
compressed sensing (CS) 75, 336
conditional variational autoencoders
  (CVAEs) 118

constant false alarm rate (CFAR)
  algorithm 9, 44–6, 108, 372,
  381, 388
contactless beat-to-beat BP estimation
  using Doppler radar 338–42
continuous-wave (CW) radar 369
  coherent radar 230
  Doppler radar 323, 338
convex optimisation 77
convolutional autoencoders (CAEs)
  110, 115–16
convolutional neural network (CNN)
  110, 115, 160–5, 263, 286
co-operative fusion 208
co-operative sensor networks 192
cross-ambiguity function (CAF) 44
cylinder target 228–9

data augmentation 118
data-driven classification 138, 160
  deep supervised learning for human
    gait classification
    convolutional neural networks
      (CNNs) 160–5
    recurrent neural networks (RNNs)
      165–7
  deep unsupervised learning for
    human gait classification 167
    adversarial autoencoders 170–5
    generative adversarial networks
      (GANs) 168–70
  flowchart of data-driven
    classification approach 111
data-driven feature learning methods
  110–11
  autoencoders (AEs) 113–14
  convolutional autoencoders (CAEs)
    115–16
  convolutional neural networks
    (CNNs) 115
  genetic algorithm-optimised
    frequency-warped cepstral
    coefficients 112–13
  principal component analysis
    111–12

data-level fusion 344
data preprocessing 108
  dimensionality reduction (DR) 108–9
  mitigation of clutter, interference, and noise 109–10
DC offset calibration technique 335–7
decision-level fusion 192, 194, 206–8, 345–6
decoys 217–20, 231–2, 245–6
deep convolutional neural network (DCNN) 20–2, 82, 91–3, 98, 160, 162
deep learning 17, 19, 41, 139, 262
deep neural networks (DNNs) 103
  classification with data-driven learning 110–11
    autoencoders (AEs) 113–14
    convolutional autoencoders (CAEs) 115–16
    convolutional neural networks (CNNs) 115
    genetic algorithm-optimised frequency-warped cepstral coefficients 112–13
    principal component analysis (PCA) 111–12
  performance comparison of DNN architectures 126–8
  radar challenge 116
    synthetic signature generation from MOCAP data 117–18
    synthetic signature generation using adversarial learning 118–20
    transfer learning from optical imagery 116–17
  radar signal model and preprocessing 105–6
    2D input representations 106–8
    3D input representations 108
    data preprocessing 108–10

recurrent neural networks (RNNs) and sequential classification 128–9
  training 167
deep supervised learning for human gait classification
  convolutional neural networks (CNNs) 160–5
  recurrent neural networks (RNNs) 165–7
deep unsupervised learning for human gait classification 167
  adversarial autoencoders 170–5
  generative adversarial networks (GANs) 168–70
deramping 296
diastolic BP (DBP) 338
dictionary learning framework 79–81, 97
digital down-converter (DDC) module 327
digital-IF CW Doppler radar 323, 338, 345
  DC offset calibration technique 335–7
  healthcare sensing, applications to 338
    contactless beat-to-beat BP estimation using Doppler radar 338–42
    multi-sensor-based sleep-stage classification 342–6
  implementation of digital-IF layer 327
    decimation 332–5
    digital quadrature demodulation 329–32
    direct IF sampling 328–9
  principles of 323–6
  RF layer 326–7
dimensionality reduction (DR) 108–9
direct signal interference (DSI) effect 46
discrete-time Fourier transform (DTFT) 373

discriminator network 121, 168, 171
DivNet-15 118, 127–8
Doppler centroid 14, 188–9
Doppler effect 142, 291
Doppler frequency 5, 14, 16, 26, 48–9,
    281, 306, 313
Doppler shift 49, 56, 82, 182, 202, 232
Doppler-spectrum 274, 374

electromagnetic predictions of birds and
    UAVs 263
  electromagnetic properties, size and
    shape 263–4
  radar cross section (RCS) predictions
    264–9
explicit gait models 158

fast Fourier transform (FFT) 10, 44,
    106, 311, 373
feature extraction, micro-Doppler 44
  data alignment stage 45–6
  data pre-processing stage 46
feature extraction for individual sensors
    186
  classifiers 191
  features from radar data 187
    automatic feature extraction 188
    handcrafted features 188–90
  features from wearable sensors
    190–1
feature-level fusion 192, 194, 344–5
feature vector extraction 236
  averaged cadence velocity diagram
    (ACVD)-based feature vector
    approach 236–7
  2D signature-based feature vectors
    237
    Krawtchouk (Kr) moments-based
      feature vector approach 238
    pseudo-Zernike (pZ)
      moments-based feature vector
      approach 237–8
    2D Gabor filter-based feature
      vector approach 239
feed-forward network (FFN) 18–19

finite difference time domain (FDTD)
    258
fixed-wing UAVs 259–60, 271
forget gate 165
Fourier dictionary 76
Fourier domain 239, 371
Fourier transform 82, 107, 235, 373
frequency-modulated continuous wave
    (FMCW) radar 105, 142, 182,
    198, 208, 280, 291, 369–70
  antenna and antenna array 298
    beamforming 299–304
    two-way pattern and MIMO
      305–7
  applications of micro-Doppler effects
    314
    fall detection 316–18
    gesture recognition 314–16
    human activity categorising 319
  fundamentals 292–5
  radar link budget analysis 308–9
  radar transceiver 296
    receiver 297–8
    transmitter 296–7
  signal processing 309
    micro-Doppler 313–14
    range-Doppler processing 311–13
    range processing 310–11
frequency modulation (FM) 151–2
Fresnel zone 41, 58
F-score 196–7
fully convolutional network (FCNs)
    263
fuzzy logic 195, 207

Gabor dictionary 76, 82, 84–5, 98
gated recurrent units (GRUs) 129
Gaussian mixture model (GMM)
  in conjunction with a hidden Markov
    model (GMM-HMM) 376
  GM-PHD filter 373
generative adversarial networks (GANs)
    118, 168–70
generator network 168

genetic algorithm-optimised frequency-warped cepstral coefficients 112–13
GPS disciplined oscillators (GPSDO) 6
greedy algorithms 78–9
ground-based surveillance radar 137–9, 142, 175

Hamming window 161, 241, 334
hard fusion 195
Hausdorff distances 84
   nearest neighbour classifier based on modified Hausdorff distance 88–9
helicopters 259, 282
Hempel filtering 58
high-resolution quadratic time-frequency distributions 184
Hughes' finding 341
human gait 139
   Boulic–Thalmann model 141
   human running model 141
   human walking model 140–1
   implementation 153–4
   measuring the signature of 141
     micro-Doppler signature 142–4
human motion 37–40, 57, 95, 97–8, 139, 141
   motion capturing activities on 145–7
human running model 141
human walking model 140–1

image entropy 189
ImageNet Large Scale Visual Recognition Challenge (ILSVRC) 103
independent component analysis (ICA) 139
industrial, scientific and medical (ISM)–band CW 145, 156
inertial measurement unit (IMU) 185
inverse synthetic aperture radar (ISAR) 260

isometric feature mapping (Isomap) 139
iterative hard thresholding (IHT) 79

joint angle and delay estimation (JADE) 56
Joint Probabilistic Data Association (JPDA) 373

Kalman filter 152, 372–3
   -based multi-target tracking block 370
Kinect sensor 117–18
K-means algorithm 84, 86–9
K-nearest neighbours (KNN) 191, 240, 345–6
Krawtchouk (Kr) moments-based feature vector approach 238, 243–4, 249, 252–3

L1-norm principal component and discriminant analyses 351
   L1-LDA-based activity classification 359
     classifier 360–1
     illustrative example 361–2
     L1-LDA algorithm 360
     problem formulation 359–60
   L1-PCA-based classification 355
     classifier 356–7
     illustrative example 357–9
     L1-norm PCA 355–6
     L1-PCA through bit flipping 356
   radar signal model 353–4
label consistent KSVD (LC-KSVD) algorithm 80, 98
linear discriminant analysis (LDA) 139, 351
locally linear embedding (LLE) 139
logarithmic opinion poll (LOGP) 195
long short-term memory (LSTM) RNNs 129, 165–6, 286
low-slow-small (LSS)-targets 278–9, 285

magnetic sensors 198, 202–5
  magnetic Hall Effect sensors 186
  magnetoresistance sensors 186
maximum CPI 272–3
mean arterial pressure (MBP) 340
Mel-frequency cepstral coefficients
    (MFCCs) 112, 219
Mel-frequency filter bank 112, 185
micro-Doppler analysis 263, 273–7
  on polarimetric data 277–8
micro-Doppler effect 313–14
micro-Doppler signature 9–11, 138,
    142–4, 260
  analysis of 48–50
  visualisation of 144
model-driven classification 138, 147–9
  classification and parameters
      estimation results 154–9
  human gait implementation 153–4
  particle-filter 152–3
  results of 149–52
Moens–Kortewegs formula 340–1
Monte Carlo approach 241, 252
motion capture (MOCAP)
    data 117
  activities on human motion 145–7
moving target indication (MTI) 371–2
multicopters 259–60, 276
multi-hypothesis tracker (MHT) 373
multilayer perceptrons (MLPs) 263
multimodal analysis 194
multimodal information fusion 208–10
multiple inputs, multiple output
    (MIMO) functionality 42, 304,
    306–7
multiple-signal classification (MUSIC)
    algorithm 55–6
multi-sensor-based sleep-stage
    classification 342–6
multi-sensor fusion
  feature selection 196–8
  principles and approaches 192–6
multi-sensor information fusion 344
multistatic radar micro-Doppler 1

bistatic and multistatic radar
    properties 4–6
multistatic radar data, analysis of 9
  feature diversity for multistatic
      radar systems 12–14
  micro-Doppler signatures 9–11
  multiple personnel for
      unarmed/armed classification
      14–17
  multistatic drone micro-Doppler
      classification 22–31
  personnel recognition using
      multistatic radar data 19–22
  simple neural networks for
      multistatic radar data 17–19
  radars and experimental setup 6–9
multi-view analysis 194

Naïve Bayes algorithm 191, 195–6, 345
nearest subspace classifier (NSC) 357
NetRAD 1, 6–7, 22
neural network (NNs) 262
NeXtRAD 1, 22
non-cooperative target recognition
    (NCTR) 263
Nyquist sampling theorem 327–8

orthogonal frequency division
    multiplexing (OFDM) 40
orthogonal matching pursuit (OMP)
    algorithm 47, 75, 78, 84, 86

particle-filter 152–3
passive bistatic radar (PBR) 2
passive radar approaches for healthcare
    35
  ambient assisted living context 36–7
  for human activity monitoring 38–42
  human activity recognition using
      Wi-Fi passive radars 42–59
  sensor technologies in healthcare
      37–8
passive Wi-Fi radars (PWR) 43
Peak Sidelobe Level (PSL) ratio 237
phase-locked loop (PLL) 296, 326

photoplethysmograph (PPG) device 341
polysomnography (PSG) 342
power spectral density (PSD) 190
principal component analysis (PCA) 12, 44, 82, 108, 138–9, 188, 262, 351, 369
probability hypothesis density (PHD) tracking method 370, 373, 381
pseudo-Zernike (pZ) moments-based feature vector approach 237–8, 242–3
pulsed radar 6, 142, 280
pulse repetition frequency (PRF) 6, 143, 189, 281–3

quadratic TF distributions (QTFDs) 107–8

radar challenge 116
   synthetic signature generation from MOCAP data 117–18
   synthetic signature generation using adversarial learning 118–20
   transfer learning from optical imagery 116–17
radar classification approaches
   feature-based approach 138
   model-based approaches 138
radar cross section (RCS) 232, 257–8, 261, 266, 268, 278
radar measurements, target analysis using 269
   body RCS 269–70
   maximum CPI 272–3
   micro-Doppler analysis 273–7
      on polarimetric data 277–8
   rotor RCS 270–2
radar power budget 143
radar return model at radio frequency 219
   cone 224
      cone plus fins 224–8
      cylinder 228–9
      sphere 229–30

radar system considerations 278
   bandwidth 279–80
   carrier frequency and polarisation 279
   pencil-beam or ubiquitous radar 283–5
   pulse repetition frequency (PRF) 281–3
   waveform and coherency 280
radial basis function (RBF) 379, 385
radio frequency, radar return model at 219
   cone 224
      cone plus fins 224–8
      cylinder target 228–9
      sphere 229–30
range and angle resolution of the radar 143
range-Doppler imaging methods 316
range-Doppler processing 294, 311–13
range processing 310–11
range-time-Doppler cube 394
Rayleigh region 277
Rayleigh scatterer 278
Recall combiner 195–6
recognition accuracy
   versus size of training dataset 91–2
   versus sparsity 90–1
   for unknown personnel targets 92–3
rectified linear unit (ReLU) activation functions 120
recurrent neural networks (RNNs) 128–9, 165–7, 263, 286
Relief-*F* 197
residual neural network (ResNet) 103, 127
root mean square deviation (RMSD) 374–5
rotary-wing UAVs 259
rotor RCS 270–2

S-Band BMD radar system 231
scattering matrix 278
sensor fusion algorithms 193
SFS (sequential feature selection) 197

short-time Fourier transform (STFT)
9–10, 106, 182, 185, 232, 274,
354, 369
signal-level fusion 192–3
signal-to-noise ratio (SNR) 262, 344,
368
signatures of small drones and birds as
emerging targets 257
classification methods 285–7
electromagnetic predictions of birds
and UAVs 263
electromagnetic properties, size
and shape 263–4
radar cross section (RCS)
predictions 264–9
radar system considerations 278
bandwidth 279–80
carrier frequency and polarisation
279
pencil-beam/ubiquitous radar
283–5
pulse repetition frequency (PRF)
281–3
waveform and coherency 280
target analysis using radar
measurements 269
body RCS 269–70
maximum CPI 272–3
micro-Doppler analysis 273–7
micro-Doppler analysis on
polarimetric data 277–8
rotor RCS 270–2
unmanned aerial vehicles (UAVs),
classes and configuration of
258–60
singular value decomposition (SVD)
12, 188, 219, 369, 377
sleep-stage classification,
multi-sensor-based 342–6
smart tracking algorithms 138
soft fusion 194–5
software-defined radio (SDR) technique
47, 327
sparse representation classifier (SRC)
47

sparse signal recovery 76
dictionary learning 79–81
signal model 76–7
typical algorithms 77
Bayesian sparse signal recovery
77–8
convex optimisation 77
greedy algorithms 78–9
sparsity-based dictionary learning of
human micro-Dopplers 94–5
measurement data collection 95
single channel source separation of
micro-Dopplers from multiple
movers 95–7
target classification based on
dictionaries 97–8
sparsity-driven micro-Doppler feature
extraction 81–2
clustering for central time–frequency
trajectory 86–8
computational costs considerations
93–4
extracting time–frequency trajectory
85–6
measurement data collection 82–4
nearest neighbour classifier based on
modified Hausdorff distance
88–9
recognition accuracy
versus size of training dataset
91–2
versus sparsity 90–1
for unknown personnel targets
92–3
spectral entropy 191
spectrogram 10–11, 21–2, 82–4, 86,
91, 95–6, 109–10, 144, 146–51,
153–66, 168–73, 186–90, 200,
219, 232, 241, 285–6
spectrogram processing 374–5
track-based spectrogram extraction
373–4
support-vector machine (SVMs) 41,
57, 191, 262, 370, 378–9

synthesis dictionary learning (SDL) 79, 97
synthetic aperture radar (SAR) 118
synthetic signature generation
  from MOCAP data 117–18
  using adversarial learning 118–20
    auxiliary conditional GANs (case study) 120–1
    diversity of ACGAN-synthesised signatures 124–5
    kinematic fidelity of ACGAN-synthesised signatures 121–4
    PCA-based kinematic sifting algorithm 125–6
systolic BP (SBP) estimation 338, 341

target detection techniques 371–2
time-division multiplexing (TDM) 306
time–frequency analysis 369
time–frequency distribution (TFD) 219
time–frequency processing 354
time–frequency representations 185
time–frequency trajectory, extracting 85–6
track-based spectrogram extraction 373–4
tracking techniques 372–3
transceiver/receiver module (TRM) 285
transfer learning from optical imagery 116–17
t-SNE 171–2
2D Gabor filter-based feature vector approach 239, 245, 252–3
2D input representations 106–8
2D signature-based feature vectors 237
  Krawtchouk (Kr) moments-based feature vector approach 238
  pseudo-Zernike (pZ) moments-based feature vector approach 237–8
  2D Gabor filter-based feature vector approach 239

ubiquitous radar 262, 283–5
ultra-wideband (UWB) radar 37, 291
unarmed/armed classification, multiple personnel for 14–17
unmanned aerial vehicles (UAVs) 257, 260, 262
  classes and configuration of 258–60

visualisation of micro-Doppler signature 144

waveform and coherency 280
wearable sensors 185–6, 190–1
Wiener-filter algorithm 344
Wi-Fi CSI informatics to monitor human activities 36, 51
  from micro-Doppler (μ-D) to channel state information (CSI) 51–3
  Wi-Fi CSI-enabled healthcare applications 57–9
  Wi-Fi CSI signal capture and parameter extraction 53–6
Wi-Fi passive radars, human activity recognition using 42
  fundamentals 42–4
  micro-Doppler activity recognition 44–50
Wigner distribution 107
Wigner–Ville distribution (WVD) 107
wing beat frequency (WBF) 275
wrapper methods 13, 196

X-band continuous wave radar 262

zero Doppler line 5, 46, 49